Lecture Notes in Physics

W0055312

Springer-Verlag Berlin Heidelberg GmbH

The Editorial Policy for Proceedings

The series Lecture Notes in Physics reports new developments in physical research and teaching – quickly, informally, and at a high level. The proceedings to be considered for publication in this series should be limited to only a few areas of research, and these should be closely related to each other. The contributions should be of a high standard and should avoid lengthy redraftings of papers already published or about to be published elsewhere. As a whole, the proceedings should aim for a balanced presentation of the theme of the conference including a description of the techniques used and enough motivation for a broad readership. It should not be assumed that the published proceedings must reflect the conference in its entirety. (A listing or abstracts of papers presented at the meeting but not included in the proceedings could be added as an appendix.)

When applying for publication in the series Lecture Notes in Physics the volume's editor(s) should submit sufficient material to enable the series editors and their referees to make a fairly accurate evaluation (e.g. a complete list of speakers and titles of papers to be presented and abstracts). If, based on this information, the proceedings are (tentatively) accepted, the volume's editor(s), whose name(s) will appear on the title pages, should select the papers suitable for publication and have them refereed (as for a journal) when appropriate. As a rule discussions will not be accepted. The series editors and Springer-Verlag will normally not interfere with the detailed editing except in fairly obvious cases or on technical matters.

Final acceptance is expressed by the series editor in charge, in consultation with Springer-Verlag only after receiving the complete manuscript. It might help to send a copy of the authors' manuscripts in advance to the editor in charge to discuss possible revisions with him. As a general rule, the series editor will confirm his tentative acceptance if the final manuscript corresponds to the original concept discussed, if the quality of the contribution meets the requirements of the series, and if the final size of the manuscript does not greatly exceed the number of pages originally agreed upon. The manuscript should be forwarded to Springer-Verlag shortly after the meeting. In cases of extreme delay (more than six months after the conference) the series editors will check once more the timeliness of the papers. Therefore, the volume's editor(s) should establish strict deadlines, or collect the articles during the conference and have them revised on the spot. If a delay is unavoidable, one should encourage the authors to update their contributions if appropriate. The editors of proceedings are strongly advised to inform contributors about these points at an early stage.

The final manuscript should contain a table of contents and an informative introduction accessible also to readers not particularly familiar with the topic of the conference. The contributions should be in English. The volume's editor(s) should check the contributions for the correct use of language. At Springer-Verlag only the prefaces will be checked by a copy-editor for language and style. Grave linguistic or technical shortcomings may lead to the rejection of contributions by the series editors. A conference report should not exceed a total of 500 pages. Keeping the size within this bound should be achieved by a stricter selection of articles and not by imposing an upper limit to the length of the individual papers. Editors receive jointly 30 complimentary copies of their book. They are entitled to purchase further copies of their book at a reduced rate. As a rule no reprints of individual contributions can be supplied. No royalty is paid on Lecture Notes in Physics volumes. Commitment to publish is made by letter of interest rather than by signing a formal contract. Springer-Verlag secures the copyright for each volume.

The Production Process

The books are hardbound, and the publisher will select quality paper appropriate to the needs of the author(s). Publication time is about ten weeks. More than twenty years of experience guarantee authors the best possible service. To reach the goal of rapid publication at a low price the technique of photographic reproduction from a camera-ready manuscript was chosen. This process shifts the main responsibility for the technical quality considerably from the publisher to the authors. We therefore urge all authors and editors of proceedings to observe very carefully the essentials for the preparation of camera-ready manuscripts, which we will supply on request. This applies especially to the quality of figures and halftones submitted for publication. In addition, it might be useful to look at some of the volumes already published. As a special service, we offer free of charge LaTeX and TeX macro packages to format the text according to Springer-Verlag's quality requirements. We strongly recommend that you make use of this offer, since the result will be a book of considerably improved technical quality. To avoid mistakes and time-consuming correspondence during the production period the conference editors should request special instructions from the publisher well before the beginning of the conference. Manuscripts not meeting the technical standard of the series will have to be returned for improvement.

For further information please contact Springer-Verlag, Physics Editorial Department II, Tiergartenstrasse 17, D-69121 Heidelberg, Germany

Emmi Meyer-Hofmeister   Henk Spruit (Eds.)

Accretion Disks – New Aspects

Proceedings of the EARA Workshop
Held in Garching, Germany,
21–23 October 1996

 Springer

Editors

Emmi Meyer-Hofmeister
Henk Spruit
Max-Planck-Institut für Astrophysik
D-85740 Garching, Germany

Scientific Organising Committee:
H. Spruit, E. Meyer-Hofmeister, F. Meyer, R. Sunyaev

Conference Secretary:
P. Berkemeyer

Cataloging-in-Publication Data applied for.

Die Deutsche Bibliothek - CIP-Einheitsaufnahme

Accretion disks - new aspects : proceedings of the EARA workshop, held in Garching, Germany, 21 - 23 October 1996 / Emmi Meyer-Hofmeister ; Henk Spruit (ed.).

(Lecture notes in physics ; Vol. 487)
ISBN 978-3-662-14142-7   ISBN 978-3-540-68715-3 (eBook)
DOI 10.1007/978-3-540-68715-3

ISSN 0075-8450
ISBN 978-3-662-14142-7

Typesetting: Camera-ready by the authors/editors
Cover design: *design & production* GmbH, Heidelberg
SPIN: 10550722   55/3144-543210 - Printed on acid-free paper

Preface

This book grew out of a meeting of the EARA held October 21–23 at the Max-Planck-Institut für Astrophysik (MPA) in Garching. EARA, the European Association for Research in Astronomy (EARA[1]) is a consortium of European research institutes consisting of the Institut d'Astrophysique de Paris, the Institute of Astronomy of the University of Cambridge, the Sterrewacht Leiden, and the MPA.

The meeting brought together the expertise on accretion disks at the EARA institutes, with additional keynote speakers invited from outside the EARA circle. The result has been a workshop in the tradition of the 'Theory of accretion disks' meetings held in Garching in 1989 and 1993.

The focus of the meeting this time has been on a number of topics at the center of the current debates about accretion disks. The contributions to the workshop cover the puzzles presented by the X-UV spectra of AGN and their variability, the recent numerical simulations of magnetic fields in disks, the remarkable behavior of the superluminal source 1915+105 and the 'bursting pulsar' 1744-28, to mention a few of the topics.

Professor Osaki's talk on a unification model for the outburst behavior of non-magnetic cataclysmic variables was a lecture given on 23 October, at the occasion of Friedrich Meyer's official retirement.

Due to its relatively small size, the meeting was quite interactive. We have been fortunate in securing excellent texts from essentially all participants. Special thanks are due to the conference secretary Petra Berkemeyer, who also provided competent assistance in preparing the camera-ready text.

Garching
April 1997

Scientific Organising Committee
H. Spruit
E. Meyer-Hofmeister
F. Meyer
R. Sunyaev

---

[1] http://www.ast.cam.ac.uk/IOA/EARA/EARA.html

Contents

List of Participants

Marek Abramowicz
   University of Göteborg, Sweden
Ulrich Anzer
   MPI für Astrophysik, Garching, Germany
Axel Brandenburg
   University of Newcastle, UK
Max Camenzind
   Landessternwarte Königstuhl, Heidelberg, Germany
Chris Campbell
   University of Newcastle, UK
Susy Collin
   Institue d'Astrophysique de Paris, France
Tiziana Di Matteo
   Institute of Astronomy, Cambridge, UK
Thomas Dörrer
   Institut für Astronomie u. Astrophysik, Tübingen,Germany
Cornelis P. Dullemond
   Sterrewacht Leiden, The Netherlands
Wolfgang J. Duschl
   Inst. für Theoretische Astrophysik, Heidelberg, Germany
Andrew Fabian
   Institute of Astronomy, Cambridge, UK
Marat Gilfanov
   MPI für Astrophysik, Garching, Germany
Jochen Greiner
   MPI für Extraterrestrische Physik, Garching, Germany
Wolfgang Hummel
   Universitätssternwarte Muenchen, Germany
Andrew King
   University of Leicester, UK
Willy Kley
   MPG AG-Gravitationstheorie, Jena, Germany
Stefanie Komossa
   MPI für Extraterrestrische Physik, Garching, Germany

Sergey Kuznetsov
    MPI für Astrophysik, Garching, Germany
Bifang Liu
    MPI für Astrophysik, Garching, Germany
Friedrich Meyer
    MPI für Astrophysik, Garching, Germany
Emmi Meyer-Hofmeister
    MPI für Astrophysik, Garching, Germany
Gordon Ogilvie
    Institute of Astronomy, Cambridge, UK
Yoji Osaki
    University of Tokyo, Japan
John Papaloizou
    Queen Mary & Westfield Colledge, London, UK
Jochen Peitz
    Landessternwarte Königstuhl, Heidelberg, Germany
P.O.Petrucci
    Observatoire de Grenoble, France
Hans Ritter
    MPI für Astrophysik, Garching, Germany
Günther Rüdiger
    Astrophysikalisches Institut, Potsdam, Germany
Bruce Sams
    Garching, Germany
Sergey Sazonov
    MPI für Astrophysik, Garching, Germany
Susanne Schandl
    Institut für Astronomie u. Astrophysik, Tübingen,Germany
Steinn Sigurdsson
    Institute of Astronomy, Cambridge, UK
Henk Spruit
    MPI für Astrophysik, Garching, Germany
Rudolf Stehle
    MPI für Astrophysik, Garching, Germany
Rashid Sunyaev
    MPI für Astrophysik, Garching, Germany
David Syer
    MPI für Astrophysik, Garching, Germany
Yasuo Tanaka
    Sterrekundig Inst., Universiteit Utrecht, The Netherlands
Hans-Christoph Thomas
    MPI für Astrophysik, Garching, Germany
Ulf Torkelsson
    Institute of Astronomy, Cambridge, UK

Sergey Trudolyubov
  MPI für Astrophysik, Garching, Germany
Marie-Hélène Ulrich
  ESO, Garching, Germany
Jan van Paradijs
  Sterrenkundig Inst. "Anton Pannekoek", Amsterdam, The Netherlands

X-ray spectrum of low-mass X-ray binaries

Y. Tanaka[1,2]

[1] Astronomical Institute, University of Utrecht, Utrecht, The Netherlands
[2] Institute of Space and Astronautical Science, Sagamihara, Kanagawa-ken, Japan

Abstract: This paper gives an overview of the properties of the observed X-ray spectra of low-mass X-ray binaries containing a neutron star (neutron-star LMXBs) and those of X-ray binaries containing a black hole (black-hole XBs). The X-ray spectra of these systems change their characteristics with the mass accretion rate. At high X-ray luminosities, typically $L_x > 10^{37}$ erg sec^{-1}, soft thermal emission from the accretion disk dominates. There is a distinct difference in the X-ray spectrum between neutron-star LMXBs and black-hole XBs. When the luminosity decreases below a certain level ($L_x \sim 10^{37}$ erg sec^{-1}), the X-ray spectra of both systems undergo a transition to a hard power-law form. At further low luminosities ($L_x < 10^{33-32}$ erg sec^{-1}), the spectra become very soft. The characteristics of these X-ray spectra at different X-ray luminosities (different accretion rates) are described and the implications are discussed.

1 Introduction

A wealth of observational results on the X-ray spectrum of binary X-ray sources has become available from various X-ray astronomy satellites. The results show that most of the observed X-ray spectra fall into either of the following four distinctly different types that are displayed in Fig. 1 (Tanaka 1992). These are the incident photon spectra derived from the observed count-rate spectra by correcting for the instrument response. In this paper, we show incident photon spectra unless otherwise stated, instead of raw count-rate spectra that are dependent on the response of the specific instrument used. (One can recognize an incident photon spectrum by the vertical scale given in units of photons cm^{-2} s^{-1} keV^{-1}, whereas the one with a vertical scale in units of counts s^{-1} keV^{-1} is a count-rate spectrum without correction for the instrument response.)

In the case of X-ray binaries containing a neutron star, the shapes of the X-ray spectra are very different depending on whether the magnetic fields of the neutron star is strong or weak. This is generally related to whether the companion is high mass or low mass. Binary X-ray pulsars contain a highly-magnetized neutron star with the surface magnetic fields of the order of 10^{12} Gauss. Most of them are Be binaries. Binary X-ray pulsars typically show a spectrum characterized by a hard power-law function with a sharp cut-off above a few tens of keV, such as shown in Fig. 1(a). On the other hand, non-pulsating X-ray binaries composed of a weakly-magnetized neutron star and a low-mass companion (neutron-star LMXBs) typically exhibit a spectrum such as shown in Fig. 1(b), when they are luminous (at high accretion rates). Qualitatively,

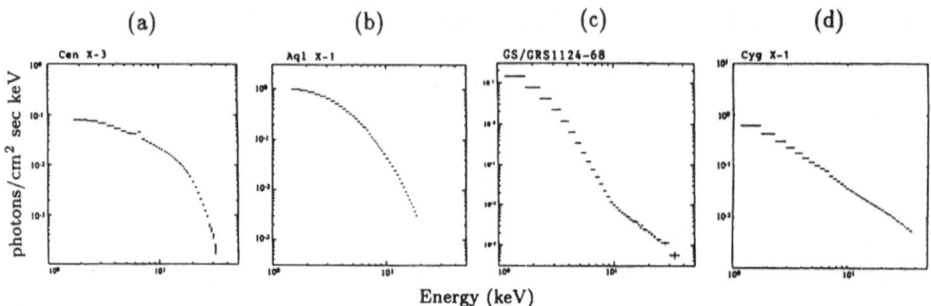

Fig. 1. Four characteristic types of X-ray spectra of X-ray binaries. These are the incident photon spectra, corrected for the instrument response.

this looks like a thermal bremsstrahlung spectrum. However, a detailed study reveals more complex properties, as discussed in section 2.1. The magnetic fields of the neutron stars in these systems are considered to be weak enough (10^9 Gauss or less) so that the influence of the magnetic fields on mass accretion can be ignored.

Other than these neutron-star LMXBs, there exist a group of sources that exhibit a qualitatively different type of spectrum characterized by a soft thermal component accompanied by a hard tail, such as shwon in Fig. 1(c). From examinations of these sources in section 2.2, it is shown that this type of spectrum is most probably a signature of an accreting black hole at high accretion rates. In the case of black-hole binaries, the X-ray spectrum is not related to the mass of the companion (either high mass or low mass). This is because in either case magnetic fields that may influence accretion are absent in black holes.

The fourth type of the spectrum is an approximately single power-law form, as shown in Fig. 1(d). Many X-ray binaries at low luminosities are found to have this type of spectrum. Both neutron-star LMXBs and black-hole XBs change their spectral shapes according to the mass accretion rate. The most outstanding change is a transition between a spectrum of thermal nature (Fig. 1(b) and 1(c)) at high luminosities and a spectrum of a single power-law form (Fig. 1(d)) at low luminosities. This phenomenon is discussed in section 3.1. It is shown that this transition occurs around a certain accretion rate regardless of whether the compact object is a weakly-magnetized neutron star or a black hole. Therefore, it is considered to be a fundamental property of accretion disks.

Many low-mass X-ray binaries are transient sources. Of about 120 LMXBs known in our galaxy (see van Paradijs 1995), about 100 are neutron-star systems of which more than 80% are persistent sources. In contrast, most, if not all, of $\sim$ 20 black-hole LMXBs known to date are transient sources. This remarkable difference between neutron-star LMXBs and black-hole LMXBs is yet to be explained (see e.g. King et al. in these Proceedings). The transient sources undergo spectacular X-ray outbursts due to a sudden commencement of mass accretion onto the compact object. In these transients, the accretion rate changes many orders of magnitude from the outburst peak through the following decay until the source returns to the quiescent state. Therefore, transient sources are very useful for the study of various accretion phenomena over a wide range of accretion rate. For a review of transient LMXBs, see e.g. Tanaka & Shibazaki (1996).

In recent years, owing to a dramatic increase of sensitivity of the X-ray instruments, X-rays from these transients in the quiescent state have become observable. The X-ray spectra at extremely low luminosities are found to be much softer than observed at higher luminosities. Study of the X-ray emission during quiescnce is a whole new subject. As discussed in section 3.2, the origin of these soft X-rays is yet to be understood.

2   X-ray spectra at high luminosities

2.1   Neutron-star LMXBs

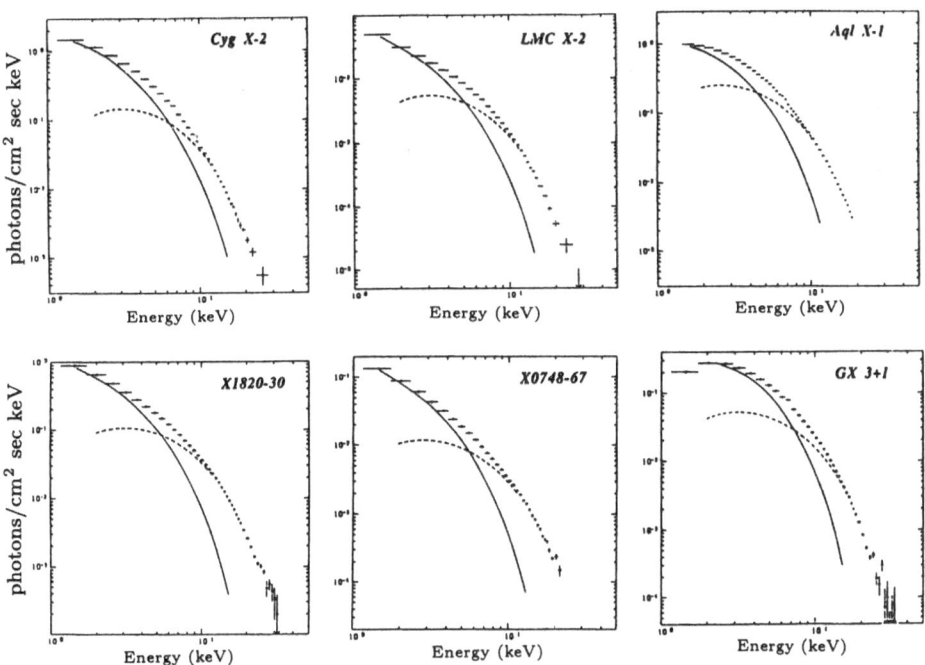

Fig. 2. Examples of the observed X-ray photon spectra of luminous ($L_x > 10^{37}$ erg sec^{-1}) neutron-star LMXBs, each comprising a soft component (solid curve) and a blackbody component (dashed curve). See text for the two components.

In the case of neutron-star LMXBs, X-rays are considered to be emitted from two possible sites: (1) an accretion disc in which thermal energy is released by viscous heating, and (2) the neutron star envelope (the boundary layer) where the kinetic energy of accreting matter is eventually deposited. In fact, when the sources are luminous, the emission spectra from these two sites can be determined respectively as described below. In what follows, the X-ray luminosity L_x is in the energy range $1-30$ keV unless otherwise specified. L_x is estimated as $4\pi d^2 I_x$ for a source distance d and a measured flux I_x, hence subject to possible anisotropy of emission and the uncertainty of the distance.

When the X-ray luminosity L_x is well above 10^{37} erg sec^{-1}, the observed spectra are shown to consist of two separate components: a soft component and a hard component, as shown in Fig. 2. Each of these components can be independently determined from the analysis of changes of the spectral shape with intensity. For example, such an exercise for the spectrum of LMC X-2 is given below (Tanaka 1992).

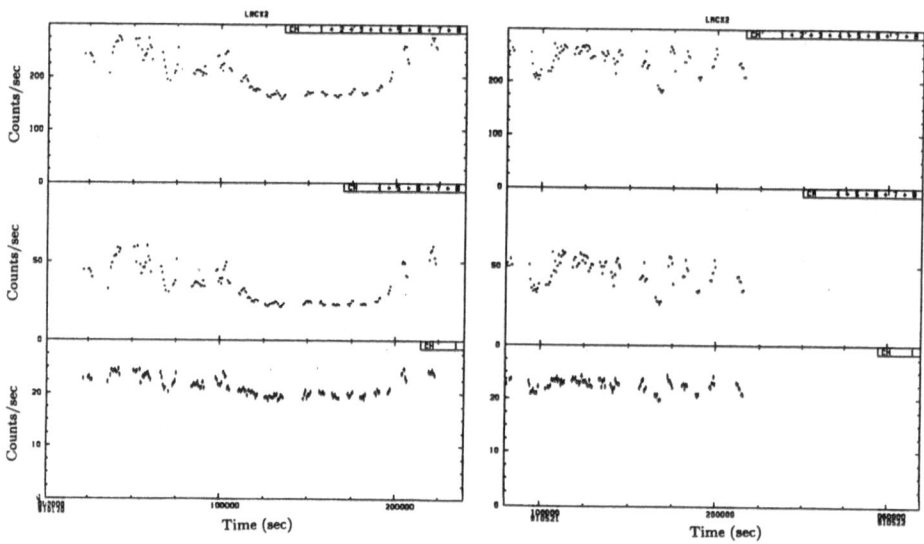

Fig. 3. The light curves of LMC X-2 during an observation with Ginga for the ranges $1-30$ keV (top), $7.0-30$ keV (middle), and $1-2.3$ keV (bottom).

Figure 3 shows the light curves of LMC X-2 in an interval of an observation with Ginga in the ranges (a) $1-30$ keV (total), (b) $7.0-30$ keV, and (c) $1-2.3$ keV, respectively. One notices that the amplitude of variation is much larger in the higher energy band. During this interval, the hardness ratio $I(4.7-23 \text{ keV})/I(1-4.7 \text{ keV})$ is found to be linearly related to the observed flux $I(1-23 \text{ keV})$, as shown in Fig. 4. (This feature is sometimes referred to as a "normal branch" on the hardness vs. intensity diagram).

We divide the data into six groups, a through f in the increasing order of

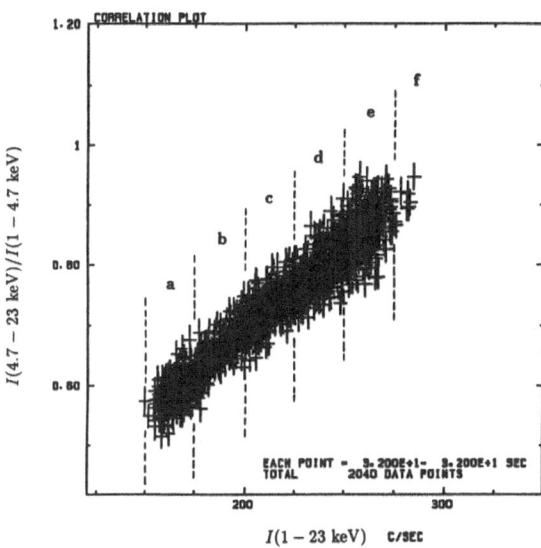

Fig. 4. The relation between the observed flux $I(1 - 23 \text{ keV})$ and the hardness ratio $I(4.7 - 23 \text{ keV})/I(1 - 4.7 \text{ keV})$. Data points for every 32 s are plotted.

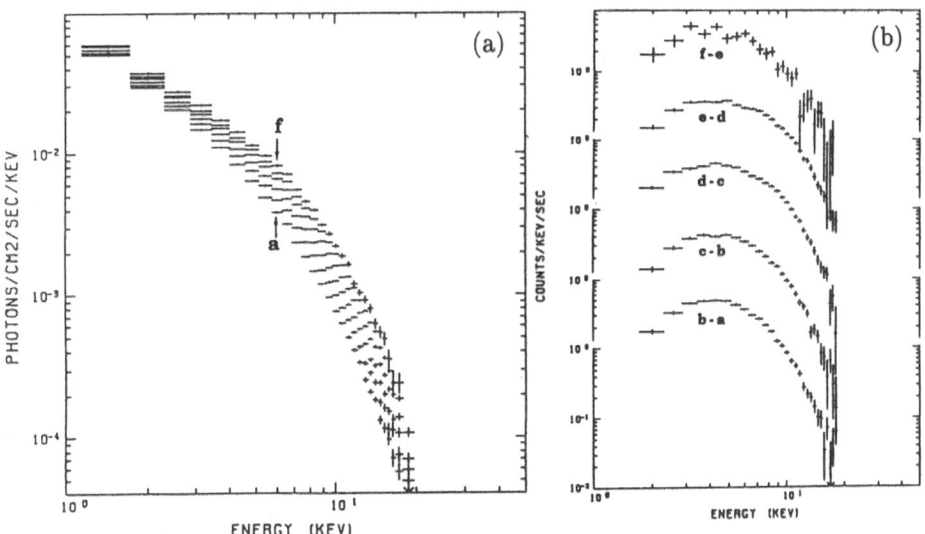

Fig. 5. (a) X-ray photon spectra of LMC X-2 at six equally separated flux levels, a through f (see Fig. 4). (b) The differences between every two spectra of adjacent flux levels (count-rate spectra, not corrected for the instrument response). Note that the vertical scales in (b) are successively shifted by a factor of 10.

I, with an equal flux ineterval ΔI as indicated in Fig. 4, and construct the spectrum for each group, as shown in Fig. 5(a). One can see in Fig. 5(a) that the flux changes very little at the lowest energy channel and that the spectra at different flux levels approach the same slope (become parallel to each other) towards the high energy end.

If we take the difference of the spectra $\Delta F(E) = F(I + \Delta I; E) - F(I; E)$ between two adjacent groups, we find that $\Delta F(E)$ is essentially identical independent of I as seen in Fig. 5(b). Therefore, it is written as

$$\Delta F(E) = B(E) \cdot \Delta I, \tag{1}$$

where $B(E)$ is a normalized function. Hence, by integrating it,

$$F(E) = S(E) + B(E) \cdot I. \tag{2}$$

$S(E)$ is an integration constant independent of I, and can be obtained by subtracting out the component of the form $B(E)$ from the observed spectrum. Thus determined $S(E)$ is shown in Fig. 6, together with the average photon spectrum of the $B(E)$ component and the sum (average photon spectrum of LMC X-2).

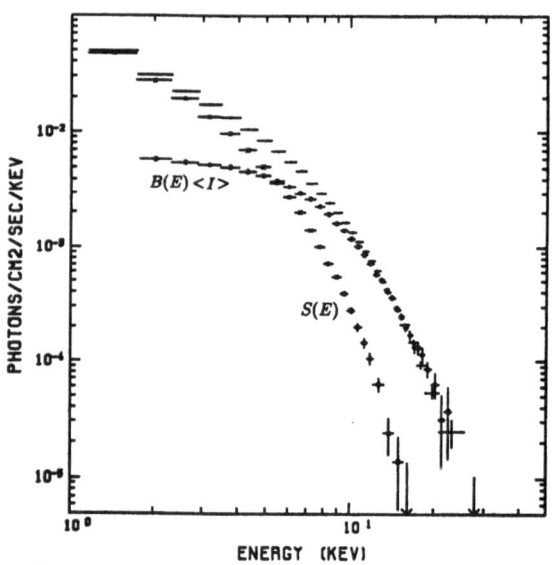

Fig. 6. The average photon spectrum of LMC X-2. Photon pectra of the soft component $S(E)$ (see text) and the average of the $B(E)$ component are separately shown.

This soft component $S(E)$ is very well expressed by a multicolor blackbody spectrum expected for the emission from an optically-thick accretion disk (Mitsuda et al. 1984) based on the standard Shakura-Sunyaev disk model (1973).

This multicolor blackbody spectrum is derived on the assumption that the energy generated by viscous heating is dissipated locally in blackbody radiation. This model includes only two free parameters, i.e. $S(E) = S(r_{in}, T_{in}; E)$, where r_{in} is the innermost disk radius and T_{in} is the color temperature at r_{in}. The excellent agreement of the observed soft component with this model makes it certain that the soft component is the emission from an optically-thick accretion disc. Hence, the luminosity of the soft component L_s (not the total luminosity L_x) represents the mass accretion rate $\dot{M}$. (This means that $\dot{M}$ was constant during this particular interval of observation of LMC X-2, despite that the total flux varied with time.) The ASCA observations of several luminous LMXBs do not reveal emission lines (except for a weak iron line at 6.7 keV), which supports that the disk is indeed optically thick. The observed color temperature T_{in} increases with L_s (or $\dot{M}$), and is typically $1.4-1.5$ keV for $L_s \sim 10^{38}$ erg sec^{-1}.

The hard component $B(E) \cdot I$ is best expressed by a modified blackbody spectrum $B_M(T_c; E)$, where T_c is a color temperature. This spectrum is very similar to that of X-ray bursts near the burst peak, with nearly the same color temperature T_c. This component is most probably the emission from the neutron star envelope (an optically-thick boundary layer). The observed T_c is typically $\sim 2.3 - 2.5$ keV. This value corresponds to an effective temperature T_{eff} of a $1.4 M_\odot$ neutron star with a radius of ~ 10 km at a luminosity near the Eddington limit, when corrected for a significant effect of electron scattering.

The intensity of this blackbody component varies by a large factor on time scales of 10 minutes to an hour, but without changing T_c. The ratio of the luminosities L_b/L_s varies typically in the range 0.2−1 (Mitsuda et al. 1984), where L_b is the luminosity of the blackbody component. It is to be emphasized that the variation of the blackbody component is unrelated to the soft component, hence not due to changes in $\dot{M}$. The observed fact seems to indicate that, for some unknown reasons, the mass flow from the innermost disk onto the neutron star is unsteady. It might be due to stagnation of accretion flow near the inner edge of the disk caused by unstable angular momentum transport, or could possibly be due to occasional mass ejection, when the luminosity is near the Eddington limit. However, these are mere speculations.

2.2   Black-hole binaries

Thus far, ten X-ray binaries are known to contain compact objects that are certainly more massive than $3 M_\odot$, established from the optically determined mass functions. Since the theoretical upper limt for a stable neutron star is $3 M_\odot$, these compact objects are most probably black holes. They are listed in Table 1 (Tanaka & Shibazaki 1996).

Eight of these ten black-hole X-ray binaries show X-ray spectra of a common characteristic shape when $L_x > 10^{37}$ erg sec^{-1}: a soft component accompanied by a hard power-law tail, such as shown in Fig. 7. The power-law tail extends beyond 100 keV (see e.g. Sunyaev et al. 1994). The other two sources show an approximately single power-law spectrum.

Table 1. Black-hole binaries established from the mass functions

Name		Spectrum[a]	Companion	F(M) $(M_\odot)$	BH mass $(M_\odot)$	Ref.
Cyg X-1	persistent	US+PL	O 9.7 Iab	0.241±0.013	~ 16 (>7)	[1]
LMC X-3	persistent	US+PL	B 3 V	2.3±0.3	> 7	[2]
LMC X-1	persistent	US+PL	O 7–9 III	0.14±0.05	~ 6(?)	[3]
J0422+32	Nova Per	PL	M 2 V	1.21±0.06	> 3.2	[4]
0620−003	Nova Mon	US+PL	K 5 V	3.18±0.16	> 7.3	[5]
1124−684	Nova Mus	US+PL	K 0–4 V	3.1±0.4	~ 6	[6]
J1655−40	Nova Sco	US+PL	F − G	3.16±0.15	4–5	[7]
1705−250	Nova Oph'77	US+PL	K ~3 V	4.0±0.8	~ 6	[8]
2000+251	Nova Vul	US+PL	early K	4.97±0.10	6–7.5	[9]
2023+338	Nova Cyg	PL	K 0 IV	6.26±0.31	8–15.5	[10]

[a] US+PL: ultrasoft + power law   PL: power law

Ref:1. Gies & Bolton (1982)   5. McClintock & Remillard (1986)   9. Filippenko et al. (1995b)
    2. Cowley et al. (1983)   6. McClintock et al. (1992)   10. Casares et al. (1992)
    3. Hutchings et al. (1987)   7. Bailyn et al. (1995)
    4. Filippenko et al. (1995a) 8. Remillard et al. (1996)

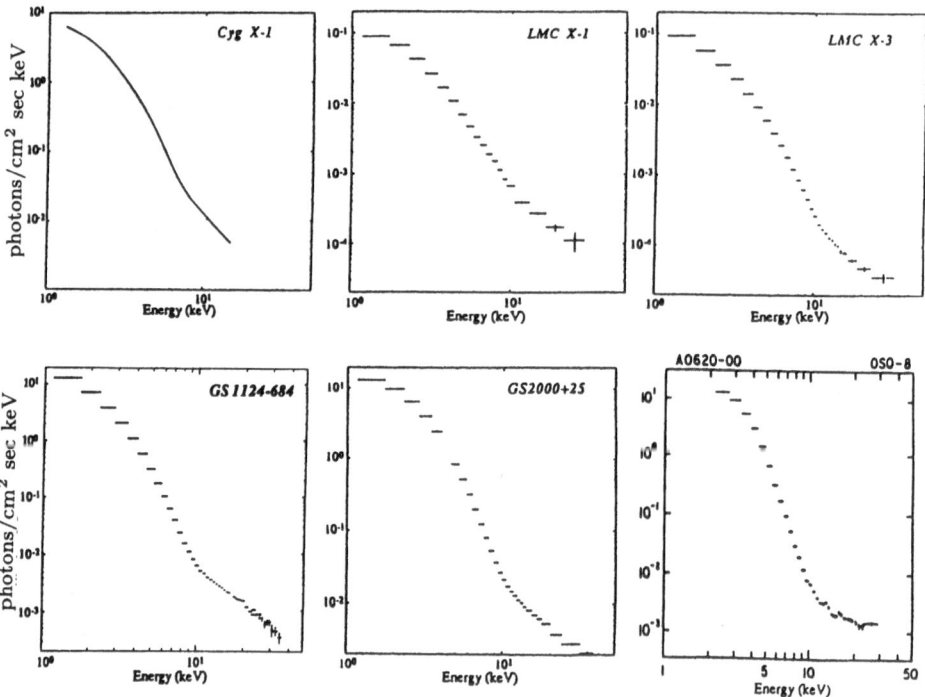

Fig. 7. X-ray photon spectra of black-hole X-ray binaries. Those of Cyg X-1 and A0620−00 are from Coe et al. (1976) and White et al. (1984), respectively.

The soft component is well described by a multicolor blackbody spectrum of the same functional form $S(r_{in}, T_{in}; E)$ as that describes the soft components of neutron-star LMXBs. This indicates that the soft component of a black-hole XB is also the emission from an optically-thick accretion disk. However, an important difference is that kT_{in} for black-hole binaries is always found to be less than 1.2 keV, significantly lower temperature than that for neutron-star LMXBs at similar luminosities. Hence, White et al. (1984) called it ultrasoft. This difference is understood in terms of the difference in the mass M_x of the compact object. If the magnetic fields are weak enough (for a neutron star) or absent (in a black hole), the accretion disk can extend to the innermost stable Keplerian orbit at a radius of $\sim 3r_g$ ($r_g = 2GM_x/c^2$). (In fact, it is probably the case, as shown below.) Then, T_{in} scales as $M_x^{-1/2}$ for a multicolor blackbody disk at a given disk luminosity (= the luminosity of the soft compoent L_s).

No blackbody component is present in the X-ray spectra of black-hole X-ray binaries. This is consistent with the absence of a solid surface in a black hole. Thus, this form of spectrum characterized by an ultrasoft component and a power-law tail is believed to be a signature of an accreting black hole.

There are about a dozen more sources showing this characterisitic spectral shape. They are also considered to be black-hole binaries. So far, not a single X-ray burst has been detected from these sources despite a long integrated observing time over a wide range of luminosity. This fact is also in support of the absence of a solid surface in the compact object of these sources in addition to the absence of a blackbody component in their spectra.

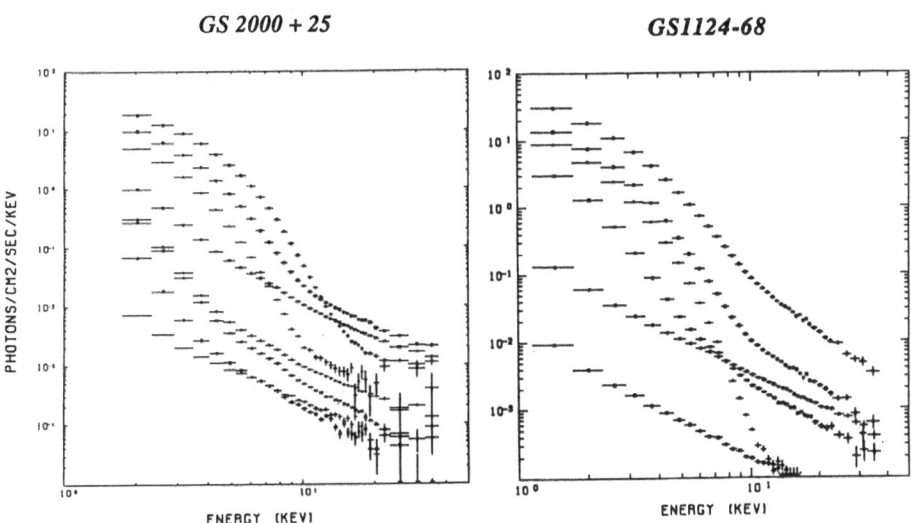

Fig. 8. Evolution of the X-ray photon spectrum of two transient black hole X-ray binaries, GS2000+25 and GS/GRS1124−68.

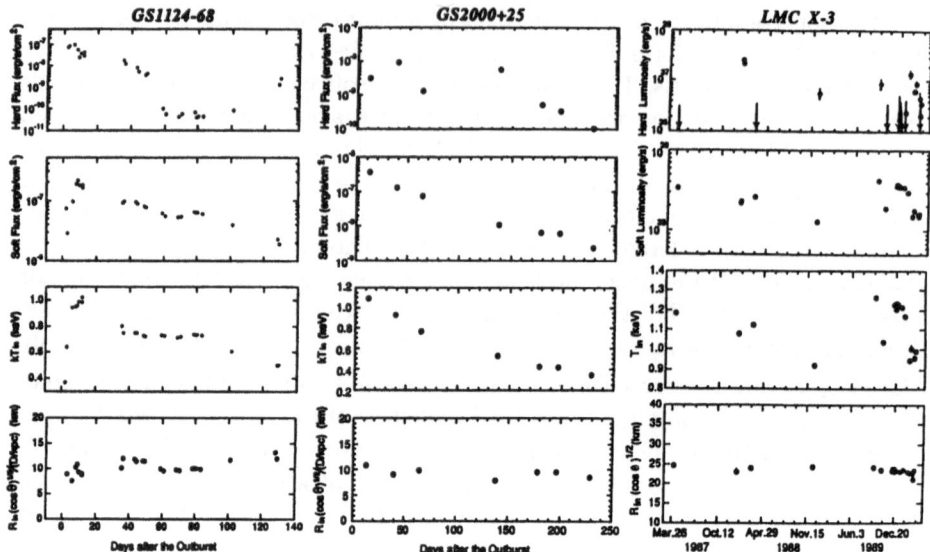

Fig. 9. Time histories of the luminosities of the hard and soft components, and the best-fit parameter values of the multicolor blackbody disk model, for three black-hole X-ray binaries, GS/GRS1124−68, GS2000+25 and LMC X-3. From the top to the bottom, the power-law flux (2-30 keV), the bolometric flux of the soft component, kT_{in}, and $r_{in}\cos^{1/2}\theta$ (see text) are shown. Note that the values of $r_{in}\cos^{1/2}\theta$ are for a normalized distance at 1 kpc for GS/GRS1124−68 and GS2000+25, and at 50 kpc for LMC X-3.

The color temperature T_{in} changes with L_s. As L_s decreases, T_{in} becomes lower (the spectrum softens). This is clearly noticeable in Fig. 8, in which the X-ray spectra of two transient black-hole binaries, GS/GRS1124−68 and GS2000+25, at various intensity levels are shown. This dependence is also evident in the kT_{in}-diagram in Fig. 9. Figure 9 represents the time histories of the luminosities of the hard and soft components as well as the best-fit spectral parameter values for the multicolor blackbody disk model, for three black-hole X-ray binaries, GS/GRS1124−68, GS2000+25 and LMC X-3 (see Tanaka & Lewin 1995, and references therein).

Remarkably, while L_s changes by a large factor, sometimes by more than an order of magnitude, the value of r_{in} remains essentially constant, as seen in Fig. 9. This implies that an optically-thick disk extends to a fixed distance from the compact object independnet of accretion rate. This fact is probably indicating that r_{in} corresponds to the radius of the innermost stable Keplerian orbit, $\sim 3r_g$.

Table 2 lists the values of $r_{in}\cos^{1/2}\theta$ determined from the observed soft components of various sources for which distance estimates are available (Tanaka & Lewin 1995), where θ denotes the inclination angle of the disk. In fact, the values for neutron-star LMXBs are not inconsistent with $3r_g$ for a $1.4M_\odot$ neutron star

Table 2. Estimated values of $r_{in}\cos^{1/2}\theta$

Black-hole binaries			Neutron-star LMXBs		
Source Name	Distance (kpc)	$r_{in}\cos^{1/2}\theta$ (km)	Source Name	Distance (kpc)	$r_{in}\cos^{1/2}\theta$ (km)
LMC X-1	50	$40^{a)}$	1608−52	$4.1^{g)}$	$6.2^{h)}$
LMC X-3	50	$24^{a)}$	1636−53	$6.9^{g)}$	8.0
A0620−00	~ 1	$25\text{-}30^{b)}$	1820−30	~ 8.5	6.1
GS2000+25	~ 2.5	$25^{c)}$	Sco X-1	~ 0.7	$4.1^{h)}$
GS/GRS1124-68	~ 3	$30^{d)}$	LMC X-2	50	10
Cyg X-1	~ 2.5	$33^{e)}$	Cyg X-2	~ 8	10
GX 339−4	~ 4	$17\text{-}22^{f)}$	Aql X-1	~ 2.5	4

[a] Ebisawa (1991)
[b] The spectrum in White et al.(1984) used.
[c] Takizawa (1991)
[d] Ebisawa et al. (1994)
[e] Dotani (1996), private communication
[f] Ebisawa (1991)

[g] Estimated from the burst peak luminosity.
    (Ebisawa et al. 1991)
[h] Mitsuda et al. (1984)

(≈ 12 km), if the effects of electron scattering ($T_c > T_{eff}$) and the gravitational redshift are corrected for (Ebisawa et al. 1991). On the other hand, the values for black-hole binaries are all larger by a factor of three to four than those for neutron-star LMXBs, showing that these compact objects are indeed more massive than $3M_\odot$ (since $r_{in} \propto M_x$). Thus, this provides a method of estimating the mass of the compact object from the X-ray spectrum alone, when the distance of the source is known.

The hard tail shows a power-law form. The observed photon indices from various sources fall within a range 2.2−2.5 (see Table 3). The intensity of the hard tail varies in a wide range more than an order of magnitude, unrelated to that of the ultrasoft component, as seen in Fig. 8 and 9. This behavior is similar to that of the blackbody component in neutron-star LMXBs. Despite large intensity changes of the power-law component, the photon index remains remarkably constant as noticeable in Fig. 8.

The site of the emission for the hard power-law component is still unknown. Possibly, it is produced by Comptonization of soft photons within $3r_g$ where matter is optically thin and hot. This would also explain why the power-law tail is absent in the luminous neutron-star LMXBs for which the optically-thick disk extends close to the neutron star surface.

3   Changes of X-ray spectrum with accretion rate

3.1   Transition between a soft state and a hard state

Black-hole binaries exhibit a dramatic change in the spectral shape between a soft thermal state (high state) at high luminosities and a hard power-law state (low state) at low luminosities. As the luminosity decreases below a certain level, the spectrum changes from the shape described in section 2.2 into an approximately single power-law form, and vice versa. This phenomenon was first observed from Cyg X-1, when Cyg X-1 was the only probable black-hole source. In addition, Cyg X-1 in the hard state shows large-amplitude intensity fluctuations (flickering) on all time scales down to 1 ms or shorter. For some time, this distinct behavior of Cyg X-1 had been considered to be a possible black hole signature. Such a transition has been observed later from at least three more black-hole binaries, GS2000+25, GS/GRS1124−68, and GX339-4. (The transitions for GS2000+25 and GS/GRS1124−68 are apparent in Fig. 8). Flickering builds up when the source goes into the power-law state (see Ebisawa et al. 1994).

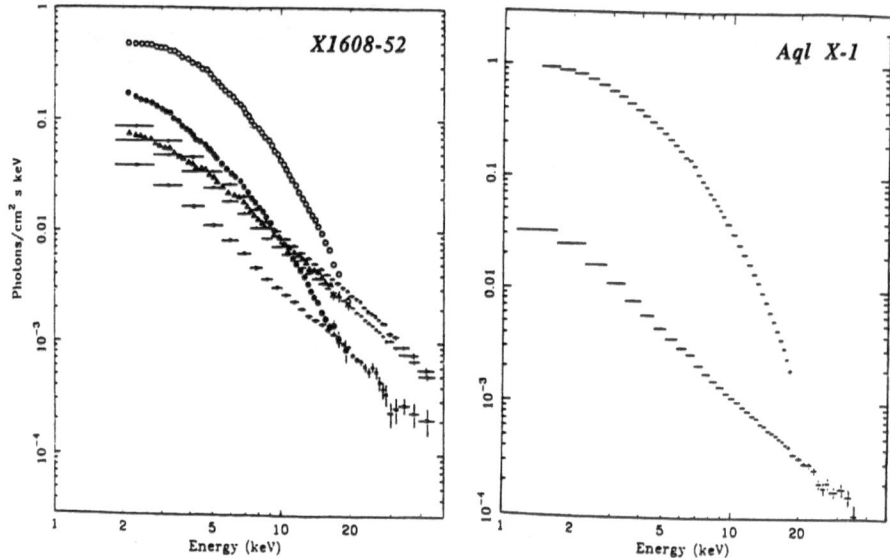

Fig. 10. X-ray photon spectra of two neutron-star LMXBs, 1608−52 (Mitsuda et al. 1989; Yoshida et al. 1993) and Aql X-1, in the soft thermal state and the hard power-law state.

However, these properties are not unique to black-hole binaries. Similar transitions have been observed from several neutron-star LMXBs. Such examples are shown in Fig. 10. In particular, the progress of such a transition was observed in detail for 1608−52, a well-known burster (Mitsuda et al. 1989). The transition occurred at around a luminosity of $\sim 10^{37}$ erg sec^{-1}, corresponding to a mass accretion rate of $\sim 10^{17}$ g s^{-1}. As the luminosity decreased to this level, first a hard tail began to show up, and then the whole spectrum went into a single power-law form within a farily small decrease in luminosity. Flckering was also observed in this state (Yoshida et al. 1993). Thus, such a transition between a soft thermal state and a hard power-law state is considered to be a fundamental property of accretion phenomenon regardless of whether the compact object is a black hole or a neutron star. The available data suggest that transitions occur across a mass accretion rate around 10^{17} g s^{-1}, but this value might vary from source to source and even one transiton to another (see Tanaka & Shibazaki 1996, and references therein).

Table 3. Photon indices of the power-law component

Source name	Single power-law state	Ultrasoft+power-law state
Black-hole XB		
Cyg X-1	1.5−1.7	
GS2023+33	1.4−1.7	
G2000+25	∼ 1.6	2.0−2.3
GS1124−68	1.6−1.7	∼ 2.5
GX339−4	∼ 1.7	∼ 2.5
LMC X-3		∼ 2.2
GS1354−64		∼ 2.3
Neutron-star LMXB		
1608−52	∼ 1.8	
Aql X-1	∼ 1.8	

The observed photon indices in the hard power-law state are in the range 1.6−1.9 for both neutron-star LMXBs and black-hole XBs as listed in Table 3. The power-law spectrum in this state is substantially harder than the hard tail of black-hole XBs when an ultrasoft component is present (see section 2.2). The photon index in this state also remains remarkably constant against large changes in luminosity.

Transition from a soft state to a hard state has been considered as due to a change in the disk structure. There is clear evidence that a sudden enhancement of an optically-thin hot plasma (with an electron temperature kT_e of the order of several tens keV or higher) occurs in the accretion disk when a source goes into a hard state. For instance, when X1608-52 went into a power-law state, the spectrum of X-ray bursts also showed a significant hard tail that had not existed before the spectral transition (Nakamura et al. 1989). This feature is shown in Fig. 11. Since X-rays of bursts are intrinsically a blackbody emission from an optically-thick neutron star atmosphere, the hard tail implies that these blackbody photons were Comptonized in passing through a hot plasma. This clearly indicates new appearance of a hot plasma associated with a transition into the power-law state. Yet, the mechanism that causes such a sudden change in the disk structure still remains to be understood.

It is important to note that, despite a drastic change in the spectral shape, the transtion between the two spectral states does not seem to cause a big jump in the total luminosity (if the power-law component is integrated to high enough

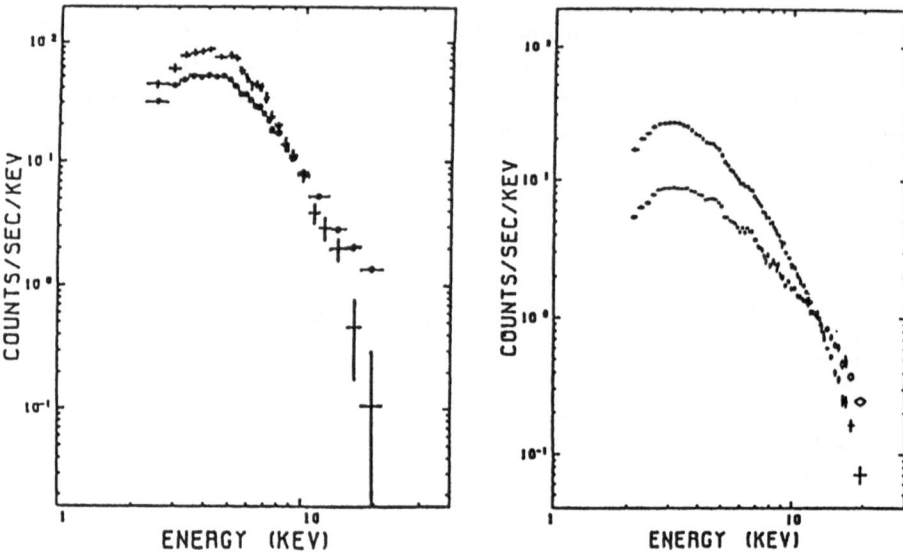

Fig. 11. Observed X-ray spectra (count-rate spectra, not corrected for the instrument response) of 1608−52 for the X-ray bursts (left panel) and the persistent emission (right panel), before a spectral transition (softer ones) and after the transition (harder ones) (Nakamura et al. 1989).

energies) before and after transition. Hence, the efficiency of radiation ($L_x/\dot{M}$) remains essentially unchanged by the change in the disk structure. Zhang et al. (1996) show that, in the recent transitions (hard-to-soft & soft-to-hard) of Cyg X-1, the increase was less than 15% in L_x in the range 1.3−200 keV and less than 70% in the bolometric luminosity in the soft state.

There is evidence that the outer disk remains cool and optically thick against Thomson scattering in the hard power-law state. For instance, Cyg X-1 in the hard state shows a 6.4-keV fluorescnce line and a significant K-absorption edge of iron near 7.1 keV, indicating that the ionization states are lower than Fe XVII (Ebisawa et al. 1996). Similar feature has been observed commonly from the sources in the hard power-law state, and is interpreted as due to a significant contribution of the X-rays reflected by Thomson scattering in a cool disk (see e.g. Tanaka 1991).

A cut-off in the power-law spectrum (an exponential steepening above a certain energy) is often observed (e.g. Sunyaev & Trümper 1979; Sunyaev et al. 1994; Harmon et al. 1994). The cut-off energy (the characteristic energy of the exponential spectrum) of various sources range from several tens keV to more than 100 keV. There is an indication that the cut-off shifts towards lower energies when the luminosity increases (Sunyaev et al. 1994; Inoue 1994), which can be interpreted as due to enhanced Compton cooling of electrons. Such a power law spectrum with a cut-off is well reproduced by Comptonization of soft photons (Sunyaev & Titarchuk 1980). In the Comptonization model, the photon index and the cut-off energy are determined by the electron temperature kT_e and the Thomson optical depth τ, or the Comptonization y-parameter: $y = (4kT_e/m_ec^2)\mathrm{Max}(\tau, \tau^2)$. Yet, the exact reasons for the constancy of the photon index against large intensity changes and for the particular photon-index values before and after a transition remain to be explained.

It is worth noting that the properties of X-ray binaries in the power-law state are strikingly similar to those of AGNs in the following points:
(1) AGNs commonly show power law spectra with the photon indices distributed within a fiarly small renge around 1.7, very similar to the values for X-ray binaries in the hard power-law state.
(2) The photon index remains constant against large changes in luminosity in both systems.
(3) A high-energy cut-off in the power-law spectrum around several tens keV is also observed commonly in Seyfert galaxies (Zdziarski et al. 1995).
(4) Both systems exhibit high time-variabilities (flickering) on various time scales down to the shortest Keplerian periods.

These similarities suggest that the basic process of accretion is essentially the same in both systems, despite huge differences in the scale and power.

3.2 X-ray spectrum in quiescence

A drastic change in the accretion flow seems to occur when the accretion rate further decreases below 10^{16} g s^{-1}(corresponding to a luminosity of $\sim 10^{36}$ erg sec^{-1}) as discussed below. The light curves of the transient black-hole LMXBs A0620−00, GS2000+25 and GS/GRS1124−68 show a fairly smooth exponential decay as shown in Fig. 12, except for a secondary and a tertiary humps (see Tanaka & Shibazaki 1995 for references). Another transient black-hole LMXB GRO J0422+32 also shows a similar exponential decay (Harmon et al. 1994). However, in the final stage of the decay, they all decline rather abruptly from the level around 10^{36} erg sec^{-1}, as noticed in Fig. 12. Similar behavior of quick turn-off around 10^{36} erg sec^{-1}was also observed from Cen X-4 and 1608-522 (see Tanaka & Shibazaki 1996 for references). This feature has been considered to indicate that the accretion flow is choked when the accretion rate falls below $\sim 10^{16}$ g s^{-1}.

On the other hand, optical observations of the transient sources after returning to the quiescent state invariably show a significant emission from the disk, indicating that mass transfer from the companion still continues. McClintock et al. (1995) estimate the mass transfer rate onto the outer disk to be of the order of 10^{16} g s^{-1}for A0620-00 even at an X-ray luminosity of $\sim 10^{31}$ erg sec^{-1}. If the X-rays come from an optically-thick inner disk, this X-ray luminosity corresponds to an accretion rate of $\sim 10^{11}$ g s^{-1}. This result suggests that very little of the transferred mass goes onto the compact object and that matter will steadily accumulate in the disk. This is in support of a disk instability model for transient X-ray outbursts (e.g. Osaki 1974; Meyer & Meyer-Hofmeister 1981).

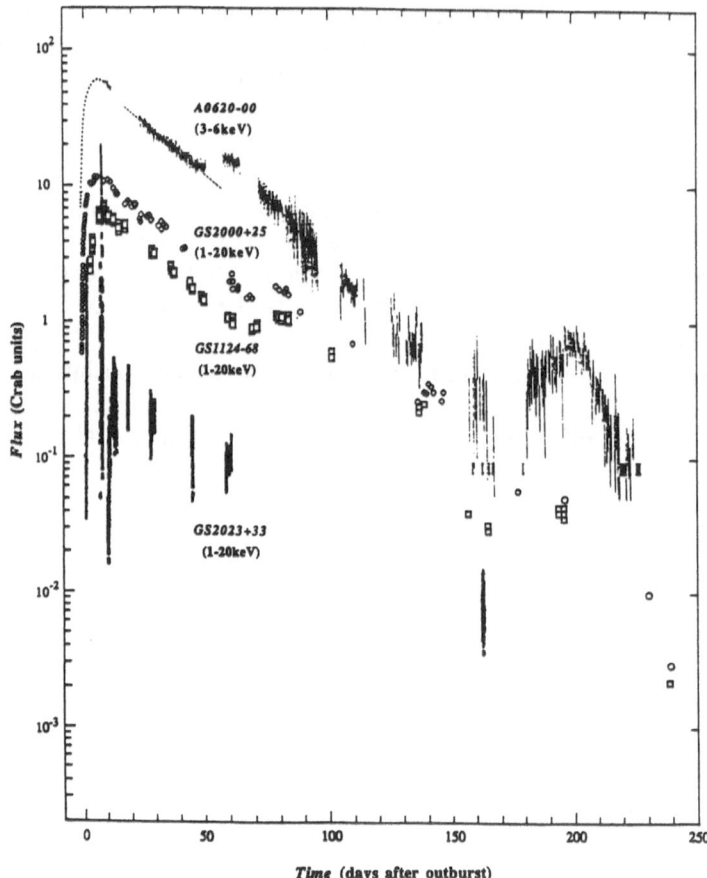

Fig. 12. X-ray light curves of four bright transient black-hole X-ray binaries, A0620−00, GS2000+25, GS2023+33 and GS/GRS1124−68. The observed fluxes are shown in units of the Crab Nebula intensity in the energy band indicated for each source. (see Tanaka & Shibazaki 1996 for references.)

On the other hand, Narayan et al. (1996a) propose an advection-dominated disk model for low accretion rates. In this model, the inner part of the disk becomes optically thin, so that the thermal energy released by viscous heating is mostly advected into the compact object. In such a disk, the efficiency of radiation is expected to be extremely small ($\sim 10^{-3} - 10^{-4}$ of the energy released). This model can also explain an abrupt drop of X-ray luminosity below a certain accretion rate as due to a transition from a radiation-dominated disk into an advection-dominated disk. They show that this model can reproduce the observed spectrum of the quiescent black-hole systems in the optical, UV and X-ray bands. However, this advection-dominated disk is against a disk instability model for transient outbursts, since matter will not accumulate in the disk.

Table 4. Results of X-ray observations of transient sources in quiescence

Source name	Class[a]	Observed range (keV)	Luminosity (10^{32} erg sec^{-1})	kT_b[b] (keV)	Ref.
Cen X-4	NS	0.5−4.5	4−8		[1]
		0.5−10	2.4	0.2	[2]
				+ hard tail	
Aql X-1	NS	0.2−2.4	4.4	0.3	[3]
1608−522	NS	0.5−10	6	0.3	[2]
0620−003	BH	0.2−2.4	0.06	0.16	[4]
2023+338	BH	0.2−2.4	80	0.2	[5]
		0.5−10	12	power law (not blackbody) photon index ~ 1.9	[6,7]
2000+251	BH	0.2−2.4	< 0.1		[3]
		0.5−10	< 0.1		[7]
1124−684	BH	0.5−10	< 0.1		[7]
GX339−4	BH	0.5−10	< 0.1		[7]

[a] NS: neutron-star LMXBs. BH: black-hole LMXBs.
[b] for a blackbody spectrum.

Ref.: 1. van Paradijs et al. (1987)   5. Wagner et al. (1994)
    2. Asai et al. (1996)   6. Narayan et al. (1996b)
    3. Verbunt et al. (1994)   7. ASCA results, unpublished
    4. McClintock et al. (1995)

With the recent high-sensitivity X-ray telescopes, in particular ROSAT and ASCA, X-rays from several transient sources after returning to the quiescent state have been detected in the luminosity range from 10^{33} erg sec^{-1} to even lower than 10^{31} erg sec^{-1}. The lowest of the positive detections is A0620-00 at 6×10^{30} erg sec^{-1} (McClintock et al. 1995). For some others, upper limits have been obtained at a level of 10^{31} erg sec^{-1}. The observed results are listed in Table 4.

The X-ray spectra in these low luminosity levels are commonly found to be very soft, except that GS2023+33 showed a hard spectrum in one occasion (see Table 4). If the spectrum is fitted with a blackbody spectrum, kT_b is around $0.2-0.3$ keV for all, much softer than a power-law spectrum observed when L_x is in the range 10^{36-37} erg sec^{-1}. If it is indeed blackbody emission, the emitting area turns out to be only $1-10$ km^2 (Verbunt et al. 1994; McClintock et al. 1995; Asai et al. 1996), much smaller than the area of the inner accretion disk. Note, however, that the precise forms of the spectra are still quite uncertain. A steep power law or a bremsstrahlung spectrum can slso give an acceptable fit. Figure 13 shows examples of X-ray spectra of transient sources (Cen X-4 & 1608−52) in quiescence observed with ASCA (Asai et al. 1996). Note that Cen X-4 has a significant hard tail. Whether the change into such a soft spectrum occurs gradually or abruptly (like a transition) with respect to the change in L_x is not yet known.

At present, where these soft X-rays come from is yet unclear. The observed

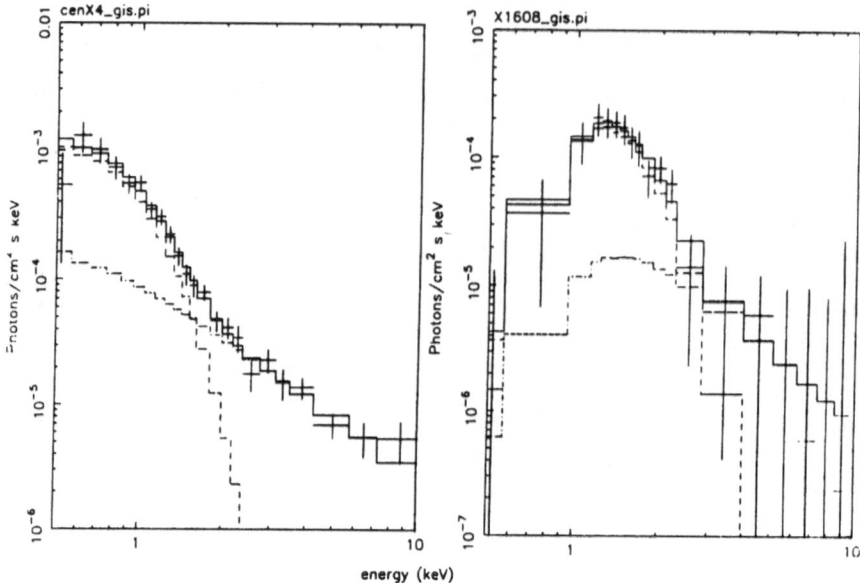

energy (keV)

Fig. 13. X-ray photon spectra of Cen X-4 (left panel) and 1608−52 (right panel) during the quiescent state (Asai et al. 1996).

results in the optical and X-ray bands show close similarities between neutron-star LMXBs and black-hole LMXBs in quiescnce. This would argue for a common origin of the observed soft X-rays for both systems.

An imortant question to be resolved is whether the accretion flow is choked or instead advectively goes into the compact object. Suppose the accretion flow is dominated by advection, it is certain that the mass flow at a rate inferred from the optical disk luminosity ($\sim 10^{16}$ g s^{-1}) does not reach the neutron star. Otherwise, the neutron star would emit X-rays at a level of $\sim 10^{36}$ erg sec^{-1}, absolutely in contradiction with the observed luminosity. It is possible that the accretion flow is stopped at some distance from the neutron star, expelled by the magnetic fields of a rapidly-spinning neutron star (Illarionov & Sunyaev 1975). On the other hand, if that were the case, a significant difference in the spectral shape would be expected bewteen neutron-star systems and black-hole systems, e.g. the characteristic temperature of the spectrum for neutron-star systems would be lower than that for black-hole systems because of a systematically larger disk radius ($\gg 3r_g$) at which the accretion flow is intercepted, hence a less amount of graviational energy available. However, the observed results so far obtained do not show such a systematic difference between the two systems.

Mass accretion at such an extremely low luminosity is an entirely new subject. Obviuosly, more detailed observations are required to understand the disk structure and the origin of the soft X-rays from the sources in quiescence. These problems are not only crucial for the physics of accretion but are also related to various other important issues (see discussions in Tanaka & Shibazaki 1996).

Acknowledgement: This work has been done while the author stayed at the Astronomical Institute, University of Utrecht. He is grateful for the support of the University and for the kind hospitality of the staff of the Institute.

References

Asai K., Dotani T., Mitsuda K., Hoshi R., Vaughan B. et al., 1996, Publ. Astron. Soc. Japan 48, 257

Bailyn C.D., Orosz J.A., Mcclintock J.E, Remillard R.A., 1995, Nature 378, 157

Casares J., Charles P.A., Naylor T., 1992, Nature 355, 614

Coe M. J., Engel A.R., Quenby J.J., 1976, Nature 259, 544

Cowley A.P., Crampton D., Hutchings J.B., Remillard R., Penfold J.E., 1983, Ap. J. 272, 118

Ebisawa K., Mitsuda K., Hanawa T., 1991, Ap. J. 367, 213

Ebisawa K., 1991, PhD. Thesis, Univ. of Tokyo

Ebisawa K., Ogawa M., Aoki T., Dotani T., Takizawa M. et al., 1994, Publ. Astron. Soc. Japan 46, 375

Ebisawa K.,   et al., 1996, Ap. J. 467, 419.

Filippenko A.V., Matheson T., Ho L.C., 1995a, Ap. J. 455, 614

Filippenko A.V., Matheson T., Barth A.J, 1995b, Ap. J. Lett. 455, L139

Gies D.R., Bolton C.T., 1982, Ap. J. 260, 240

Harmon B.A., Zhang S.N., Wilson C.A., Rubin B.C., Fishman G.J., 1994, *The Second Compton Symposium*, C.E. Fichtel, G. Gehrels, & J.P. Norris (eds.), p. 210. AIP

Hutchings J.B., Crampton D., Cowley A.P., Bianchi L., Thompson I.B., 1987, Astron. J. 94, 340

Illarionov A.F., Sunyaev R.A., 1975, Astron. Astrophys. 39, 185

Inoue H., 1994, *Multi-Wavelength Continuum Emission of AGN, Proc. IAU Symp. 159, Geneva*, T.J.-L. Courvoisier & A. Blecha (eds.), p.73. Kluwer

King A.R., 1996, these Proceedings

McClintock J.E., Remillard R.A., 1986, Ap. J. 308, 110

McClintock J.E., Bailyn C.D., Remillard R.A., 1992, IAU Circ. No. 5499

McClintock J.E., Horne K., Remillard R.A., 1995, Ap. J. 442, 358

Meyer F., Meyer-Hofmeister E., 1981, Astron. Astrophys. 104, L10

Mitsuda K., Inoue H., Koyama K., Makishima K., Matsuoka M. et al., 1984, Publ. Astron. Soc. Japan 36, 741

Mitsuda K., Inoue H., Nakamura N., Tanaka Y., 1989, Publ. Astron. Soc. Japan 41, 97

Nakamura N., Dotani T., Inoue H., Mitsuda K., Tanaka Y. et al., 1989, Publ. Astron. Soc. Japan 41, 617

Narayan R., McClintock J.E., Yi I., 1996a, Ap. J. 457, 821

Narayan R., Barret D., McClintock J.E., 1996b, preprint. To appear in Ap. J.

Osaki Y., 1974, Publ. Astron. Soc. Japan 26, 429

Remillard R.A., Orosz J.A., McClintock J.E., Bailyn C.D., 1996, Ap. J. 459, 226

Shakura N.I., Sunyaev R.A., 1973, Astron. Astrophys. 24, 337

Sunyaev R.A., Trümper J., Nature 279, 506

Sunyaev R.A., Titarchuk L.G., 1980, Astron. Astrophys. 86, 121

Sunyaev R., Churazov E, Gilfanov M., Vikhlinin A., Markevich M. et al., 1994, *New Horizon of X-Ray Astronomy*, F. Makino, T. Ohashi (eds.), p211. Universal Academy

Takizawa M., 1991, M.Sc. Thesis, Univ. Tokyo

Tanaka Y., 1991, *Iron Line Diagnostics in X-Ray Sources*, A. Treves, G.C. Perola & L. Stella (eds.), Lecture Notes in Physics 385, 98. Springer Berlin

Tanaka Y., 1992, *Ginga Memorial Symposium, ISAS*, F. Makino & F. Nagase (eds.), p.19

Tanaka Y., Lewin W.H.G., 1995, *X-Ray Binaries*, W.H.G. Lewin, J. van Paradijs, E.P.J. van den Heuvel (eds.), p.126. Cambridge University Press

Tanaka Y., Shibazaki N., 1996, Ann. Rev. Astron. Astrophys. 34, 607

van Paradijs J., Verbunt F., Shafer R.A., Arnaud K.A., 1987, Astron. Astrophys. 182, 47

van Paradijs J., 1995, *X-Ray Binaries*, W.H.G. Lewin, J. van Paradijs, E.P.J. van den Heuvel (eds.), p.536. Cambridge University Press

Verbunt F., Belloni T., Johnston H.M., van der Klis M., Lewin W.H.G., 1994, Astron. Astrophys. 285, 903

Wagner R.M., Starrfield S.G., Hjellming R.M., Howell S.B., Kreidl T.J., 1994, Ap. J. Lett. 429, L25

White N.E., Kaluzienski L.J., Swank J.L., 1984, *High-Energy Transients in Astrophysics*, S. Woosley (ed), p.31. AIP

Yoshida K., Mitsuda K., Ebisawa K., Ueda Y., Fujimoto R. et al., 1993, Publ. Astron. Soc. Japan 45, 605

Zdziarski A.A., Johnson W.N., Done C., et al., 1995, Ap. J. 438, L63

Zhang S.N., Cui W., Harmon B.A., Paciesas W.S. Remillard R.E. et al., 1996, Ap. J. in press

Black holes in X-ray binaries

J. van Paradijs[1,2]

[1] Astronomical Institute 'Anton Pannekoek', University of Amsterdam & Center for
High-Energy Astrophysics
[2] Department of Physics, University of Alabama in Huntsville

Abstract: In this paper I provide a brief review of the following aspects of
research in black-hole X-ray binaries: X-ray spectra, mass determinations, and
source states. In addition, I discuss several recent developments resulting from
observations of the transient GRS 1915+105.

1 Introduction

Cyg X-1 was the first X-ray source which was shown to be a member of a binary star. This discovery (Webster & Murdin 1972; Bolton 1972) followed the
determination of an accurate ($\sim 1'$) error box by Rappaport et al. (1971), which
contained a radio source (Braes & Miley 1971; Hjellming & Wade 1972) coinci-
dent with the 8th magnitude known supergiant HD 226868. Optical spectroscopy
of this star revealed a 5.6 day periodic radial-velocity variation with an amplitude $K_{opt} = 64$ km/s, and a corresponding mass function $f_{opt}(M) = 0.25$ M$_\odot$.
Under the assumption that the supergiant has a 'normal' mass of $\gtrsim 15$ M$_\odot$ these
results led to the conclusion that the mass of the compact star in Cyg X-1 was
higher than 3 M$_\odot$, which exceeds the maximum possible mass of a neutron star.

The identification of the radio source with Cyg X-1 was confirmed when its
brightness showed a large increase correlated with a major hardening of the
2-10 keV spectrum of Cyg X-1 (Tananbaum et al. 1972). We now know that
this spectral hardening is caused by the disappearance of an 'ultra-soft' spectral
component in the X-ray spectrum, signalling a transition from a 'high' (or 'soft')
state to a 'low' (or 'hard') state (see Sect. 4).

Following the discovery of the binary X-ray pulsar Cen X-3 (Schreier et al.
1972) and many other similar systems, and of X-ray bursters (Grindlay et al.
1976; Belian et al. 1976), research in X-ray binaries in the 1970's was dominated
by systems in which the accreting compact object is a neutron star.

The discovery of strong rapid variability of the X-ray flux of Cyg X-1 (Oda
et al. 1971; see also Oda 1976, for a review of early work on Cyg X-1) led to
the idea that such variability is a tell-tale sign of an accreting black hole, which
might be used to distinguish them from accreting neutron stars. On the basis of
this idea Cir X-1 was long considered a black-hole candidate. However, neutron
star X-ray binaries can also show rapid variability, as was strikingly illustrated
by the transient V 0332+53, which was initially suggested as a possible black-
hole system, but later shown to be an X-ray pulsar (Stella et al. 1985), whose
pulse amplitude happened to be relatively weak compared to that of the red

noise component in the power density spectrum (PDS). Also Cir X-1 was shown to be a neutron star when it emitted type I X-ray bursts (Tennant et al. 1986).

Tananbaum (1973) distinguished two groups of galactic X-ray sources on the basis of their 2-10 keV X-ray spectra, i.e., sources with relatively hard X-ray spectra, several of which were binary X-ray pulsars, and sources with relatively soft X-ray spectra, similar to Sco X-1, none of which were then known to be binary systems. We now know that these two groups represent accreting neutron stars with strong ($B \sim 10^{12}$ G) and weak ($B \lesssim 10^{10}$ G) magnetic fields, respectively (see various chapters in Lewin, Van Paradijs & Van den Heuvel 1995, for reviews of these objects).

In 1977 Ostriker extended this result by suggesting that black-hole X-ray binaries might be distinguished by the shape of their X-ray spectra. This idea was put on a firm footing by White & Marshall (1984) who showed that in an X-ray colour-colour diagram, derived from the HEAO-1 A-2 sky survey the two sources, then known to contain black holes (Cyg X-1, in its 'soft' state, and LMC X-3) occupied the extreme upper-left corner, together with several other sources, some of which were transient. A few years later McClintock & Remillard (1986) found that the mass function of the transient A 0620–00, obtained from the radial-velocity curve of the low-mass secondary star (which became detectable in quiescence, after the X-ray luminosity had decreased by a very large factor) equals 3.18 ± 0.16 M$_\odot$. This immediately showed that the compact star in this system is too massive to be a neutron star, and gave confidence in the idea that X-ray spectra are an efficient way to select X-ray binaries with black holes.

Currently, ten X-ray binaries are known to contain black holes on the basis of a dynamical mass determination; seven of these are transient low-mass X-ray binaries. Another 17 systems are suspected to be black hole X-ray binaries on the basis of their X-ray spectra (see Tables 1 and 2, from White & Van Paradijs 1996). According to a recent analysis the total number of transient black-hole X-ray binaries in the Galaxy is ~ 500 (White & Van Paradijs 1996).

In Sections 2 and 3 I will discuss the X-ray spectra of black-hole X-ray binaries, and their mass determinations, respectively. In Section 4 I discuss the concept of source states. Several recent results are presented in Section 5.

2   X-ray spectra of black-hole X-ray binaries

Fig. 1 shows typical examples of the X-ray spectra of accreting neutron stars with low and high magnetic fields, respectively, and of an accreting black hole. The spectra of weak-field neutron stars (e.g., Sco X-1, X-ray burst sources) can be approximately described by a thermal-bremsstrahlung model, with $kT_{\rm TB} \sim$ 5 keV. The spectra of strong-field neutron stars (i.e., X-ray pulsars) are power laws (photon indices ~ 1) with a high-energy cut off at several tens of keV.

In the X-ray spectra of accreting black holes one finds two components, whose relative strengths can vary by a large factor. One is a power law, with photon indices in the range ~ 1.5 to ~ 2.5, which dominates the high-energy part

($\gtrsim 10$ keV) of the spectrum, and is occasionally detected at energies of several hundreds of keV. The other component is limited to photon energies below 10 keV, and is called the 'ultra-soft' component. It is roughly described by a Planck function, with $kT_{bb} < 1$ keV. For extensive reviews of the X-ray spectra of black-hole X-ray binaries I refer to Gilfanov et al. (1995) and Tanaka & Lewin (1995).

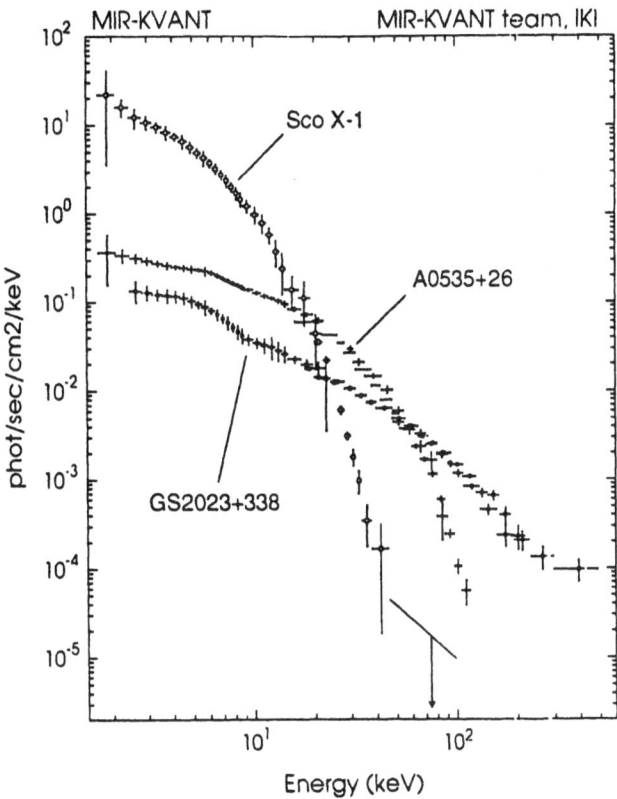

Fig. 1. Spectra obtained with MIR-KVANT of three types of galactic X-ray bina-ries with different compact objects: Sco X-1 (weak-field neutron star), A 0535+26 (strong-field neutron star), and GS 2023+338 (a black hole). (From Gilfanov et al. 1995).

The ultra-soft component is generally interpreted as the emission from an optically thick, geometrically thin, accretion disk. For a standard accretion disk the temperature distribution $T(r)$ is given by

$$T^4(r) = 3GM_X\dot{M}_d/8\pi\sigma r^3 \tag{1}$$

where r is the radial distance from the center, M_X is the mass of the accreting object, and $\dot{M}_d$ is the mass transfer rate through the optically thick disk. Note

24   J. van Paradijs

that $\dot{M}_d$ is not necessarily equal to the total mass accretion rate, as part of the flow may pass through a geometrically thick very hot advective flow (Rees et al. 1982; Narayan 1996), or leave the system after having passed through the disk (e.g., in a jet ejected from the near vicinity of the compact star).

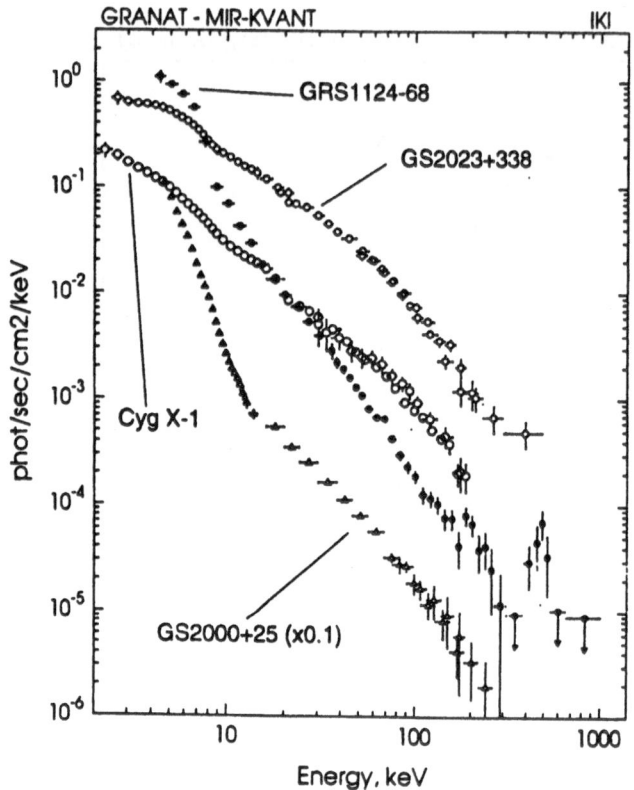

Fig. 2. X-ray spectra of several black-hole X-ray binaries, showing various combinations of ultra-soft and power law components (from Gilfanov et al. 1995).

Mitsuda et al. (1984) assumed that the local emission from the disk is Planckian, and derived the following expression ('multi-temperature disk blackbody') for the flux observed from the disk:

$$f(E) = \frac{8\pi R_{in}^2 \cos i \, T_{in}^{8/3}}{3d^2} \int_{T_{out}}^{T_{in}} T^{-11/3} \, B(E,T) \, dT \tag{2}$$

here i is the inclination angle of the disk, R_{in} is the inner radius of the disk, d is the source distance, $B(E,T)$ is the Planck function, and T_{in} and T_{out} are the disk temperatures at the inner and outer disk radii. Note that the disk is assumed to be perfectly flat; therefore, at very high inclination angles the model may not be applicable, e.g., because of self-occultation of the disk. In this model, if general-relativistic effects are ignored, the total disk luminosity L_d is given by (Makishima et al. 1986) $L_d = 4\pi R_{in}^2 \sigma T_{in}^4$, i.e., formally the expression is the same as for a spherical uniform blackbody emitter with radius R_{in} and temperature T_{in}, although all disk emission originates outside R_{in}.

The parameter T_{in} is determined from the shape of the observed spectrum; R_{in} is a factor in the normalization of the fit, and can be obtained if the distance and inclination angle are known. Together, R_{in} and T_{in} determine the mass flow rate through the disk, according to

$$\dot{M}_d = 8\pi\sigma R_{in}^3 T_{in}^4 / (3GM_X) \tag{3}$$

Usually, it is assumed that the inner disk radius is located at three times the Schwarzschild radius, inside of which stable orbits around a non-rotating black hole do not exist, i.e., $R_{in} = 6GM_X/c^2$. However, around a maximally spinning Kerr black hole the inner disk radius can be as small as the Schwarzschild radius.

Spectral fits with this model have been made to X-ray spectra obtained with *Ginga* throughout the outbursts of several BHXT. Remarkably, R_{in} remained constant, while the disk luminosity changed by more than an order of magnitude (see Tanaka & Lewin 1995). This has given some confidence in the applicability of the model. The values of R_{in} obtained from the fits are consistent with three Schwarzschild radii for stellar-mass black holes.

The power law component in the X-ray spectra of accreting black holes (both in X-ray binaries and in active galactic nuclei) is generally interpreted as the result of Compton up-scattering of low-energy photons in a very hot medium, generally associated with a disk corona, or a geometrically thick inner disk. Approximating this spectral component as a power law with an exponential cut off at high energies, the photon index, Γ, of the power law is given by $\Gamma \approx -1/2 + \sqrt{9/4 + \pi^2/3y}$, where y is the Compton parameter $y = 4kT\tau^2/m_ec^2$ (T and τ are the temperature and the scattering optical depth of the hot electron gas).

The nature of the very hot electron gas is not immediately obvious. The heating mechanism may be related to magnetic processes on the disk surface, analogous to coronal heating in late-type stars. The hot scattering medium may be a by-product of an advective flow in which the ion temperature is of order the virial temperature (Rees et al. 1982; Narayan 1996); the electron temperature is then determined by a balance between heating due to Coulomb interactions with the ions, and cooling due to upscattering of low-energy photons (see the contribution by H. Spruit to this Volume). It has been suggested that scattering may occur on a converging bulk flow in the near vicinity of a black hole (Blandford & Payne 1981; Payne & Blandford 1981; Chakrabarthy & Titarchuk 1995); the expected photon index in this case is ~ 2.5.

3   Mass determinations

In determining the mass of an X-ray source, using Newtonian effects only, the fundamental quantity is the mass function $f_{opt}(M)$ which is determined from the orbital period, P_{orb}, and the amplitude, K_{opt}, of the radial-velocity variations of the mass donor by

$$f_{opt}(M) \equiv M_X^3 \sin^3 i/(M_X + M_2)^2 = \frac{K_{opt}^3 P_{orb}}{2\pi G} \qquad (4)$$

The corresponding quantity $f_X(M)$ can be determined for binary X-ray pulsars:

$$f_X(M) \equiv M_2^3 \sin^3 i/(M_X + M_2)^2 = \frac{4\pi^2 (a_X \sin i)^3}{G P_{orb}^2} \qquad (5)$$

(The connection to observational parameters is written differently, since in the case of X-ray pulsars the observed quantities are usually pulse arrival times, whereas from optical spectra one measures radial velocities.)

If both mass functions can be measured, their ratio immediately gives the mass ratio $q \equiv M_X/M_2 = f_{opt}(M)/f_X(M)$, and both masses are then determined separately, by up to a factor $\sin^3 i$. The orbital inclination can be estimated from the duration of an X-ray eclipse (if it occurs), and from the amplitude of an ellipsoidal optical light curve, which reflects the tidal and rotational distortion of the (nearly or fully) Roche-lobe filling secondary star. The latter approach, of course, requires that the light of the secondary dominates the optical emission from the binary (see Van Paradijs 1990, for a review of optical light curves of X-ray binaries).

In the case of black-hole X-ray binaries $f_X(M)$ cannot be determined, but nonetheless important information can be inferred from $f_{opt}(M)$, in conjunction with other observables. In the first place, $f_{opt}(M)$ provides a firm lower limit to M_X, since $\sin i$ cannot exceed unity and $M_2 > 0$. The observed mass functions of BHXT are, with one exception, larger than ~ 3 M$_\odot$ (see Table 1), and this has provided the strongest evidence yet for the existence of black holes with masses in the stellar range.

For systems with a high-mass companion (Cyg X-1, LMC X-1 and LMC X-3) the mass functions by themselves do not show that the compact object is a black hole. In these cases one has to either assume that the companion is not extremely undermassive for its luminosity (see, e.g., Gies and Bolton 1986, for a discussion of this issue), or use additional observational constraints on the system. Examples of such observational constraints are the following (see Van Paradijs 1983; Gies and Bolton 1986; Charles 1996, for reviews):

(i) The amplitude of an ellipsoidal optical light curve, if necessary corrected for the effect of an accretion disk, provide a relation between q, i and a Roche lobe (volume) filling factor f. (ii) The duration (or absence) of eclipses provides another such relation. (iii) The apparent magnitude and spectral type (or colour index) of the secondary star determine the ratio R_2/d of secondary radius R_2

to distance d. If the source distance, and therefore R_2, is known this constrains the relation between M_X and M_{opt}. (iv) Since both K_{opt} and the observed rotational velocity $V_{rot} \sin i$ (measurable from broadening of spectral lines in the secondary's spectrum) are projected by the same factor $\sin i$, their ratio is given by $V_{rot} \sin i / K_{opt} = (q^{-1} + 1) f^{1/3} r_2(q)$, where r_2 is the ratio of the (volume averaged) radius of the secondary. Useful expressions for r_2 have been given by Paczynski (1971) and Eggleton (1983).

A summary of the mass determinations of black holes in X-ray binaries is given in Table 1 (from White & Van Paradijs 1996).

4   Source states

During the last decade our understanding of accreting weak-field neutron stars has much improved by the introduction of the concept of 'source states', defined by both temporal and spectral properties of these X-ray sources; this concept has also proven fruitful in studying accreting black holes (see Van der Klis 1994, 1995 for recent reviews, and Lewin et al. 1988 for early developments).

Weak-field neutron stars are found in low-mass X-ray binaries, and we now recognize two groups of such systems, which are called Z sources and atoll sources after the shapes of source loci in X-ray colour-colour diagrams (CD; Hasinger & Van der Klis 1989; see Fig. 3). The Z sources have X-ray luminosities close to the Eddington limit (L_{Edd}), and the neutron stars in these systems have surface dipolar fields of order 10^{10} G. They show three source states related to each of the three branches of the Z-shaped track in the X-ray CD, i.e., the horizontal branch, the normal branch and the flaring branch; the mass accretion rate increases along the Z track in this order. The atoll sources have systematically much lower X-ray luminosities (mainly between $\sim 10^{-2}$ and 0.5 L_{Edd}), and $B \lesssim 10^9$ G (Hasinger & Van der Klis 1989). Two source states have been recognized in the atoll sources, the 'island state' and 'banana state' (see Fig. 3).

The banana state is characterized by a slightly curved track in the CD along which the sources move on a time scale of hours. In the upper part of the banana branch their PDS show only a power law shaped very-low frequency component. In the lower part of the banana, which connects to the island in the CD, a high-frequency component appears in addition to the VLFN. In the island state the X-ray spectral shape doesnt't change very much on time scales of order a day. In the PDS the high-frequency component is very strong, roughly shaped as a power law with a low-frequency cut off. The island state PDS of several atoll sources are very similar to the PDS observed for black-hole X-ray binaries in the low state (see below). Between the island state and the banana state the mass accretion rate increases (Hasinger & Van der Klis 1989).

Also in the case of black holes the simultaneous analysis of the spectra and the fast variability has led to the distinction of source states, as follows (see Fig. 4). In the 'low state' (LS) the ultra-soft component is absent, so the X-ray spectrum shows only the hard power law component. The PDS then shows a strong broad-band noise component, which at high frequencies is a power law with slope ~ -2 with a variable low-frequency cut off ν_{co}. Belloni & Hasinger (1990) found that as ν_{co} changes the high-frequency part of the PDS remains the same; thus the lower ν_{co} is, the higher the fractional variations in the X-ray flux. From the long-term monitoring with BATSE Crary et al. (1996) found that the r.m.s. fractional hard X-ray variability of Cyg X-1 is correlated with the photon index of the power law component in the X-ray spectrum.

In the high state (HS) the X-ray spectrum shows the ultra-soft component, which in some cases completely dominates the emission. The amplitude of X-ray flux variations is then much smaller than in the low state, in the PDS they are represented by a weak power law.

TABLE 1

BLACK HOLE BINARY CANDIDATES FROM RADIAL VELOCITY MEASUREMENTS

Name	ID	Date (yr)	Transient	Companion	P_{orb} (days)	$f(M)$ ($M_\odot$)	M_X ($M_\odot$)	Reference
HMXRB:								
Cyg X-1..................	HD226868	1972	No	O9.7 Iab	5.6	0.25	≥ 7	1, 2
LMC X-3.................		1982	No	B3V	1.7	2.3	≥ 7	3
LMC X-1.................		1983	No	O7–9 III(?)	4.2	0.12	≥ 2.6	4
LMXRB:								
A 0620−00	V616 Mon	1986	Yes	K5 V	0.32	2.91	≥ 2.9	5
GS 2023+338	V404 Cyg	1992	Yes	G9 V–K0 III	6.47	6.26	≥ 6.1	6
GS/GRS 1124−68	XN Mus 92	1992	Yes	K0 V–K4 V	0.43	3.1	≥ 2.9	7
GRO J0422+32..........	V518 Per	1995	Yes	M2	0.21	1.2	≥ 2.6	8, 9
GRO J1655−40	XN Sco 95	1995	Yes	F3–F6	2.60	3.16	≥ 3.2	10
GS 2000+25..........	QZ Vul	1995	Yes	K5 V	0.35	5.0	≥ 6.4	11, 12
H1705−25................	XN Oph 77	1996	Yes	K3	0.52 (0.70)	4.0	≥ 3.4	13, 14

REFERENCES.— 1. Webster & Murdin 1972; 2. Bolton 1972; 3. Cowley et al. 1983; 4. Hutchings et al. 1987; 5. McClintock & Remillard 1986; 6. Casares et. al. 1992; 7. McClintock, Bailyn, &Remillard 1992; 8. Filippenko, Matheson, & Ho 1995; 9. Casares et. al. 1995b; 10. Bailyn et al. 1995; 11. Casares, Charles, & Marsh 1995a; 12. Filippenko, Matheson, & Barth 1995; 13. Remillard et al. 1996; 14. Martin et al. 1995.

TABLE 2

BLACK HOLE CANDIDATES IN LOW-MASS X-RAY BINARIES

Name	Type	Signature	l^{II}	b^{II}	d (kpc)	z (pc)	d Reference
GRO J0422+32 (V518 Per).................	Transient	D/UH	165.9	−11.9	2.5	−516	1
A 0620−00 (V616 Mon)....................	Transient	D/US	210.0	−6.5	0.8	−91	2
GRS 1009−45 (XN Vel 93).................	Transient	UH	275.9	9.3	...	...	
GRS 1124−68 (GU Mus)	Transient	D/US/UH	295.3	−7.1	3.9	−482	3
GS 1354−645 (Cen X-2)	Transient	US/UH	310.0	−2.8	...	...	
A 1524−62 (TrA X-1)......................	Transient	US/UH	320.3	−4.4	...	...	
4U 1543−47.................................	Transient	US/UH	330.9	5.4	4	376	4
4U 1630−47 (Nor X-1).....................	Transient	US/UH	336.9	0.3	...	...	
GRO J1655−40 (XN Sco 94)	Transient	D/US/UH/J	345.0	2.5	3.2	140	5
GX 339−4 (4U1658-48)	Variable	US/UH	338.9	−4.3	4	−300	6
H 1705−25 (V2107 Oph)...................	Transient	US/UH	358.6	9.1	6	950	7,8
GRO J1719−24 (GRS 1716−249)	Transient	UH	0.1	7.0	...	...	
KS 1730−312	Transient	UH	356.7	1.0	...	...	
1E 1740.7−2942...........................	Variable	UH	359.1	−0.1	8.5	−15	6
H 1743−32.................................	Transient	US/UH	357.1	−1.6	...	...	
SLX 1746−331.............................	Transient	US	356.8	−3.0	...	...	
4U 1755−338	Persistent	US	357.3	−4.9	...	...	
GRS 1758−258	Variable	UH	4.5	−1.4	...	...	
GS 1826−24...............................	Variable	UH	9.3	−6.1	...	...	
EXO 1846−031.............................	Transient	US/UH	30.0	−0.9	10	−157	9
GRS 1915+105	Variable	UH?/J	45.3	−0.9	12.5	−196	10
4U 1957+11................................	Persistent	US	51.3	−9.3	...	...	
GS 2000+25 (QZ Vul)	Transient	D/US/UH	63.4	−3.0	2.5	−131	6
GS 2023+33 (V404 Cyg)	Transient	D/UH	73.2	−2.1	8	−293	11

REFERENCES.— 1. Shrader et al. 1994; 2. Oke 1977; 3. West 1991; 4. Chevalier 1989; 5. Hjellming & Rupen 1995; 6. Tanaka & Lewin 1995; 7. Griffiths et al. 1978; 8. Martin et al. 1995; 9. Parmar et al 1993; 10. Mirabel & Rodriguez 1994; 11. Wagner et al. 1992.

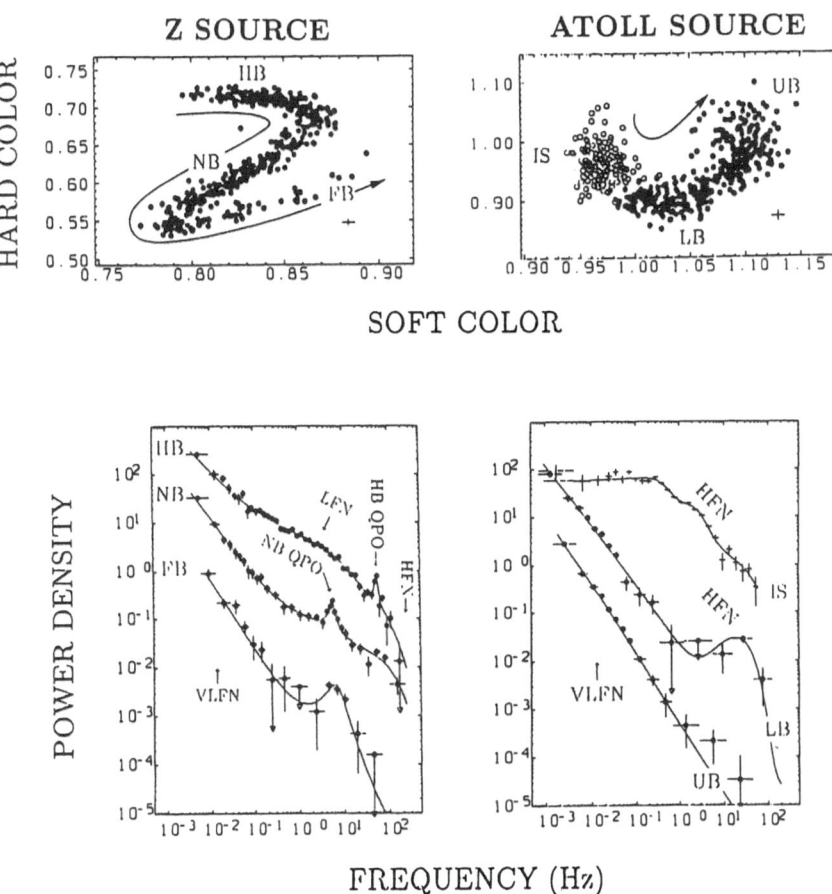

Fig. 3. X-ray colour-colour diagrams and power density spectra typical of Z sources and atoll sources (Van der Klis 1995).

In the very high state (VHS) the X-ray spectrum contains a strong US component, and a relatively steep power law component at high energies. The PDS shows, in addition to a highly variable broad-band noise component, strong quasi-periodic oscillations with frequencies of order 10 Hz and substantial harmonic content. In N Mus 1991 the VHS was observed at the peak of the outburst; afterwards this source went into the HS which, in turn, during the decay of the outburst was followed by the LS. On the basis of this temporal order Van der Klis (1994) concluded that the sequence LS - HS - VHS is one of increasing mass accretion rate.

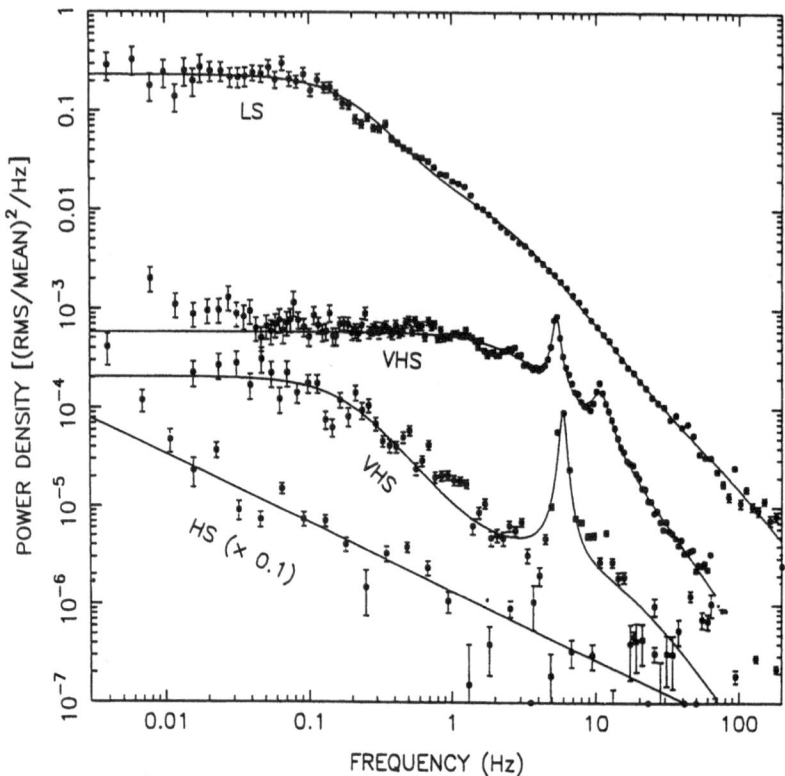

Fig. 4. Power spectra from *Ginga* data of black-hole X-ray binaries in the low state (Cyg X-1), high and very-high states (GS 1124–68) (from Van der Klis 1995).

Recently, this pleasingly simple global picture of black-hole states was affected by the discovery that the PDS of N Mus 1991 showed VHS characteristics when the source moved from the HS to the LS. This possible 'intermediate state' has also been found in Cyg X-1 (Belloni et al. 1996).

Work by the Granat/Sigma group (see, e.g., Barret & Vedrenne 1994) showed that at low luminosities the X-ray spectra of some LMXB with neutron stars are relatively hard power laws. Van Paradijs & Van der Klis (1994) showed that there is a general anti-correlation between spectral hardness in the 13-80 keV range and X-ray luminosity, and that the X-ray spectra of NS-LMXB with $L_X \sim 10^{-2}L_{Edd}$ are as hard as those of black-hole binaries (see also the review of Gilfanov et al. 1995). Currently available data are consistent with the idea that only black holes can show the combination of a hard power law X-ray spectrum and a high X-ray luminosity ($\gtrsim 10^{37}$ erg/s) (for a recent summary of this issue I refer to Barret et al. 1996).

The LS power spectra of accreting black holes are strikingly similar to those of atoll sources in the island state (see Fig 5), not only with respect to the shape of the broad-band noise component, but als with respect to the invariance of its high-frequency part as the low-frequency cut off changes (Belloni-Hasinger effect). Together with the spectral similarities this suggests that at low accretion rates the dominant factors in the emission processes near black holes in the LS and near weakly magnetized neutron stars in the island state are the same; in particular, the presence of a hard surface in the case of accreting neutron stars, and possibly of a significant magnetic field, do not appear to have much effect on the X-ray spectral and temporal properties.

Another example of strong similarity between accreting weak-field neutron stars and black holes is provided by Cir X-1, whose power spectrum, when it is very bright, is very similar to those of accreting black holes (see Van der Klis 1995), in particular by the presence of a high-frequency broad-band noise component and QPO. The black-hole power spectra show higher harmonics of these QPO; this may be a distinguishing property of these objects. This is consistent with the idea (Hasinger & Van der Klis 1989) that Cir X-1 is an atoll source (i.e., the neutron star magnetic field is very weak) accreting near the Eddington limit.

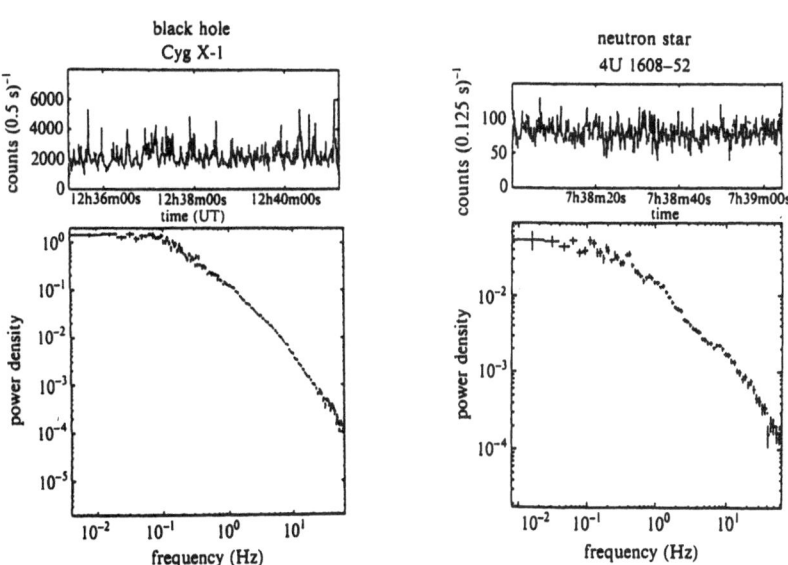

Fig. 5. Light curves and power spectra of a black hole (Cyg X-1) in the low state and an atoll source (4U 1608–52) in the island state (from Van der Klis 1995).

5   GRS 1915+105: some recent developments

In this Section I briefly discuss some recent results which extend the above picture of accreting black holes, based on radio and X-ray observations of the transient GRS 1915+105.

5.1   Relativistic jets

Mirabel & Rodriguez (1994) found that the X-ray transient GRS 1915+105 ejected radio emitting 'blobs' in opposite directions, with an angular speed that appeared superluminal at the distance $D = 12.5$ kpc they estimated from 21 cm absorption line observations. Using the expression describing relativistic proper motion ($\mu_{1,2}$) of symmetrically ejected emitters

$$\mu_{1,2} = \frac{\beta \sin \theta}{1 \pm \beta \cos \theta} \frac{c}{D} \tag{6}$$

they derived from $\mu_1 = 17.6$ mas per day, and $\mu_2 = 9.0$ mas per day that $\beta = V/c = 0.92$.

Jets and two-sided ejection had been found before in the X-ray binaries SS 433 (see Margon 1984), Cyg X-3 (Schalinski et al. 1995), 1E 0236+610 (Massi et al. 1993), GRS 1758–258 (Rodriguez et al. 1993) and 1E 1740.7–2942 (Mirabel et al. 1992). The nature of the compact objects in the first three sources is not known; the last two objects likely contain black holes on the basis of their hard power law X-ray spectra and high X-ray luminosity (see Sect. 4). The connection between accreting black holes and relativistic jets was strengthened by the discovery (Hjellming & Rupen 1995) that also the X-ray transient GRO J1655–40 showed superluminal expansion of double-sided radio 'blobs' ($\beta = 0.92$ for this system as well). The mass function for this system (Bailyn et al. 1995) indicates that it contains a black hole (see Table 1).

These results provide a strong link between galactic black-hole X-ray binaries and active galactic nuclei (AGNs), a subset of which eject superluminal radio jets (see Antonucci 1993, for a recent review of AGNs), and give strong support to the relativistic interpretation of superluminal motion in AGN. The link is reinforced by the recent finding of Sams et al. (1996) that both BH X-ray binaries and AGN with relativistic jets follow one relation between the size and surface brightness of the jets and the accretion rate onto the black hole (see also the contribution to this Volume by Sams et al.).

5.2   Extreme variability in GRS 1915+105

After its first outburst in 1994 the transient GRS 1915+105 has remained very active (see Harmon et al. 1994). The Rossi XTE detected one of its recurrent outbursts, in April 1994, and has continued observations on a weeky basis. On several occasions GRS 1915+105 has shown extremely complicated patterns of variability, in which a time interval of relatively low X-ray flux with little variability is followed by what is best described as an outburst which, in turn, is followed by an interval of rapid switching between low and high X-ray fluxes. After this the source returns to the low-flux state and the above pattern is repeated. On October 6, 1996 the duration of one such cycle of variability was ~ 20 minutes (see Fig. 6).

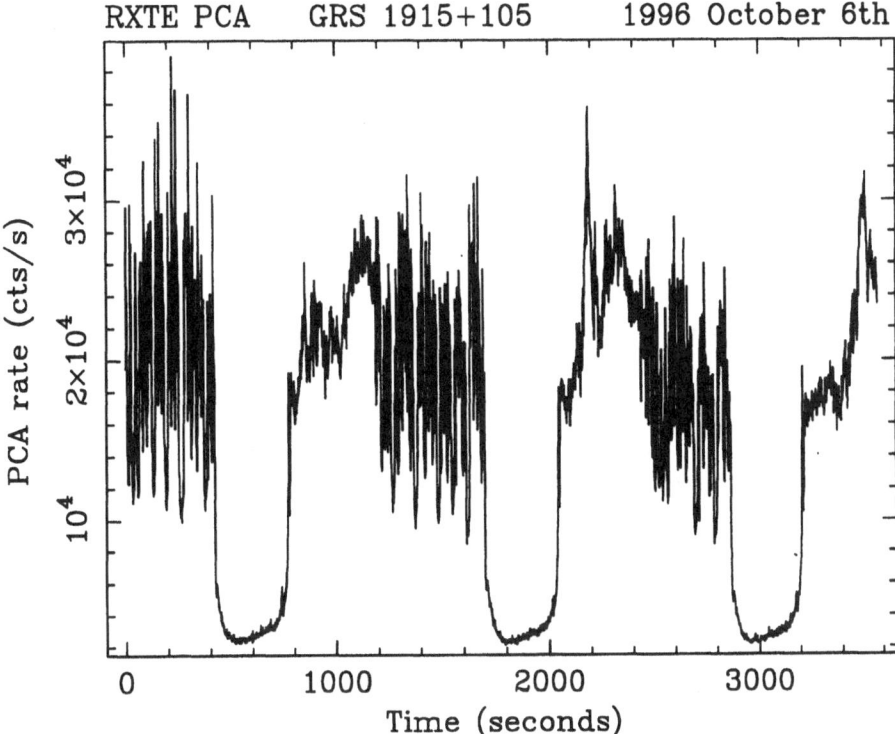

Fig. 6. X-ray light curve of GRS 1915+105 observed with the Rossi XTE on October 6, 1996 (from Belloni et al. 1997).

From a time resolved spectral analysis, using a combination of a power law plus a multi-color disk blackbody (see Sect. 2), Belloni et al. (1997) found that during the quiescent state in this variability the inner disk radius, R_{in}, was quite large (several hundred km). During the outbursts R_{in} decreased to ~ 20 km. From the values of R_{in} and T_{in} during quiescence they inferred that then the mass transfer rate through the disk is extremely high, in the range $7\ 10^{-7}$ M$_\odot$/yr to $6\ 10^{-6}$ M$_\odot$/yr. Only a very small fraction of this mass flow is emitted in the form of X rays during the outbursts. Belloni et al. concluded that most of the internal energy in the disk is advected into the black hole (a fraction may be ejected from the system, perhaps in a collimated fashion).

If the spectral inrepretation of Belloni et al. is correct one can draw the conclusion that the compact object in GRS 1915+105 does not have a hard surface. Of course, this is the standard picture of a black hole, following their description in general relativity. It is gratifying that the observations of GRS 1915+105 support this description.

Acknowledgement: I thank Tomaso Belloni, Michiel van der Klis and Henk Spruit for their comments on early versions of this paper. This work has been supported by NASA under conract NAG5-nnnn

References

Antonucci, R. 1993, ARA&A 31, 473
Bailyn, C.D. et al. 1995, Nat 374, 701
Bailyn, C.D., Orosz, J.A., McClintock, J.E., Remillard, R.A. 1995, Nat 378, 157
Barret, D., Vedrenne, G. 1994, ApJS 92, 505
Barret, D. et al. 1996, ApJ (in press)
Belian, R.D., Conner, J.P., Evans, W.D. 1976, ApJ 206, L135
Belloni, T., Hasinger, G. 1990, A&A 227, L33
Belloni, T. et al. 1996, ApJ 472, L107
Belloni, T. et al. 1997, ApJ (submitted)
Bolton, C.T. 1972, Nat 235, 271
Blandford, R., Payne, 1981, MNRAS 194, 1033
Braes, L.L.E., Miley, G.K. 1971, Nat 232, 246
Casares, J., Charles, P.A., Naylor, T. 1992, Nat 355, 614
Casares, J., Charles, P.A., Marsh, T.R. 1995a, MNRAS 277, L45
Casares, J. et al. 1995b, MNRAS 276, L35
Chakrabarthy, S., Titarchuk, L. 1995, ApJ 455, 623
Charles, P.A. 1996, in *Compact Stars in Binaries*, IAU Symp. 165, J. van Paradijs, E.P.J. van den Heuvel, E. Kuulkers (Eds), p. 341
Chevalier, C. 1989, in *Proceedings 23rd ESLAB Symposium*, ESA SP-296, p. 341
Cowley, A.P. et al. 1983, ApJ 272, 118
Crary, D.J. et al. 1996, ApJ 462, L71
Eggleton, P.P. 1983, ApJ 268, 368
Filippenko, A.V., Matheson, T., Barth, J. 1995, ApJ 455, L139
Filippenko, A.V., Matheson, T., Ho, L.C. 1995, ApJ 455, 139
Gies, D., Bolton, C.T. 1986, ApJ 304, 371
Gilfanov, M. et al. 1995, in *The Lives of the Neutron Stars*, M.A. Alpar, U. Koziloglu, J. van Paradijs (Eds), Kluwer Academic Publishers, p. 331
Griffiths, R.E. et al. 1978, ApJ 221, L63
Grindlay, J.E. et al. 1976, ApJ 205, L127
Harmon, B.A. et al. 1994, Nat 374, 703
Hasinger, G., Van der Klis, M. 1989, A&A 225, 79
Hjellming, R.M., Wade, C.M. 1971, ApJ 168, L21
Hjellming, R.M., Rupen, M.P. 1995, Nat 375, 464
Hutchings, J.B. et al. 1987, AJ 94, 340
Lewin, W.H.G., Van Paradijs, J., Van der Klis, M. 1988, SSR 46, 273

Lewin, W.H.G., Van Paradijs, J., Van den Heuvel, E.P.J. (Editors) 1995, *X-ray Binaries*, Cambridge University Press

Makishima, K. et al. 1986, ApJ 308, 635

Margon, B. 1984, ARA&A 22, 507

Martin, A.C. 1995, MNRAS 274, L46

Massi, M., Paredes, J.M., Estalella, R., Fellin M. 1993, A&A 269, 249

McClintock. J.E., Remillard, R.E. 1986, ApJ 308, 110

McClintock, J.E., Bailyn, C., Remillard, R. 1992, ApJ 399, L145

Mirabel, I.F., Rodriguez, L.F. 1994, Nat 371, 46

Mirabel, I.F. et al. 1992, Nat 358, 215

Mitsuda, K. et al. 1984, PASJ 36, 741

Narayan, R. 1996, ApJ 462, 136

Oda, M. 1976, SSR 20, 757

Oda, M. et al. 1971, ApJ 166, L1

Oke, J.B. 1977, ApJ 217, 181

Ostriker, J. 1977, Ann. New York Acad. Sci. 302, 229

Paczynski, B. 1971, ARA&A 9, 183

Parmar, A.N., Angelini, L., Roche, P., White, N.E. 1993, A&A 279 179

Payne, and Blandford, R. 1981, MNRAS 196, 781

Rappaport, S., Zaumen, W., Doxsey, R. 1971, ApJ 168, L17

Rees, M.J. et al. 1982, Nat 282, 17

Remillard, R.A., Orosz, J.A., McClintock, J.E., and Bailyn, C.D. 1996, ApJ 459, 226

Rodriguez, L.F., Mirabel, I.F., Marti, J. 1993, ApJ 401, L15

Sams, B.J., Eckart, A., Sunyaev, R. 1996, Nat 382, 47

Schalinski, C.J. et al. 1995, ApJ 447, 752

Schreier, E. et al.1972, ApJ 172, L12

Shrader, C.R. et al. 1994, ApJ 434, 698

Stella, L. et al. 1985, ApJ 288, L45

Tanaka, Y., Lewin, W.H.G., 1995, in *X-ray Binaries*, Lewin, W.H.G., Van Paradijs, J., Van den Heuvel, E.P.J., Cambridge University Press, p. 126

Tananbaum, H. 1973, IAU Symposium 55, 9

Tananbaum, H. et al. 1972, ApJ 177, L5

Tennant, A. et al. 1986, MNRAS 221, 27P

Van der Klis, M. 1994, ApJS 92, 511

Van der Klis, M. 1995, in *X-ray Binaries*, Lewin, W.H.G., Van Paradijs, J., Van den Heuvel, E.P.J., Cambridge University Press, p. 252

Van Paradijs, J. 1983, in *Accretion Driven Stellar X-ray Sources*, W.H.G. Lewin, E.P.J. van den Heuvel (Eds), Cambridge University Press, p. 189

Van Paradijs, J. 1990, in *Neutron Stars, Theory and Observations*, D. Pines, J. Ventura (Eds), Kluwer Academic Publishers, p. 289

Van Paradijs, J., Van der Klis, M. 1994, A&A 281, L17

Wagner, R.M., Kreidl, T.J., Howell, S.B., Starrfield, S.G. 1992, ApJ 401, L97

Webster, L., Murdin, P. 1972, Nat 235, 37

West, R.M., 1991, in *Proc. Workshop on Nova Mus 91*, Lyngby, Ed. S. Brandt, p. 143.

White, N.E., Marshall, F. 1984, ApJ 281, 354

White, N.E., Van Paradijs, J. 1996, ApJ 473, L25

GX 339-4: Hard X-ray observations; possible mechanism for transient outbursts

S. Trudolyubov[1,2], M. Gilfanov[1,2], E. Churazov[1,2], R. Sunyaev[1,2],
SIGMA team[3,4]

[1] Space Research Institute,Russian Academy of Sciences, Profsoyuznaya 84/32, 117810 Moscow, Russia
[2] Max – Plank – Institut für Astrophysik, Karl – Schwarzschild – Str. 1, 85748 Garching, Germany
[3] Service d'Astrophysique, DAPNIA/DSM, Bt 709, CEA Saclay, 91191 Gif sur Yvette Cedex, France
[4] Centre d'Etude Spatiale des Rayonnements 9, avenu du Colonel Roche, BP 4346, 31029 Toulouse Cedex, France

Abstract: The results of hard X-ray/soft gamma-ray observations of GX 339-4 with GRANAT/SIGMA are reported. The spectral and temporal properties of the source during its four successive hard X-ray outbursts in 1990-1994 were studied in details. We suggest that the mechanism of GX 339-4 outbursts is triggering of the irradiation – driven instability in the low mass binary system.

1 Observations

GX 339-4 has been observed by SIGMA during five sets of observations in 1990-1994. During four observational periods activity of the source in hard X-ray energy domain was detected with typical 35-150 keV flux $\approx 200 - 400$ mCrab.

The results of 1990 – 1991 SIGMA observations were reported by Bouchet et al. (1993). In the present work we report results of 1992 – 1994 observations and discuss the possible origin of source outbursts in the framework of the mass transfer instability model (MTI; Hameury, King, & Lasota (1986), (1988), Hameury et al. (1990)).

1.1 1992 Flare Observations

During the February-March 1992 SIGMA and ART-P observations the source was found in the *off* state with 3σ upper limits on the average 35-150 keV flux 18 mCrab (*see also* Grebenev et al. (1993)).

The source was on again during October 1992 observations. On Oct., 12-14, 1992 it was detected by SIGMA in its *hard (low)* state with the average 35-150 keV flux ≈ 220 mCrab. According to the BATSE data (Harmon et al. (1994)) the lightcurve and spectral evolution of GX 339-4 in the hard X-ray – soft gamma ray (≥ 20 keV) energy band was very similar to that of 1991, Fall flare. The average source spectrum is clearly seen up to 300 keV and it's shape is well described by the comptonization model with electron temperature

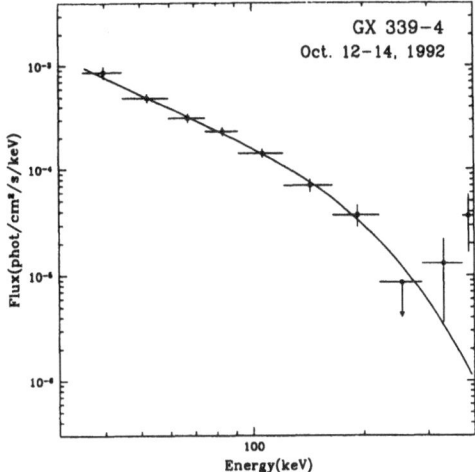

Fig. 1. The 40 – 400 keV spectra of GX 339-4 obtained by SIGMA in Fall 1992.

$kT_e \approx 40$ keV and Thomson optical depth $\tau \approx 3.1$ or by an optically thin thermal bremsstrahlung with the characteristic temperature ≈ 160 keV in the $40 - 300$ keV energy range (Figure 1). It should be noted that the source spectrum at that time was somewhat harder than usual hard state outburst spectra.

1.2 1994 Flare Observations

GX 339-4 was the target of four SIGMA observations in February – March, 1994. The source has been found in *hard* (*low*) state with spectral and temporal characteristics close to those for the 1991 hard flare (Figure 2).

According to the SIGMA data the averaged hard X-ray spectrum has parameters very close to that of previously observed hard state outbursts; the best fit parameters of comptonized spectrum are $kT_e \approx 33 \pm 4$ keV, $\tau \approx 2.95 \pm 0.50$ (assuming spherical geometry).

2 Discussion

According to BATSE results, three successive hard X-ray outbursts of GX 339-4 with $\approx 440 \pm 30$ days cycle duration occured during 1991-1994 (Fishman et al. (1991); Harmon et al. (1992); Harmon et al. (1994a)). It was pointed out that the spectral and temporal properties of the source are very similar for these outbursts (Harmon et al. (1994)), which is fully consistent with the results of SIGMA observations.

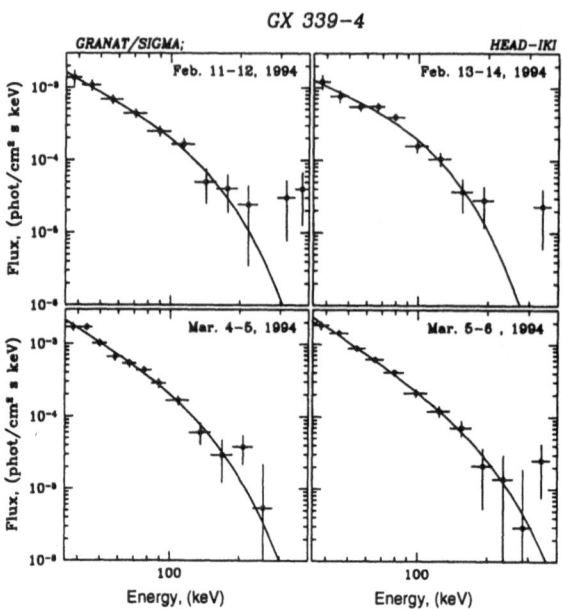

Fig. 2. The 40 – 400 keV spectra of GX 339-4 obtained by SIGMA in 1994.

2.1   Hard X-ray Spectral Evolution During the Outbursts

In Figure 3 (*middle panel*), the best-fit bremsstrahlung temperature is shown as a function of time since the start of the outbursts for the 1991, 1992, 1994 SIGMA observations (all SIGMA observations when the statistically significant flux from GX339-4 was detected) The outbursts start times were taken from the (Harmon et al. (1994a), Harmon et al. (1994)). The upper panel in Fig. 3 shows the 40–300 keV flux from the source. While some outburst–to–outburst variations in the flux history are present, the SIGMA data indicates tentatively that the spectral evolution during all three outbursts was quite similar and may be described as a gradual softening of the source high energy spectrum.

It has been recently shown for various black hole candidates (BHC) that the hardness of the high energy part of the spectrum is often correlated with the X-ray luminosity (Kuznetsov et al. (1996), Revnivtsev et al. (1996)). In particular, the data acquired from observations of several X-ray Novae indicate, that the hardness of the spectrum is generally anti-correlated with the mass accretion rate. Assuming, that the hard spectral component production mechanism in GX339-4 is essentially the same we may tentatively suggest, that the gradual softening of the GX339-4 spectrum during outburst corresponds to monotonic increase of the mass accretion rate.

Furthermore, comparing the simultaneous results of BATSE (Harmon et al.

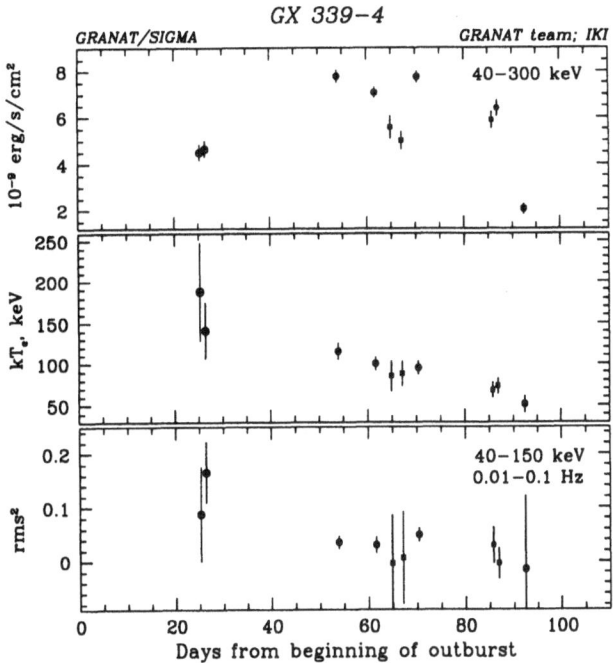

Fig. 3. The evolution of the characteristics of hard X-ray radiation from GX 339-4 with time since the beginning of the 1991, 1992, 1994 hard state outbursts: 40 – 300 keV energy flux – *upper panel*; hardness of the 40 – 300 keV spectrum vs. fitted OTTB electron temperature – *middle panel*; (0.01 – 0.1 Hz) fractional *rms²* variation of 40 – 150 keV flux – *lower panel*. Solid circles, open circles and solid squares in each pannel correspond to the 1991, 1992 and 1994 SIGMA data respectively. The starting dates of outbursts were taken from Harmon et al. (1994a), Harmon et al. (1994b).

(1994a)) and ART – P and SIGMA (Grebenev et al. (1993), Bouchet et al. (1993)) on the 1991 GX 339-4 outburst, we can conclude that the beginning of the *hard-to-soft* state transition coincided with the peak of the hard X-ray (≥ 20 keV) luminosity. On the other hand it was demonstrated that the soft spectral state corresponds to higher mass accretion rate than the hard state (Trudolyubov et al. (1996)). Therefore, the peak of the mass accretion rate is reached well after the maximum of the hard X-ray luminosity – the observed decrease of the hard X-ray luminosity after it's maximum corresponds to the transition to soft spectral state and doesn't necessarily reflect decrease of the accretion rate.

According to the EXOSAT/ME (Ilovaisky et al. (1986)) and GINGA observations (Ueda et al. (1994)) the *off* state spectrum of GX339-4 in the 2–20 keV band is extremely hard with the photon index ~ 1.7. This value is close to that for the hard (*low*) state (Grebenev et al. (1993), Bouchet et al. (1993), this paper), while the X-ray luminosity in off state is ~ 100 times lower than during the hard outburst. This is yet another example of surprisingly weak dependence of the slope ($\sim 2 - 20$ keV) of comptonized spectrum upon luminosity (Gilfanov et al. 1995).

2.2   Relation between spectral hardness and level of short–term variability

We have searched for possible correlation between short term fluctuations of the hard X-ray flux and the hardness of the source spectrum. The relation between the fractional rms (in the $10^{-2} - 10^{-1}$ Hz frequency range) of the $40 - 150$ keV flux fluctuations and best-fit bremsstrahlung temperature is shown in Figure 4. It is seen from Fig. 4 that softening of the source spectrum is accompanied with decrease of the fractional rms. This behaviour resembles that of Cygnus X-1 (Kuznetsov et al. (1996)) and Nova Persei 1992 (GRO J0422+32). This fact may hint on the general property of the hard spectral component production mechanism.

2.3   On the Origin of the Outbursts

The optical studies (Cowley, Crampton, & Hutchings (1987), Callanan et al. (1992)) identified GX 339-4 with a binary system containing a compact object with mass $1M_\odot \leq M_c \leq 2M_\odot$ and a probably evolved low mass secondary with luminosity $L_s \leq L_\odot$. In addition, the 14.8-hr binary period has been reported recently (Callanan et al. (1992)).

The BATSE observations have demonstrated that GX 339-4 hard X–ray (≥ 20 keV) light curve during outburst is characterized by an initial rise of the flux during ~ 1 month followed by slower increase up to the peak value during ~ 2 months and relatively rapid drop within ~ 20 days (Harmon et al. (1993), Harmon et al. (1994)). The evolution of the X–ray spectrum suggests that the maximum of the accretion rate is reached some time after the maximum of the hard X–ray luminosity, i.e. more than ~ 3 month after the start of the outburst. This type of behavior is opposite to that of X–ray Novae during primary outbursts characterized by a short rise time ($\sim$ one week) and much slower decay of the X–ray luminosity on the timescale of months usually attributed to the disk instabilities.

Increase of the mass accretion rate onto compact object is thought to be the origin of the transient outbursts in the low mass X-ray binaries (LMXBs). Basing on the different mechanisms that explain this phenomena two competing models have been constructed. The mass transfer instability model (MTI; Hameury, King, & Lasota (1986), (1988), Hameury et al. (1990)) suggests that the outburst is caused by sudden increase of the mass transfer rate through the

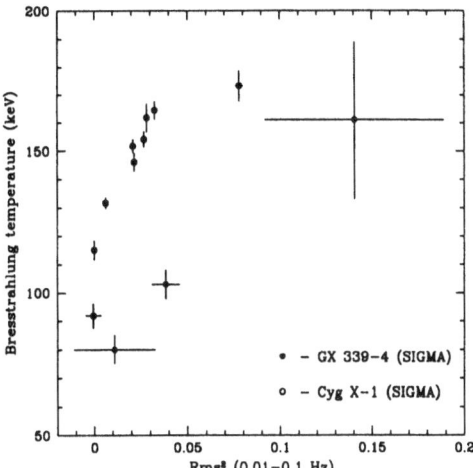

Fig. 4. GX 339-4 hard X-ray flux fluctuations vs. $10^{-2} - 10^{-1}$ Hz fractional rms^2 variation of 40 – 150 keV flux dependence on the hardness of source spectrum vs. fitted OTTB electron temperature. For comparison the Cyg X-1 results are shown (Kuznetsov et al. (1996)).

inner Lagrangian point (L_1) due to the expansion of the secondary's outer layers heated by hard X-rays generated in the vicinity of the compact object. The disk instability model (DTI; Lin & Taam (1984), Huang & Wheeler (1989), Mineshige & Wheeler (1989)) attributes the outburst to the rising of the mass transfer rate through the accretion disk itself. Let us consider GX 339-4 state transitions in the framework of these models.

In the mass transfer instability model (MTI) the transient process is governed by the change of the mass transfer rate from a companion star on a timescale of its envelope expansion (Gontikakis & Hameury (1993)) and by changing of mass transfer through the accretion disk on its diffusion timescale $\tau_{dif} \sim (r/v_r)$ $\sim (1/\alpha\Omega)(r/H)^2 \sim$ several months (Lightman (1974)) where r and H are the radius and the thickness of the disk, α and Ω are viscosity parameter and the Keplerian disk angular velocity (Shakura & Sunyaev (1973)). On the other hand, the averaged measured quiescent GX 339-4 X-ray luminosity ($\geq 10^{35} D_{4kpc}^2$ ergs s^{-1}) (Ilovaisky et al. (1986), Ueda et al. (1994)) is known to be higher than the required to initiate the expansion of the secondary's outer layers $\sim 10^{34} M_s^2$ ergs s^{-1} (Hameury, King, & Lasota (1986), Chen, Livio, & Gehrels (1993)) where

M_s is the secondary mass in the solar units. The disk thermal instability (DTI) has a characteristic timescale of order the disk heating wave propagation time $\tau_{heat} \sim (r/\alpha c_s) \sim (1/\alpha \Omega)\, (r/H) \sim$ few days (Meyer (1984)) where the c_s is the speed of sound. These facts and the absence of the fast rise in hard X-rays allow us to suppose the mass transfer instability (MTI) as an origin of the GX 339-4 hard state outbursts rather than the disk thermal instability (DTI).

Supposing that GX 339-4 outbursts are caused by the matter overflow through the inner Lagrangian point (L_1), let us define the degree of secondary Roche lobe (RL) overfill ΔR for a given system mass transfer rate. In the case of secondaries with sufficiently deep convective envelope the equation of state can be reasonably approximated by the polytropic law $p \propto \rho^{5/3}$, the relation between the mass transfer rate $\dot{M}$ the degree of RL overfill is follows: $\dot{M} \sim (M_s/P_B)(\Delta R/R_s)^3$ (Livio (1992), Lubow & Shu (1975)) where M_s and R_s are the secondary mass and radius, P_B – is the binary period. The estimated hard state binary mass transfer is $\sim 3 \times 10^{-9}\,(0.1/\eta)\ M_\odot$ year^{-1}, where η is the efficiency of X-ray production, assuming the X-ray luminosity $\sim 2 \times 10^{37}$ ergs s^{-1} (Grebenev et al. (1993)). Therefore, for the GX 339-4 hard state outbursts $\Delta R/R_s \sim 10^{-4}$, taking into account the uncertainty of the secondary mass determination. Can it be attributed to the intrinsic secondary radius fluctuations? It is known that in low mass binary systems, such as cataclysmic variables, long term secondary radius fluctuations exist, which is possibly linked to cycles in magnetic activity (Gontikakis & Hameury (1993)). In addition, solar observations show $\Delta R/R_\odot \sim 10^{-4}$ within the solar cycle (Gilliland (1981)). Although the possible influence of these fluctuations on the long term source behavior can not be completely excluded, their associated time scale (years) is too long to explain the observed transient events.

We propose that the sequence of the emission episodes during GX 339-4 outbursts is caused by the gradual increase of the mass transfer from the secondary onto the compact object due to the triggering of the irradiation-driven instability of the secondary outer layers. When the secondary outer layers expand to some extend, the mass transfer through the disk increases resulting in the *hard-to-soft* source state transition. X-ray illumination of the secondary is effective until thick accretion disk shields the L_1 region, quenching the instability (Hameury, King, & Lasota (1986)). The following contraction of the unilluminated part of the secondary causes falling of the system mass transfer rate below critical value.

Proposed transient mechanism doesn't require the strict periodicity of the source outbursts. Furthermore, some destabilizing factors such as secondary intrinsic radius fluctuations discussed above, superimposed to the main instability cycle, may cause dramatic increase of the system mass transfer, resulting in the observed *super-high* state (Miyamoto et al. (1991)) or even be able to produce out of turn outbursts.

Acknowledgement: This work was supported in part by the RBRF grant 96-02-18588 and ESO C&EE Grant A-01-103. S. Trudolyubov was partially supported by grant of the International Science Foundation.

References

Bouchet, L., Jourdain, E., Mandrou, P., Roques, J. P., 1993, ApJ 407, 739

Callanan, P. J., Charles, P. A., Honey, W. B., & Thorstensen, J. R., 1992, Mon. Not. Roy. Astr. Soc. 259, 395

Chen, W., Livio, M., Gehrels, N., 1993, ApJ 408, L5

Cowley, A. P., Crampton, D., & Hutchings, J. B., 1987, AJ 92, 195

Fishman, G. J., Harmon, B. A., Finger, M. H., Pasiesas, W. S., Brock, M., & Meegan C. A., 1991, IAU Circ. 5327

Hameury, J. M., King, A. R., Lasota, J. P., 1986, A&A 162, 71

Hameury, J. M., King, A. R., Lasota, J. P., 1988, A&A 192, 187

Hameury, J. M., King, A. R., Lasota, J. P., 1990, ApJ 353, 585

Harmon, B. A., Fishman, G. J., Paciesas, W. S., & Finger, M. H., 1992, IAU Circ. 5647

Harmon, B. A., Zhang, S. N., Wilson, C. A., Rubin, B. C., Fishman, G. J., & Paciesas, W. S., 1993, in Proc. 304 The Second Compton Symposium, eds. C. E. Fichtel, N. Gehrels, J. P. Norris (New York: AIP), 210

Harmon, B. A., Paciesas, W. S., Zhang, S. N., Fishman, G. J., & Finger, M. H., 1994, IAU Circ. 5915

Harmon, B. A., Wilson, C. A., Pasiesas, W. S., & Pendleton, G. N., 1994, ApJ 425, L17

Huang, M., & Wheeler, J. C., 1989, ApJ 343, 229

Ilovaisky, S. A., Chevalier, C., Motch, C., Chiappetti, L., 1986, A&A 164, 67

Gilfanov, M. R. et al, 1991, Soviet. Astron. Lett. 17, 437

Gilfanov, M., Churazov, E., Sunyaev, R., Vikhlinin, A., Finoguenov, A., et al, 1995, in The Lives of the Neutron Stars, NATO ASI C450, Kluwer, Dordrecht, 331

Gilliland, R. L., 1981, ApJ 248, 1144

Gontikakis, C., Hameury, J. M., 1993, A&A 271, 118

Grabelsky, D. A.,

Grebenev, S. A. et al, 1993, A&AS 97, 281

Kuznetsov, S., et al, 1996, (in preparation)

Lin, D. N. C., & Taam, R. E., 1984, in AIP Conf. Proc. 115 High Energy Transients in Astrophysics, ed. S. E. Woosley (New York: AIP), 83

Lightman, A. P., 1974, ApJ 194, 419

Livio, M., 1992, in 22d Saas Fee Advanced Course, Interacting Binaries, ed. H. Nussbaumer

Lubow, S. H., Shu, F. H., 1975, ApJ 198, 383

Makishima, K., Maejima, Y., Mitsuda, K., Bradt, H. V., Remillard, R. A., Tuohy, I. R., Hoshi, R., & Nakagawa, M., 1986, ApJ 308, 635

Meyer, F., 1984, A&A 131, 303

Mineshige, S., & Wheeler, J. C., 1989, ApJ 343, 241

Miyamoto, S., Kimura, K., Kitamoto, S., Dotani, T., & Ebisawa, K., 1991, ApJ 383, 784

Paczynski, B., 1971, Ann. Rev. A&A 9, 183

Revnivtsev, M. G. *et al*, 1996, (in preparation)

Shakura, N. I., & Sunyaev, R. A., 1973, A&A 24, 337

Sunyaev, R. A., & Titarchuk, L. G., 1980, A&A 86, 121

Trudoyubov, S., Gilfanov, M., Churazov, E., Borozdin, K., Sunyaev, R., *et al*, 1996, Astron. Letters 22, 664

Ueda, Y., Ebisawa, K., & Done, C., 1994, PASJ 46, 107

Spectral and temporal variations of the X–ray emission from black hole and neutron star binaries

M. Gilfanov[1,2], E. Churazov[1,2], R. Sunyaev[1,2]

[1] Max-Planck-Institut für Astrophysik, Karl-Schwarzschild-Str. 1, 85740 Garching bei Munchen, Germany
[2] Space Research Institute, Profsouznaya 84/32, 117810 Moscow, Russia

Abstract:
 Spectral and temporal variability of black hole and neutron star binaries in soft and hard X–ray energy domain are discussed basing on the MIR–KVANT/TTM, GRANAT/SIGMA and some ASCA observations. Along with prominent spectral changes associated with low/high state transitions the variations of spectral and short term variability characteristics of the hard X–ray emission were observed in the low spectral state of several black hole binaries. Possible relation of these variations with change of the mass accretion rate and low/high spectral state transitions is discussed.

1 Introduction

Numerous studies of black hole binaries in the X–ray/low γ–ray energy domain during the past decade revealed complicated pattern of the spectral variability. The most prominent spectral changes are associated with well known and intensively studied transition between soft/hard spectral states (e.g. Tanaka, 1989; Grebenev et al. 1993, Sunyaev et al. 1993). This transitions manifest themselves as a dramatic redistribution of bulk of the emitted energy over wavelength (e.g. Fig.1). Along with that less evident "subtle" variations of the spectral properties are often observed (e.g. Fig.2). In many cases spectral changes are accompanied with changes in the short term aperiodic variability properties (e.g. Miyamoto et al. 1995, van der Klis, 1995, Belloni et al., 1996ab). The spectral and short term variability characteristics of the emission are directly observable quantities, provided sufficient spectral and temporal resolution, sensitivity and energy coverage.

In many theoretical models the geometry and conditions in the X–ray emission generation zone are defined primarily by the mass accretion rate. Opposite to spectral and temporal characteristics the mass accretion rate isn't directly measurable quantity. In some cases it can be determined in the model dependent way. Related quantity is the bolometric luminosity which is not directly measurable either due to limited energy coverage or, more important, due to low energy interstellar absorption. The bolometric luminosity can be in principle estimated via extrapolation of the spectral model – best fit to the observed

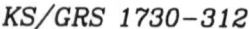

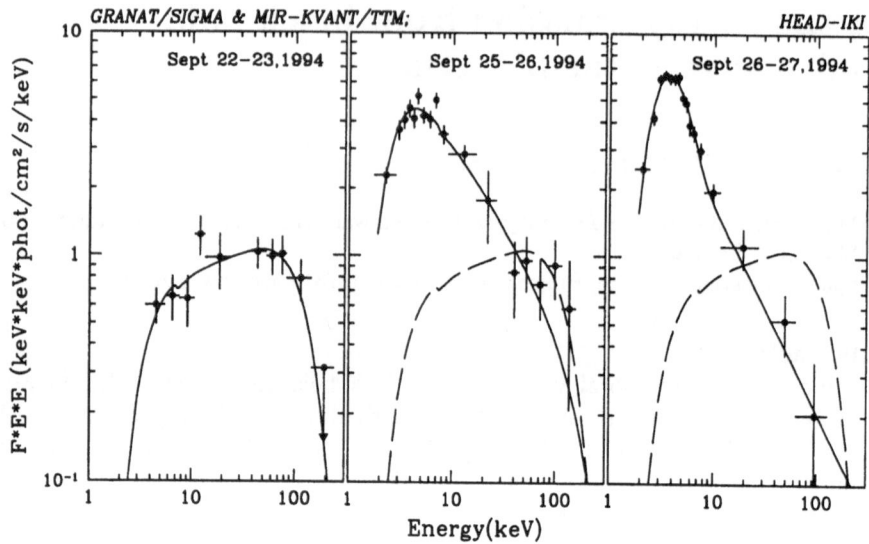

Fig. 1. The spectra of KS1730-312 in different spectral states (TTM and SIGMA data, from Trudolyubov et al. 1996 .

spectrum. That obviously requires certain assumptions to be made about spectral behavior beyond the energy range covered by the instrument, i.e. is spectral model dependent. The only directly measurable quantity is the luminosity in some restricted energy range defined by the bandpass of the instrument.

As is well known, the Comptonization theory and, in particular, the simplest approximation given by Sunyaev & Titarchuk 1980 formula was quite successful in describing individual spectra of the black hole candidates in the low spectral state (e.g. Sunyaev & Truemper, 1979; Grebenev et al. 1993). The Compton up scattering of low frequency radiation in hot optically thin part of the accretion flow is very likely a mechanism of generation of the hard spectral component. However the structure of the accretion flow and in particular origin of the hot electrons and geometry of the Comptonization region are still unclear.

Below we discuss some results on spectral/temporal variation of several black hole candidates basing on the MIR–KVANT/TTM, GRANAT/SIGMA and some ASCA observations.

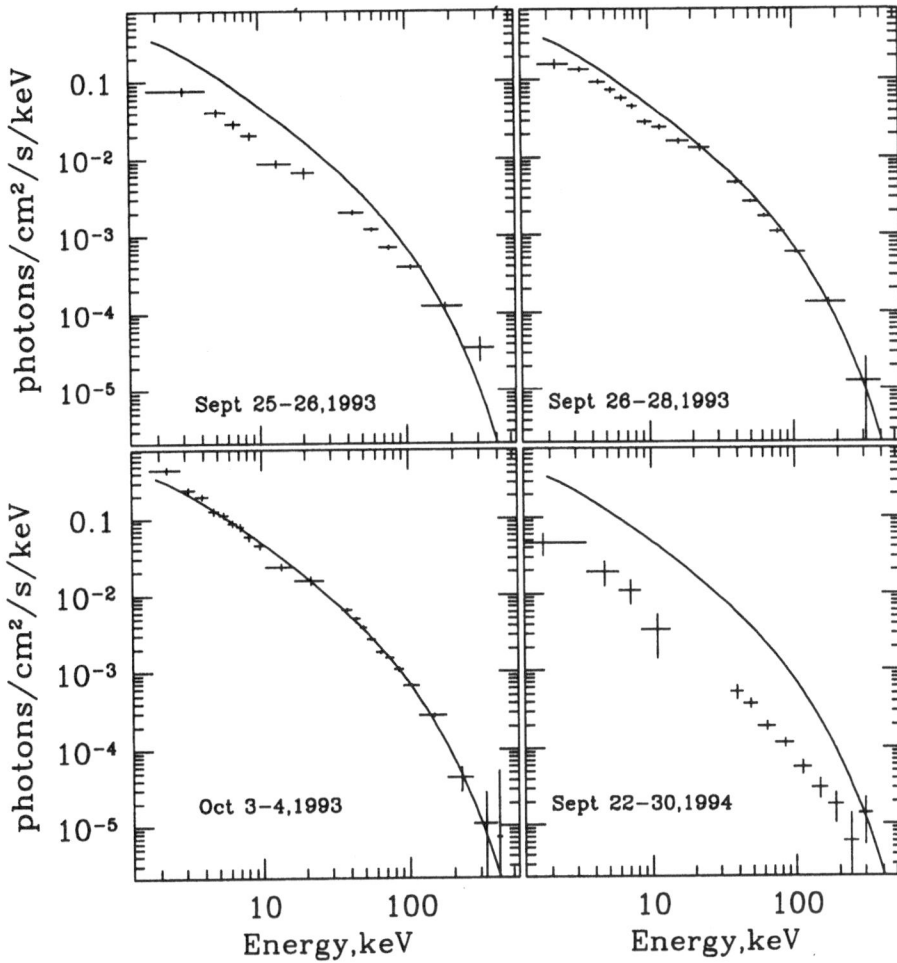

Fig. 2. The spectra of GRS1716-249 (Nova Oph 1993) at different epoch (TTM and SIGMA data, from Revnivtsev et al. 1997a).

2 Spectral states

2.1 Black hole binaries

Existence of different spectral states for black hole candidates was established more than two decades ago (e.g. Tananbaum et al. 1972, Holt et al. 1976, Ogawara et al. 1982). Nonetheless the luminosity change (and it's sign even) from the hard state to the soft was unclear. The primary purpose of this paragraph is to estimate typical luminosity levels corresponding to each spectral state for several sources. For completeness we review briefly the pattern of spectral states with emphasis on the spectral properties.

Different spectral states are roughly distinguished according to relative importance of the soft and the hard spectral components. Three spectral states are usually distinguished:

1. *low* or *hard* – the emission is dominated by the hard spectral component which shape is adequately described by the Comptonization model.
2. *high* or *soft* – The primary contribution to the emergent emission is due to the soft spectral component, the hard component being absent or extremely steep and weak.
3. *superhigh* – both the soft and the hard spectral components are present, with the major fraction of the X–ray luminosity being emitted in the soft spectral component. Properties of the hard spectral component are quite different from that in the low state.

The commonly accepted point of view is that the soft spectral component is due to emission from optically thick and geometrically thin (part of the) accretion disk of the type predicted by the standard model of Shakura & Sunyaev 1973. The hard spectral component is believed to result from Compton upscattering of soft photons in vaguely defined optically thin Comptonization region. The temporal variability properties of these two spectral components are very different (e.g. Miyamoto et al., 1991, 1994, van der Klis, 1995). Besides that, spectral and short term variability properties of the hard spectral component observed in the low and superhigh spectral states are also quite different (Miyamoto et al. 1995, van der Klis, 1995, Belloni et al., 1996ab). In particular, sufficiently steep photon index (≈ 2.5) and small fraction of the luminosity emitted in the hard power law component in the superhigh state could be easily accommodated in the disk–corona model without postulating that major part of the gravitational energy of accreting matter is dissipated in the rarefied corona (e.g. Gilfanov et al. 1991).

Since the bulk of the emitted energy might shift over the photon energy by $\sim$ two orders of magnitude (typical black body temperature of the soft component is $\sim 0.3 - 1.0$ keV, the $F_E \times E^2$ for the hard component peaks at ~ 100 keV) accurate luminosity estimate requires broad energy coverage – from $\lesssim 1$ keV to $\gtrsim$ few hundred keV. Besides that accurate estimate of luminosity of the soft spectral component ($kT_{bb} \sim 0.3 - 1$ keV) requires in many cases proper account for the interstellar absorption. The latter was proven to be important for the recently observed soft (or intermediate, Belloni et al. 1996a, Zhang et al. 1997) spectral state of Cyg X–1.

The Table 1 presents luminosity estimates for several sources. Note that the numbers given in the table do not correspond to the whole range of the luminosities for particular spectral state. They rather represent some luminosity levels at which particular source was picked up by different observatories/instruments. The high state luminosity for Cyg X–1 was calculated using the best fit parameters for the soft and hard spectral components from Dotani et al. 1996 assuming

Table 1. Luminosity in different spectral states for several black hole candidates

Spectral state	Nova Mus	KS1730-312	Cyg X–1
low	$\approx 5 \cdot 10^{35}$	$\approx 4.4 \cdot 10^{37}$	$\sim (2-4) \cdot 10^{37}$
high	$\approx 3 \cdot 10^{36}$	$\approx 1.2 \cdot 10^{38}$ (?)[1]	$\sim 8 \cdot 10^{37}$ (?)[1]
superhigh	$\approx 2 \cdot 10^{37}$		

The source distance was assumed 1 kpc for Nova Mus, 8.5 kpc for KS1730-312 and 2.4 kpc for Cyg X–1. The energy range is 1–300 keV for Nova Mus and KS1730-312 and 0.5–300 keV for Cyg X–1. No correction for interstellar absorption was applied to Nova Mus and KS1730-312 data.

[1] Most likely intermediate state between low and high spectral states (see Belloni et al. 1996a for Cyg X–1).

hydrogen column density of $6 \cdot 10^{21}$ cm^{-2}. The Table 1 demonstrates, that the luminosity increases from the low state to high and superhigh spectral states.

2.2   Neutron star binaries

The spectral study of X–ray bursters, (e.g. Langmeier et al. 1987, Mitsuda et al. 1989, Ford et al. 1996), demonstrate, that X–ray bursters (weakly magnetized neutron star binaries) show *bimodal spectral behavior similar to that observed for black hole binaries* (see also Barret & Vedrenne 1994, Tanaka & Shibazaki 1996, Revnivtsev et al. 1997b). At least two distinct spectral states of X–ray bursters may be identified – *soft/high* and *hard/low* (cf. spectral states of black hole candidates). The 4U1705-44 and 4U1608-56 are discussed below as two relatively well studied examples (Revnivtsev et al. 1997b).

The high/soft state spectrum corresponds to luminosity level of $L_X \sim (3-9) \times 10^{37}$ erg/s[1] or $L_X \sim (0.2-0.6) \times L_{Edd}$ and roughly[2] could be represented as a blackbody spectrum with temperature of $kT_{bb} \sim 1-2$ keV possibly with superimposed weak and rather steep power law tail which diminishes above $\sim 20-30$ keV. In the case of 4U1705-44 the upper limit on the luminosity in the higher energy domain during the soft state is 2×10^{36} erg/s (2σ, 35–100 keV, assuming the spectral shape similar to that observed in the low state).

At lower luminosity level ($L_X \sim (0.7-2) \times 10^{37}$ erg/s or $L_X \sim 0.1 \times L_{edd}$) the spectrum becomes considerably harder with photon index in the low energy limit of $\sim 1.6-1.8$ and exponential cut-off at $\sim 30-50$ keV. The low state spectrum could be generally described by the Comptonization model.

---

[1] The hard X–ray luminosities and upper limits for these two X–ray bursters obtained by GRANAT/SIGMA and cited in this subsection are from Revnivtsev et al., 1997b.

[2] Detailed discussion of the spectral behavior of the X–ray bursters at high luminosity is beyond scope of these paper (see e.g. White, Nagase & Parmar, 1995 for review)

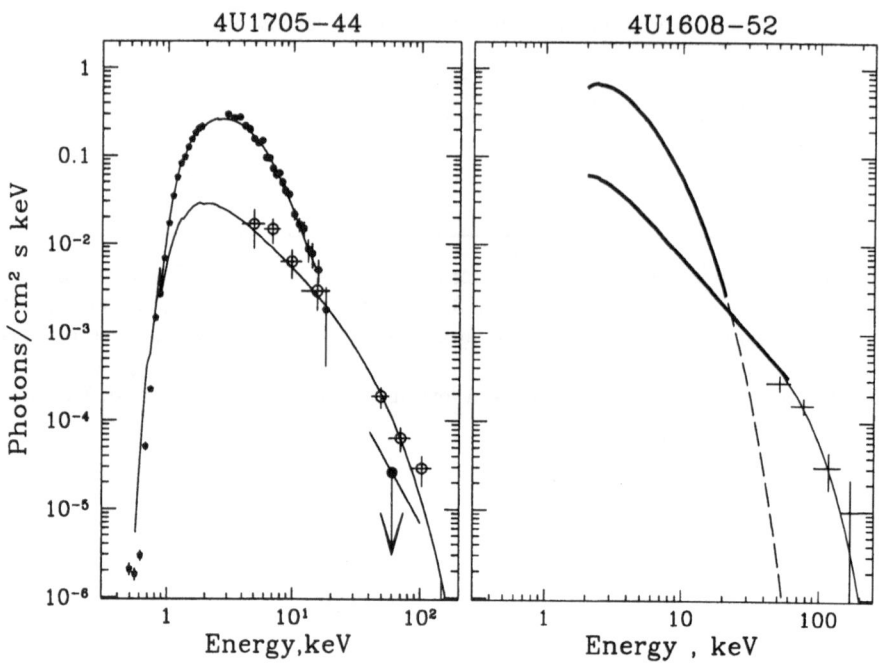

Fig. 3. The spectra of 4U1705-44 and 4U1608-52 in soft and hard spectral states (from Revnivtsev et al. 1997b).

The overall spectral shape in the soft and the hard states is generally similar to that of black hole candidates except for somewhat higher value of kT_{bb} in the soft state. Another important difference is that the X-ray bursters have $\sim 2-3$ times lower energy of the exponential cut-off of the spectrum in the hard state (Fig.4, Gilfanov et al. 1993). This difference is directly observable and opens a possibility to distinguish an accreting low magnetized neutron star from a black

Table 2. Luminosity in different spectral states for two X-ray bursters

Spectral state	4U1705-44	4U1608-52
low	$\sim 1 \cdot 10^{37}$	$\sim 0.7 \cdot 10^{37}$
high	$\sim 5-9 \cdot 10^{37}$	$\sim 3 \cdot 10^{37}$

The source distance was assumed 7.4 kpc for 4U1705-44 and 3.6 kpc for 4U1608-52. The energy range is 0.5–200 keV.

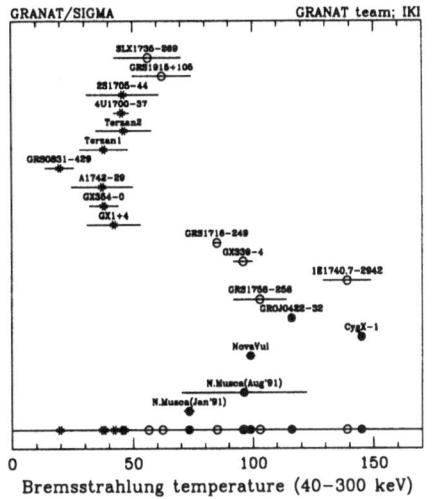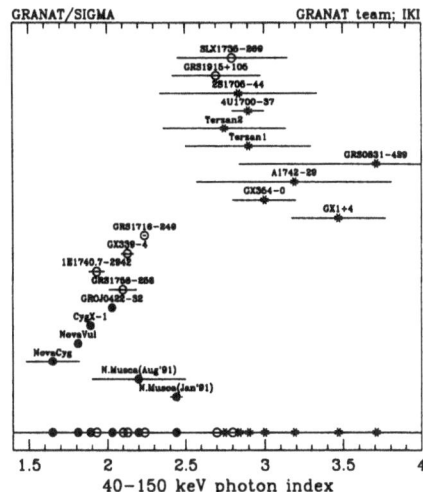

Fig. 4. The hardness of the spectra of black hole binaries (in low or superhigh spectral state) and neutron star binaries (low spectral state) as observed in the hard X-ray domain expressed in terms of best-fit bremsstrahlung temperature (left) and 40–150 keV photon index (right). The binary systems with known nature of the compact object are marked by asterisk (a neutron star) or filled circle (a black hole). The GRANAT/SIGMA data; from Gilfanov et al. 1993.

hole. On the other hand it's worth mentioning that in the hard state the spectral slope sufficiently below the cut–off energy is quite the same as for the hard state spectra of black hole candidates.

The X–ray bursters undergo spectral transitions at approximately the same luminosity level as black hole binaries (Tables 1,2) although the mass of the compact object might differ by more than the order of magnitude. In terms of critical Eddington luminosity in the case of X–ray bursters transition occurs at $L_X/L_{Edd} \sim$few$\times0.1$ while for black hole binaries the threshold luminosity corresponds to $L_X/L_{Edd} \lesssim 0.1$.

The most apparent difference between weakly magnetized neutron star binaries and black hole binaries is presence of the neutron star. Existence of relatively cool optically thick surface in vicinity of the accretion disk could affect the energy balance and/or the structure of the inner part of the accretion disk. In particular the feedback (via reprocessing of the hard Comptonized radiation) between the neutron star surface and hot electrons in the Comptonization region could be the reason for lower electron temperature observed in the low state of X–ray bursters (Sunyaev et al. 1991, Gilfanov et al. 1993; Fig.4).

If the neutron star rotation frequency is sufficiently below the break up frequency and the magnetic field is week, more than a half of the gravitational energy of the accreting matter could be released near the neutron star surface

(Sunyaev & Shakura 1986) in addition to the energy released in the accretion disk. If the neutron star radius exceeds the radius of the last marginally stable keplerian orbit $R_{NS} > 3R_g$ the viscous boundary layer will be formed. In the opposite case of $R_{NS} < 3R_g$ the accreting matter at $R < 3R_g$ should fall toward the neutron star along the trajectories close to that of free particles with the given value of the specific angular momentum and energy. The most of the kinetic energy is then released during the impact with the neutron star surface in the narrow optically thin layer. The spectrum of outgoing radiation in this case could be extremely hard opposite to the spectrum of the viscous boundary layer (Sunyaev & Shakura 1986).

The physical conditions in the boundary layer aren't well understood yet. On the other hand, in the case of optically thick boundary layer the spectrum of outgoing radiation should be rather soft with characteristic blackbody temperature $kT \sim 1-2$ keV (for typical luminosity of X-ray bursters, $\sim 10^{37} - 10^{38}$ erg/sec). However, the EXOSAT data show that at sufficiently low luminosity, $L_X \lesssim 10^{37}$ erg/sec, the contribution of the soft blackbody component with $kT_{bb} \sim 1-2$ keV couldn't exceed $\sim 10 - 15\%$ of the overall source luminosity in the low spectral state (Mitsuda et al. 1989, Revnivtsev et al. 1997b).

3   Spectral variability of the hard spectral component

In this paragraph we will consider variability of the hard spectral component at the energies $\gtrsim 40$ keV basing on observations of the GRANAT/SIGMA (Paul et al. 1991). Since often the SIGMA observations aren't complemented with observations in the standard X-ray band it is sometimes unclear if observed spectral variations are intrinsic to the low/hard spectral state or correspond to transition from the low to high spectral state (e.g. extended episodes of very low hard X-ray flux observed for Cyg X-1 and 1E1740.7-2942).

In order to quantify shape of the hard spectral component the best fit optically thin bremsstrahlung temperature is used. This approach provides simple though rather crude single parameter representation of the hardness of spectrum. The same approximation is used to estimate the energy flux or luminosity. The short term variability is characterized by relative rms in the $0.01 - 0.1$ Hz frequency range.

3.1   Cygnus X-1 and 1E1740.7-2942

The hard X-ray light curves of both Cyg X-1 and 1E1740.7-2942 have a complex structure with short term (time scales of days to weeks) variations superimposed on long term (time scale of years) intensity changes of generally larger relative amplitude (Kuznetsov et al. 1997).

The long-term light curve of Cygnus X-1 (Fig.5) recorded by SIGMA (Salotti et al. 1992, Vikhlinin et al. 1994, Ballet et al. 1996) shows variations of the 40-150 keV flux by a factor of ~ 4. During 1990 – mid 1993 (Vikhlinin et al. 1994)

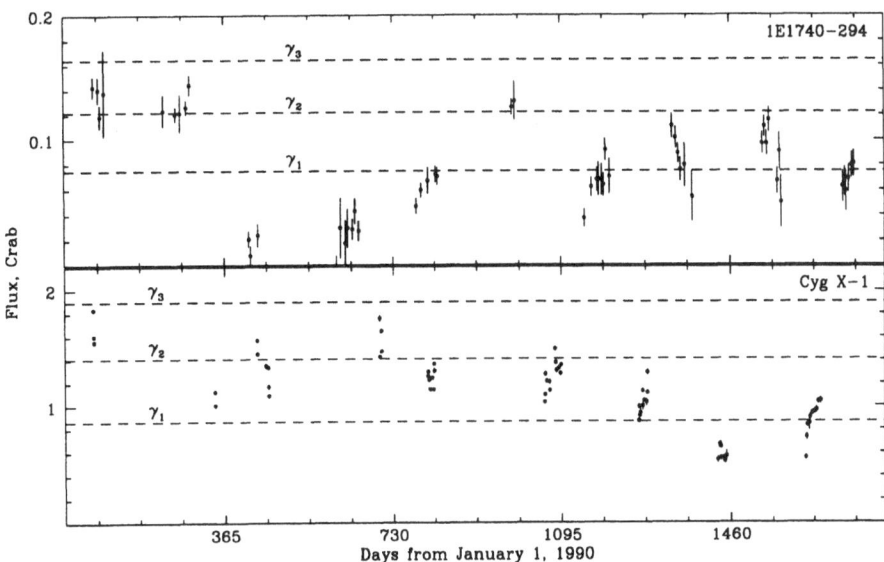

Fig. 5. The 40–150 keV light curves of Cyg X-1 (top) and 1E1740.7–2942 (bottom). Each data point represents the average over $\approx 2 - 8$ hours (Cyg X-1) and ~ 50 hours (1E1740.7–2942). The date Jan.1,1990 corresponds to MJD 47892. Approximate flux levels corresponding to the three γ–states of Cyg X-1 are shown on both panels. From Kuznetsov et al. 1997.

the source was typically detected near the γ_2 intensity level. In Dec 1993 and in the first observations in June 1994 the minimal flux was detected, ~ 0.5 Crab. According to BATSE data (Crary et al. 1996) having much better time coverage these two observational sets occurred during an extended low hard X-ray flux episode. The lowest flux from Cyg X-1 during this period was detected in Feb.1994 ($0.2\gamma_1$ – Phlips et al. 1996). During almost all SIGMA observations considerable variability on the hours–days time scale was detected – by a factor of ~ 1.5.

The light curve of 1E 1740 recorded by SIGMA (Fig5) shows a qualitatively similar pattern (Cordier et al. 1993, Churazov et al. 1993). The 40-150 keV flux from 1E 1740 changed by a factor of $\sim$10 on the time-scale of $\sim 1/2$ year with the minimal flux corresponding to the extended minimum observed during 1991 (Churazov et al. 1993). Variability by a factor of 1.5 on a days-weeks time scale was detected during most of the observational sets.

The SIGMA observations provided on one hand rather sparse time coverage – especially for Cyg X-1, and, on the other, a limited time resolution restricted by the instrument time resolution (several hours for spectral information) and, especially for 1E 1740, by the accuracy of the spectral and variability parameters

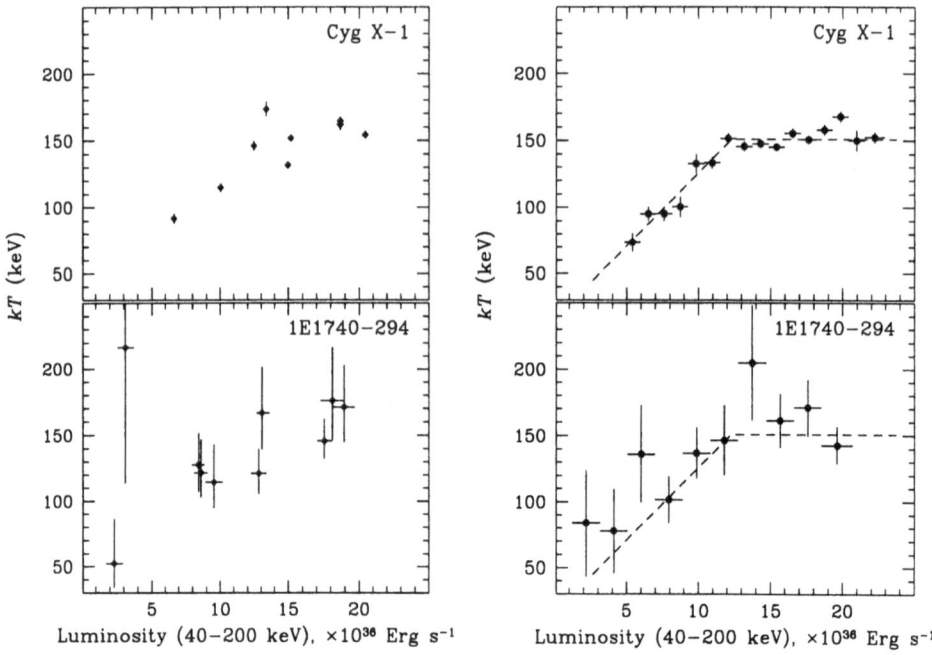

Fig. 6. The best-fit bremsstrahlung temperature plotted against the hard X-ray luminosity (40-200 keV) for Cyg X-1 (upper panels) and 1E 1740.7–2942 (lower panels). The data were averaged over 1 to 20 days of consecutive observations (left panels) and according to the source intensity (right panels). From Kuznetsov et al. 1997, see also Ballet et al. 1996.

estimation. The latter leads to the necessity of further grouping of the data. In order to verify possible effects of the data averaging two grouping methods were applied to the data and the results are shown in Fig.6 and 7 (see Ballet et al. 1996, Kuznetsov et al. 1997 for the details).

Although no strict point-to-point correlations were detected certain general tendencies are evident. For both sources an approximate correlation between kT and L_X exists. At low hard X-ray luminosity – below $\sim 10^{37}$ erg/sec – kT increases with L_X. At higher luminosity the spectral hardness depends weaker or does not depend at all on the hard X-ray luminosity. On the other hand for Cyg X–1 the spectral hardness is in general positively correlated with the relative amplitude of short-term variability. The low luminosity end of these approximate correlations (low kT and low rms) corresponds to extended episodes of very low hard X-ray flux which occurred during SIGMA observations.

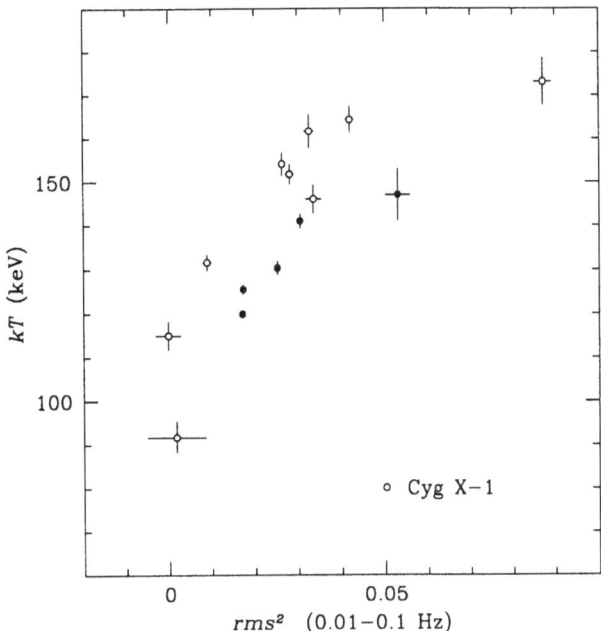

Fig. 7. The best-fit bremsstrahlung temperature plotted against the rms^2 of the short-term flux variations in the 0.01–0.1 Hz frequency range. Open circles correspond to Cygnus X-1, filled circles – to X-ray Nova Persei 1992 (GRO J0422+32). From Kuznetsov et al. 1997.

3.2   The black hole X–ray Novae

The X–ray Novae form distinct and being intensively studied class of objects (see Tanaka & Shibazaki 1996 for review). Typically they are X-ray binary systems composed of a low mass normal star and most likely a black hole. The presence of a black hole is dynamically proven in many cases – the X–ray Novae form most numerous so far class of objects known to harbor stellar mass black holes (Cowley 1992, Tanaka & Shibazaki 1996 and references therein).

Phenomenologically, from the viewpoint of the spectral evolution the X–ray Novae might be divided in to two subgroups according to the spectral state (low vs. high/superhigh) at the maximum of the X-ray light curve. The likely reason for that difference is the peak value of the dimensionless mass accretion rate $\dot{m} = \dot{M}/\dot{M}_{Crit}$. The X-ray Novae supposedly having higher $\dot{m}$ (e.g. Nova Vul 1989, Nova Mus 1991) pass through the entire range of the spectral states during their evolution, undergoing the spectral transitions similar to that observed for Cyg X–1 and GX339-4. The less luminous (referred below as "low $\dot{m}$") X–ray Novae (e.g. Nova Per 1992, Nova Oph 1993) never show soft spectral component and apparently are in the low/hard spectral state during entire outburst. The

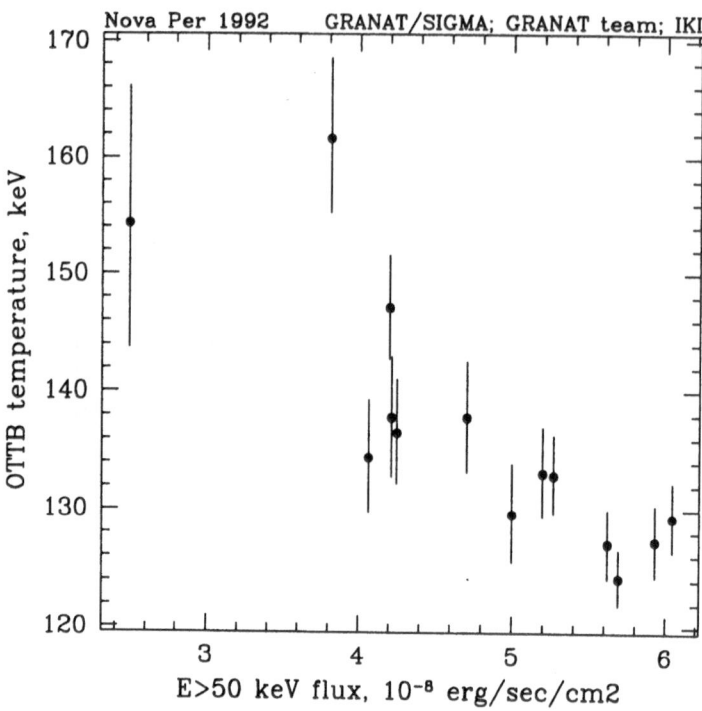

Fig. 8. The relation between spectral hardness and high energy flux for Nova Per 1992 (GRO J0422+32) (SIGMA data, from Vikhlinin et al. 1995).

spectral variability of the latter is considered below.

An example of spectral evolution (X–ray Nova Per 1992) is shown in Fig.8 – the decrease of hard X–ray flux was accompanied with hardening of the spectrum (Vikhlinin et al. 1995). Similar relation between spectral hardness and luminosity was observed for X–ray Nova Oph 1993 (e.g. Fig.2, see Revnivtsev et al. 1997a for the details). Such behavior is apparently opposite to that observed for Cyg X–1 and 1E1740.7—2942 – cf Fig.6 and 8.

3.3 Spectral variability of the hard spectral component

Study of the broad band spectra using the data of MIR–KVANT/TTM and ASCA observations close in time to the GRANAT/SIGMA observations indicates that change of the hard X–ray flux $\gtrsim 40$ keV traces change of the overall luminosity of the hard Comptonized spectral component (but not necessarily of the total X–ray luminosity). Therefore (Fig.6 and 8), the hardness of the Comptonized radiation and it's luminosity are *positively correlated* in the case of Cyg X–1 and *anti correlated* in the case of (at least some of) low $\dot{m}$ X–ray Novae.

It is very likely that in the case of the X–ray Novae the mass accretion rate is decreasing with time after the maximum of the light curve, i.e. change of the hard spectral component luminosity traces change of the mass accretion rate. For Cyg X–1 and 1E1740.7—2942 the change of the mass accretion rate is unclear and can't be directly determined. The study of the emission from the optically thick part of the disk (see below) indicates in the model dependent way that in the case of Cyg X–1 the low hard X–ray flux episode corresponded to increase of the mass accretion rate.

Thus, we may *tentatively* conclude that in the low spectral state of both low $\dot{m}$ X–ray Novae and Cyg X–1 the *increase* of the mass accretion rate leads to the *softening* of the spectrum of the Comptonized radiation. On the other hand relation between the mass accretion rate and the luminosity of Comptonized radiation is less unambiguous: the increase of the mass accretion rate might result in either increase of the Comptonized radiation luminosity (e.g. X–ray Nova Per) or it's decrease (Cyg X–1).

Regardless of the luminosity and the mass accretion rate change, the relation between the spectral hardness and the level of the short term aperiodic variability are qualitatively the same for Cyg X–1 and X–ray Nova Per (Fig.7). Similar behavior was recently found for GX339-4 (Trudolyubov et al. 1997).

4   Optically thick disk emission in the low spectral state

As was mentioned above the high/superhigh spectral states are characterized by presence of prominent soft spectral component which gives the dominant contribution to the $\lesssim 1$ keV to few hundred keV luminosity. In the low state the bulk of X–ray luminosity is emitted in the hard Comptonized spectral component. There are reasons to believe (e.g. Tanaka, 1989, Ebisawa et al. 1994) that the soft spectral component originates from geometrically thin, optically thick part of the accretion disk. The spectra observed during the superhigh spectral state show that the optically thick disk might coexist with rarefied optically thin hot region where the hard spectral component originates from. In this context the search for the soft component – emission from the optically thick part of the accretion disk in the low spectral state is of certain interest for understanding the structure of the accretion disk.

4.1   Cygnus X–1

The first indications of presence of the soft excess in the low state spectrum of Cyg X–1 were found about two decades ago (Priedhorsky et al. 1979, Balucinska-Church et al. (1995)). Most apparently it could be noticed from comparison of the values of the hydrogen column density inferred by the low energy cut-off of the X–ray spectrum, $\sim 3 \cdot 10^{21}$ cm^{-2}, with that known from 21 cm observations and interstellar reddening of the optical companion, $\sim 6 \cdot 10^{21}$ cm^{-2}. The data of the ASCA observations fully support this conclusion. That excess may be ascribed to the presence of the soft emission with the best fit blackbody temperature $T_{bb} \sim 120$ eV and luminosity of the order of $\sim 20\%$ of the overall 0.5–300 keV luminosity of the source[3] in the nominal low spectral state.

The approximation of the soft excess (ASCA observation in November, 1994) by a multicolor disk model (Shakura & Sunyaev 1973, Makishima et al. 1986) gives value of the temperature at the inner boundary of the accretion disk $kT_{in} \approx$ 136 eV and the radius of the inner boundary $R_{in} \approx 440 \pm 60$ km. The best fit value of the inner disk radius corresponds to $\approx 15R_g$ for the $10M_{\odot}$ black hole i.e. exceeds considerably the $3R_g$. That value might be interpreted as a radius at which the geometrically thin, optically thick disk approximation ceases to hold. As is well known (Shakura & Sunyaev 1976), the inner part of the standard accretion disk (Shakura & Sunyaev 1973) where the radiation pressure dominates is unstable. That inner part of the accretion flow could be responsible for generation of the hard spectral component observed in the low spectral state (Sunyaev et al. 1991; see recent discussions of the advection dominated flows by Chakrabarti & Titarchuk 1995, Narayan 1996).

Within that approximation the disk temperature depends upon the radius according to well known relation:

$$T(r) = \left(\frac{3G\dot{M}M}{8\pi\sigma r^3} \right)^{1/4} \left(1 - \left(\frac{r_0}{r} \right)^{1/2} \right)^{1/4}$$

(Shakura & Sunyaev 1973). Therefore, knowing that $T(R = 440km) = 136$ eV we can estimate the disk mass accretion rate

$$\dot{M}_{disk} \sim (2-3) \cdot 10^{17} \ g/sec$$

(assuming 2.5 kpc distance and $10M_{\odot}$ black hole). On the other hand knowing that the absorption corrected energy flux from the source at that time was $F_X(E > 300 \text{ eV}) \sim (4-5) \cdot 10^{-8}$ erg/sec/cm^2 and assuming the accretion efficiency of $\eta \sim 0.1$ we can obtain *independent* estimate of the total mass accretion rate

$$\dot{M}_{total} \sim (3-4) \cdot 10^{17} \times \left(\frac{0.1}{\eta} \right) \ g/sec,$$

which is consistent with that derived above from analysis of the low energy part of the spectrum. Note, however, that these two values estimate two essentially different quantities and could suffer from different uncertainties. The $\dot{M}_{total}$, according to the way it was derived, accounts for contribution of the matter possibly accreting via rarefied optically thin halo and, on the other hand, depends on assumed value of the accretion efficiency, i.e. importance of the advection. The $\dot{M}_{disk}$ is an estimate of accretion rate in geometrically thin keplerian disk and is not affected by these two effects, but depends on the validity of simple multicolor disk approximation.

---

[3] It should be mentioned that the luminosity estimate is rather uncertain in this case because bulk of the soft component emission is absorbed by the line–of–sight gas.

The above values of the accretion rate correspond to

$$\dot{M} \sim (0.01 - 0.03) \times \dot{M}_{crit}$$

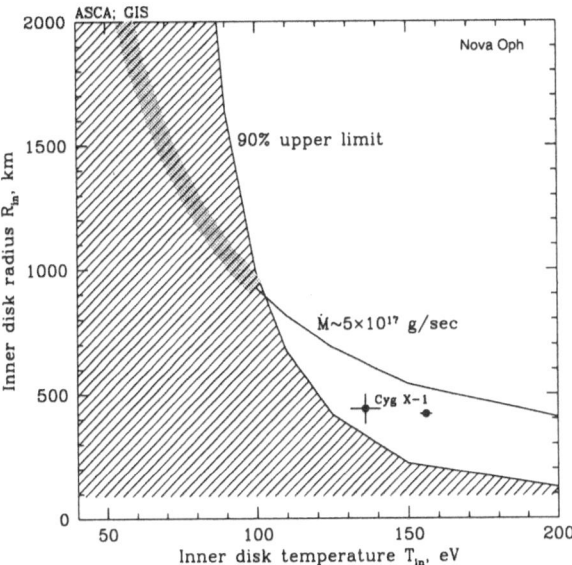

Fig. 9. Constraints on the radius and temperature at the inner boundary of the optically thick part of the disk in GRS1716–249 (X–ray Nova Oph 1993). ASCA observation on Oct.5,1993 (GIS data). The hatched area under the curve marked "90% upper limit" is a range of parameters consistent with the ASCA non detection of the soft excess in the spectrum (90% confidence, assuming $N_H = 4.5 \cdot 10^{21}$ cm^{-2}). The line marked "$\dot{M} \sim 5 \times 10^{17}$ g/sec" shows relation between R_{in} and T_{in} for that value of the disk accretion rate. Two points marked "Cyg X–1" are values measured for Cyg X–1 in 1994 (left) and 1993 (right). The source distance was assumed 2.5 kpc, $10 M_\odot$ black hole, $cos(i) = 0.5$.

4.2   GRS1716-249 (X–ray Nova Oph 1993)

The GRS1716-249 was observed by ASCA on Oct.5, 1993 near the maximum of it's X–ray luminosity. Opposite to Cyg X–1, the low frequency cut-off of the X–ray spectrum agrees quite well with the Galactic value of $N_H \approx 4.5 \cdot 10^{21}$ cm^{-2} (Della Valle et al. 1994). The upper limit on the luminosity of the soft component similar to the one observed in the spectrum of Cyg X–1 in 1994 during "standard" low spectral state (black body spectrum with $kT_{bb} \sim 100$ eV is $\sim 10^{36}$ erg/sec, assuming 2.5 kpc distance (Della Valle et al. 1994), which is less than $\sim 3\%$ of the total X-ray luminosity of the source.

The range of values of the temperature and the radius of the inner boundary of the optically thick part of the accretion disk, consistent with the ASCA GIS spectrum of the source is shown in Fig.9 (hatched area below the curve marked "90% upper limit").

The multicolor disk parameters could be restricted further, assuming that the total luminosity of the source may be used as an estimator of the disk mass accretion rate. The 0.5–300 keV source luminosity at the date of ASCA observation was $\approx 4 \cdot 10^{37}$ erg/sec (the high energy luminosity was measured by GRANAT/SIGMA – Revnivtsev et al. 1997a) which corresponds to $\dot{M}_{total} \sim 5 \times 10^{17}$ g/sec. The relation between T_{in} and R_{in} corresponding to that value of $\dot{M}_{disk}$ is shown in Fig.9 by a curve marked "$\dot{M} \sim 5 \times 10^{17}$ g/sec". Only part of that curve within the hatched area is consistent with non detection of the soft excess in the spectrum of the source by ASCA. Under this assumption, the accretion disk parameters are constrained by $T_{in} \lesssim 100 - 110$ ev, $R_{in} \gtrsim 900$ km $\approx 30 R_g$ for $10 M_\odot$ black hole.

On the other hand, assuming, that the inner disk radius is the same as for Cyg X–1, $\sim 15 R_g$, the upper limit on the inner disk temperature is $T_{in} \lesssim 120$ eV, which constraints the disk accretion rate: $\dot{M}_{disk} \lesssim 1 \cdot 10^{17}$ g/sec.

5   Broad band spectral variability of Cyg X–1

The Cyg X–1 was observed with ASCA satellite in a number of occasions during 1993–1995. Some of this observations were performed during extended low hard X-ray flux episode (although not at the minimum of the hard X-ray flux). This gives a possibility to study variations of the broad band spectrum of the source corresponding to the spectral changes observed in the high energy domain (Fig.6). We used for this purpose two ASCA observations performed in November 1993 (low hard X-ray flux episode) and one year later in November 1994 ("standard" hard spectral state of the source).

The broad band spectra of the source in 1993 and 1994 (ASCA and GRANAT/SIGMA data) are shown in Fig.10. The parameters of the spectral approximation of the ASCA data in the 0.5–10 keV energy range by the model consisting of the power law emission with reflected component and multicolor disk component (the hydrogen column density fixed at $6 \cdot 10^{21}$ cm^{-2}) are presented in Table 3.

The spectral changes observed by SIGMA in the high energy domain during low hard X-ray flux episode in 1993–1994 were accompanied with corresponding changes in the standard X-ray band (Fig.10 and second and third columns of Table 3):

1. The slope of the Comptonized radiation changes from ≈ 1.6 for standard hard state spectrum to ≈ 2.0 (low hard X-ray flux episode). As well known the steepening of the spectrum may be due to decrease of either the electron

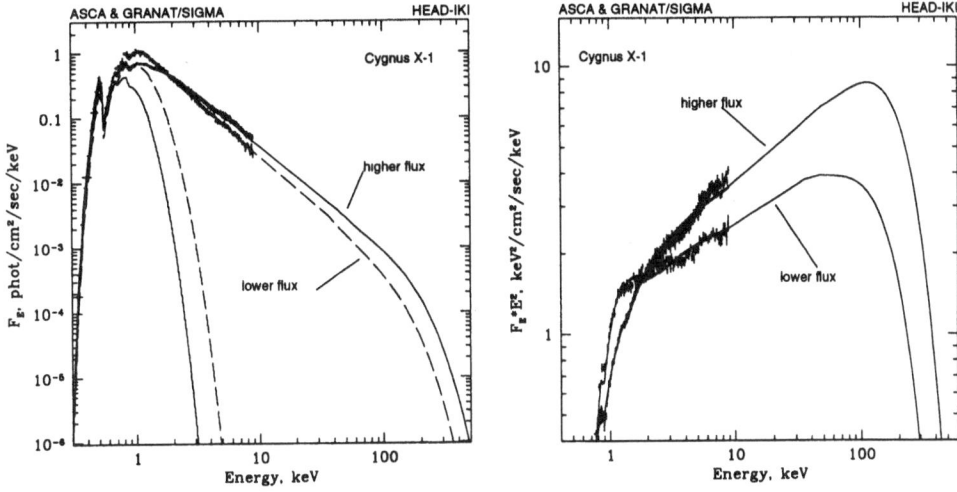

Fig. 10. The broad band spectra of Cyg X–1 in the "nominal" hard state (marked as "higher flux") and during very low hard X–ray flux episode observed in 1993 (marked "lower flux"). The behavior of the spectral flux density above 30 keV is schematically shown according to Kuznetsov et al. 1997.

temperature or the Thompson optical depth. The high energy data show that the position of the high energy cut-off of the spectrum shifts towards lower energy indicating that the electron temperature in the Comptonization region is decreasing.[4]

2. Parameters of the soft spectral component – supposedly emission from optically thick, geometrically thin outer part of the accretion disk – change as well. Qualitatively, these changes correspond to increase of the luminosity and the mean photon energy of the soft component. The relative contribution of the soft component to total X–ray luminosity increases as well. In terms of the multicolor disk approximation these changes may be attributed to increase of the disk temperature and possibly decrease of the radius of the inner boundary of the optically thick part of the accretion disk (the latter conclusion is tentative due to limited accuracy of inner radius determination, but see discussion below). Within that model these changes are caused by increase of the mass accretion rate by a factor of $\sim 1.5 - 2$.

The sign of change of the overall X–ray luminosity is rather uncertain mainly due to quite large value of interstellar absorption and the fact that the soft spectral component might contribute non negligible fraction to the total X–ray

---

[4] The ASCA and SIGMA data are not contemporaneous restricting therefore the possibility of reliable Comptonization model parameters estimate from the broad band spectral fitting

Table 3. The best fit parameters for the ASCA data approximation with the model consisting of the power law with reflection and multicolor disk emission.

Parameters [1]	25–26/11/94 "nominal" hard state	11-12/11/93 ~beginning of the low hard X–ray flux 1993 episode	30–31/05/96 [5] ~middle of the low hard X–ray flux 1996 episode
photon index, α	1.64 ± 0.01	1.99 ± 0.01	~ 2.2
T_{in}, eV	136 ± 5	156 ± 2	~ 470
R_{in}, km [3]	440 ± 60	420 ± 15	~ 110
F_{pl} [2]	0.86 (1.4)	0.75 (1.6)	3.1 (3.3)
F_{disk} [2]	0.02 (0.7)	0.06 (1.3)	0.2 (5.7)
$\dot{M}_{disk}$, 10^{17} g/sec [4]	~ 1.7	~ 2.8	~ 5

[1] The hydrogen column density was fixed at $N_H L = 6 \cdot 10^{21}$ cm^{-2}.
[2] The 0.3–9 keV energy flux of the power law and disk emission components, 10^{-8} erg/sec/cm^2 (the absorption corrected value is given in parenthesis)
[3] Assuming the source distance of 2.5 kpc, binary system inclination angle of 70°
[4] Estimated from the parameters of the soft excess using the multicolor disk approximation
[5] The multicolor disk approximation parameters were *roughly* estimated using the spectral parameters reported by Dotani et al., 1996

luminosity[5]. The observed (uncorrected for the low energy absorption) 0.5–300 keV luminosity slightly decreased from $\approx 4.1 \cdot 10^{-8}$ erg/sec/cm^2 ($\approx 3.1 \cdot 10^{37}$ erg/sec) to $\approx 2.6 \cdot 10^{-8}$ erg/sec/cm^2 ($\approx 2.0 \cdot 10^{37}$ erg/sec) with decrease of the hard X–ray flux.

Another episode of very low hard X–ray flux from Cyg X–1 occurred in May–June 1996 (Zhang et al. 1997). The source was intensively observed by GRO/BATSE in hard X–ray energy domain and by XTE and ASCA in the standard X–ray band. The accretion disk and Comptonized emission parameters roughly estimated using the spectral parameters reported by Dotani et al. (1996) are given in the fourth column of Table 3. The ASCA observation in 1993 occurred two month before the source reached the lowest value of the hard X–ray flux, whereas in 1996 Cyg X–1 was observed with ASCA nearly at the minimum of the hard X–ray flux. Correspondingly, 1996 observation found steeper power law, higher disk temperature, smaller disk inner radius and higher value of the mass accretion rate as estimated from the disk parameters (Table 3). It seems very likely, that both 1993 and 1996 events correspond to the same phenomena –

[5] Note that the for the hydrogen column density of $6 \cdot 10^{21}$ cm^2 more than $\approx 95\%$ of the disk emission with parameters given in the Table 3 is absorbed

transition of the source to the soft spectral state caused by increase of the mass accretion rate by a factor of few.

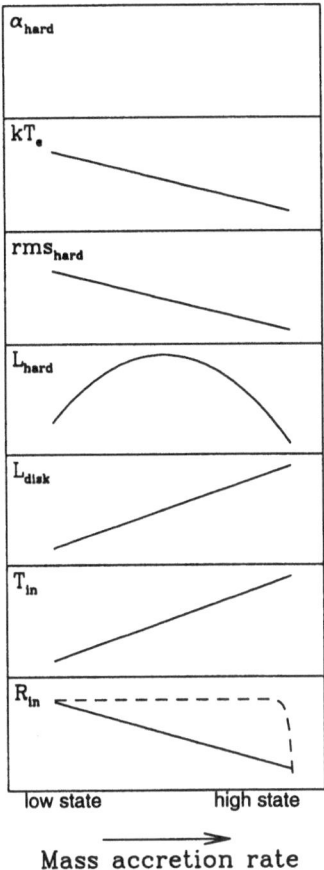

Fig. 11. Possible dependence of the hard and soft components parameters on the mass accretion rate (low through high spectral states).

6   Summary

Possible dependence of spectral characteristics of the soft and hard component upon the mass accretion rate is schematically shown in Fig.11. The range of M considered in Fig.11 covers low through high spectral states and doesn't include superhigh state supposedly corresponding to higher values of $\dot{M}$. These curves include a large degree of interpolation of the experimental data and therefore should be treated with certain caution. A fair amount of new broad band observations is needed to prove and detailize these dependences.

One of unclear points is dependence of radius of the inner boundary of the optically thick part of the accretion disk on the mass accretion rate. One possibility is that increase of the $\dot{M}$ results in smooth continuous decrease of R_{in}. The transition from the low state to the high state is consequently smooth in this case. On the other hand sudden increase of the soft X–ray luminosity observed by XTE/ASM during transition of Cyg X–1 to the soft state (Zhang et al. 1997) might indicate that two distinct accretion regimes exist: the lower $\dot{M}$ regime – the inner boundary is located sufficiently far from the compact object, at $\gtrsim (10 - 20)R_g$, and higher $\dot{M}$ regime, with the inner boundary being close to the compact object, $\sim 3R_g$. It should be mentioned though, that in any case change of the parameters of the hard spectral component (L_{hard}, α_{hard}, kT_e) is smooth (Zhang et al. 1997).

The opposite dependence of the hardness of the hard spectral component upon it's luminosity, observed for several low $\dot{m}$ X–ray Novae and, on the other hand, for Cyg X–1 and 1E1740.7-2942 could be understood assuming that these objects differ in value of $\dot{m}$. In both cases softening of the Comptonized radiation spectrum and decrease of the level of short term variability are caused by increase of the mass accretion rate, but sign of change of the Comptonized radiation luminosity is opposite at lower and higher values of the mass accretion rate (Fig.11). At low $\dot{m}$ fraction of the gravitational energy dissipated in the optically thick part of the accretion flow is small and change of the hard spectral component luminosity traces change of $\dot{M}$. Opposite to that, at higher $\dot{m}$ increase of $\dot{M}$ leads to increase of the fraction of gravitational energy dissipated in the optically thick part of the accretion flow and to decrease of the hard spectral component luminosity. This kind of non monotonic dependence of the hard component luminosity upon the mass accretion rate was probably observed for X–ray Nova Muscae after the secondary maximum of it's light curve (Paciesas et al. 1993, Ebisawa et al. 1994) and for GX339-4 (Harmon et al. 1994, Trudolyubov et al. 1997).

The consequence of non monotonic dependence of the hard X–ray luminosity upon the mass accretion rate is that at sufficiently small $\dot{M}$ Cyg X–1 and 1E1740.7-2942 should show hardness upon luminosity dependence similar to that observed for low $\dot{m}$ X–ray Novae. Such behavior was possibly observed for Cyg X–1 – the $kT - L_X$ dependence splits into two branches at $L_X \lesssim 10^{37}$ erg/sec (see Ballet et al. 1996, Kuznetsov et al. 1996).

Acknowledgment: This research has made use of data obtained through the High Energy Astrophysics Science Archive Research Center Online Service, provided by the NASA/Goddard Space Flight Center. Part of this work was supported by the RBRF grant 96-02-18588.

References

Ballet J. et al., 1996, ASCA Symp. Proc., X-Ray imaging and spectroscopy of cosmic hot Plasma. Tokyo, in press

Balucinska–Church M. et al., 1995 Astron. Astrophys., 302, 5

Barret D. & Vedrenne G., 1994, ApJS, 92, 505

Belloni T. et al., 1996a, ApJ, 472, 107

Belloni T. et al., 1996b, submitted to A&A

Chakrabarti S. & Titarchuk L., 1995, ApJ, 455, 623

Churazov E. et al., 1993, ApJ, 407, 752

Cordier B. et al., 1993, ApJ, 272, 277

Cowley A., 1992, Annu.Rev.Astron.Astrophys., 30, 287

Crary D.J. et al., 1996, ApJ, 462, L71

Della Valle M. et al., 1994, Astron. Astrophys., 290, 803

Dotani T. et al., 1996, IAUC 6415

Ebisawa K. et al., 1994, PASJ, 46, 375

Ford E. et al., 1996, ApJ 469, L37

Gilfanov M. et al., 1991, Pis'ma v Astron.Zhurn. 17, 1059

Gilfanov M. et al., 1993, in: Alpar, M., Kiziloglu, U., Van Paradijs, J., eds., AIP Conf. Proc., The lives of the neutron stars, Kluwer, Dordrecht, NATO ASI Ser., 308, p. 712

Harmon B. A. et al., 1994, ApJ, 425, L17

Holt S. et al., 1976, ApJ, 203, L63

Grebenev S. et al, 1993, A&A Suppl.Ser., 97, 281

Kuznetsov S. et al., 1996, Proc. of Röentgenstrahlung from the Universe. Garching, MPE Report 263, 157

Kuznetsov S. et al., 1997, to be submitted to M.N.R.A.S.

Langmeier A. et al., 1987, ApJ, 323, 288

Makishima K. et al., 1986, ApJ, 308, 635

Mitsuda K. et al., 1989, PASJ, 41,97

Miyamoto S. et al. 1991, ApJ, 383, 784

Miyamoto S. et al., 1994, ApJ, 435, 398

Miyamoto S. et al., 1995, ApJ, 442, 13

Narayan R., 1996, ApJ, 462, 136

Ogawara Y. et al., 1982, Nature, 295, 675

Paciesas W. et al., 1993, Alabama Univ. Preprint

Paul J. et al., 1991, Adv. Space Res., 11, 289

Phlips B. et al., ApJ, 1996, 465, 907

Priedhorsky W. et al., ApJ, 1979, 233, 350

Revnivtsev M. et al., 1997a, to be submitted to Astron.Astrophys

Revnivtsev M. et al., 1997b, to be submitted to Astron.Astrophys

Salotti L. et al., 1992, A&A, 253, 245

Shakura N. & Sunyaev R., 1973 Astron.Astrophys, 24, 337

Shakura N. & Sunyaev R., 1976 M.N.R.A.S., 175, 613

Sunyaev R. & Truemper J., 1979, Nature, 279, 506

Sunyaev R. & Titarchuk L., 1980, A&A, 86, 121

Sunyaev R. & Shakura N., 1986 Sov.Astron.Lett, 12, 117

Sunyaev R. et al., 1991, A&A, 247, L29

Sunyaev R. et al., 1993, A&A, 280, L1

Tanaka Y., 1989, *in Proceedings of* "23rd ESLAB Symposium", Bologna, Italy, ESA
   SP-296, editors: J.Hunt & B.Battrick, v.1, p.3
Tanaka Y. & Shibazaki N., 1996, Annu.Rev.Astron.Astrophys., 34, 607
Tananbaum H. et al. 1972, ApJ, 177, L5
Trudolyubov S. et al., 1996, Pisma v Astron. Zhurnal, v.22, p.740
Trudolyubov S. et al., 1997, to be submitted to Astron.Astrophys
van der Klis M., 1995, in "X–ray binaries", Eds: Lewin W., van Paradijs J. & van den
   Heuvel E., Cambridge Univ. Press, p.252
Vikhlinin A. et al., 1994, ApJ, 424, 395
Vikhlinin A. et al., 1995, ApJ, 441, 779
White N., Nagase F, & Parmar A., 1995, in "X–ray binaries", Eds: Lewin W., van
   Paradijs J. & van den Heuvel E., Cambridge Univ. Press, p.1
Zhang S. et al., 1997, submitted to ApJ Letters

X-ray spectrum of a disk illuminated by ions

H.C. Spruit

Max-Planck-Institut für Astrophysik, Postfach 1523, D-85740 Garching

Abstract: The X-ray spectrum from a cool disk inside an ion supported torus is computed. A surface layer of the disk is heated by the protons from the torus. It produces a comptonized spectrum with a shape that is only very weakly dependent on the incident energy flux and the distance from the accreting compact object, and is similar to observed spectra. Further evidence for 'ion illuminated' disks are the Li abundance in the secondaries of low mass X-ray binaries and the 450 keV lines sometimes seen in black-hole transient spectra.

1   Introduction

The X-ray spectra $F(E)$ of AGN as well as galactic black hole candidates (BHC) in their hard states are characterized by a power law of index ≈ 1 and a high energy cutoff E_c around 200 keV. Such spectra are well known to be describable by comptonization in an electron scattering layer of optical depth $\tau \approx 0.5$ and temperature $T \sim E_c$. One of the classsical problems in X-ray astronomy is to explain why τ and T should have just these values, with little variation between sources. Theoretical arguments can be given that comptonization is in fact the most important interaction between matter and radiation at temperatures of 10–100 keV, for the inferred radiation energy densities near accreting black holes, but this does not tell us what the the thickness and temperature of the interaction region are.

An optically thick accretion disk would produce spectra peaking at 1 keV and 10–100 eV for BHC and AGN, repectively. The conditions in BHC and AGN allow (at accretion rates well below Eddington) for a second form of accretion, an ion supported advection torus (Shapiro et al. 1976, Liang 1979, Rees et al. 1982, Fabian and Rees 1995, Narayan etal 1995, 1996). The ions in this flow are near their virial temperature, the electrons much cooler because of their weak interaction with the ions and their strong interaction with the radiation field. Such flows could produce, in principle, the kind of spectrum observed (Narayan et al. 1995, 1996), but again it is not clear which physics restricts the optical depth and temperature of the flow to the observed narrow ranges (cf. Haardt 1997, Maraschi and Haardt 1997).

1.1   Evaporating disks inside tori

Various geometries for the accretion flow near a black hole have been developed and are reviewed by S. Collin elsewhere in this volume. One of the possibilities is an ion supported advection torus coexisting with an optically thick accretion disk embedded in it (see fig. 1 in Collin). This possibility is attractive because the spectra of BHC in their high states show evidence of the simultaneous presence of an optically thick, thermal, accretion disk and a hotter component which produces a power law tail at higher photon energies (e.g. Tanaka, this volume).

Theoretically, one would expect exchange of both mass an energy to take place between the disk and the advection torus. Heating of the disk surface by the hot ion supported flow above would lead to an 'evaporation' of the disk surface, feeding mass into the torus. Such evaporation has been studied in detail for the case of disks in Cataclysmic Variable systems by Meyer and Meyer-Hofmeister 1994. In the inner regions of the disk the mass available in the disk is smallest, and the energy budget potentially available for evaporation largest. If the mass flow from disk into torus increases with the energy dissipation rate, and if a steady state develops, one could therefore envisage a structure consisting of three regions: an outer one in which only a geometrically thin optically thick disk is present, inside this a composite region with an evaporating disk inside a hot ion supported advection torus, and inside this a region in which only an ion supported flow exists because all disk mass has evaporated (Meyer and Meyer-Hofmeister 1994).

Depending on details of the processes of mass and energy exchange between disk and torus, the boundaries between these regions may vary. It is not necessary that the structure is steady. The model has, in principle, sufficient ingredients to allow for variability and may perhaps be developed further in the context of the various forms of variability seen in BHC.

1.2   Energy exchange between disk and torus

Energy exchange between disk and torus may be mediated by particles or by radiation. If the interaction is by radiation only, one obtains a model like that of Haardt and Maraschi (1991). To simplify the discussion, assume that the accretion takes place predominantly through the torus (this assumption can easily be relaxed). The radiation produced by the torus illuminates the disk below, which thermalizes it into an approximate blackbody spectrum. These (soft) photons are comptonized in the hot torus. In this model, approximately half the energy comes out as soft radiation and half as comptonized photons. It correctly predicts the slope of the spectrum, but still requires an ad hoc assumption about the optical depth of comptonizing region to get the right cutoff energy E_c. For further developments of this model see Haardt (1997).

As second possibility for energetic interaction are the hot protons in the torus with temperature near the virial temperature, $T_p \approx T_v \approx 400 r_g/r$ MeV. At the distance dominating the energy release, $r \approx 7r_g$, the protons thus have a temperature around 50 MeV. At this energy, they have a significant penetration depth into the cool disk. They are slowed down mainly by Coulomb interactions with the ensemble of electrons inside their Debye sphere. The 'stopping depth', expressed in terms of the corresponding Thompson optical depth, is

$$\tau_s \approx \frac{m_p}{3m_e \ln \Lambda} \frac{\beta^4}{\psi - x\psi'}, \tag{1}$$

(e.g. Ryter et al., 1970) where $\beta = v_z/c \approx (kT_p/m_pc^2)^{1/2}$ is the vertical component of the proton velocity, $\theta = kT/m_ec^2$ is a measure of the temperature of the heated layer, $\ln \Lambda \approx 20$ is a Coulomb logarithm, $x^2 = \beta^2/(2\theta)$, and ψ the error function

$$\psi = \frac{2}{\sqrt{\pi}} \int_0^x e^{-x^2} dx. \tag{2}$$

At low electron temperature $kT < m_e/m_p \, kT_p$, x is small and the factor involving the error function can be expanded. This yields

$$\tau_s \approx \frac{m_p}{m_e \ln \Lambda} \beta \theta^{3/2},$$
$$\approx \left(\frac{kT_p}{50\text{MeV}}\right)^{1/2} \left(\frac{kT_e}{50\text{keV}}\right)^{3/2}. \tag{3}$$

2   Comptonization in a layer heated by protons

2.1   Estimating the depth of the comptonizing layer

Heating by protons yields a Comptonizing layer of thickness equal to the stopping depth τ_s. This depth is a function of the electron temperature in the layer, by (2). The electron temperature on the other hand is determined by the heating and cooling processes, so that a consistent calculation of heating and cooling will yield both the electron temperature and the optical depth of the layer. With a simple estimate, we can now show that this will yield τ_s and T_s in roughly the right range.

The cooling process in the layer is the inverse Compton process, i.e. the energy loss electrons experience as they scatter the soft photons from the cool disk below. We assume that these soft photons are all (or mostly) produced by thermalization of comptonized photons from the heated layer, as in the model of Haardt and Maraschi (1991, hereafter HM). Since approximately half the Comptonized photons escape and the other half illuminates the thermalizing layer, the energy flux in the soft photons at the base of the layer must be about the same as that in the escaping comptonized photons. Such a balance is possible only if the Comptonization is sufficiently strong. In terms of the Compton y-parameter $y \approx 4\theta\tau_s$, it requires that $y \approx 1$. If the electron temperature is too low, Compton cooling of the electrons by the soft photons is too low and the layer heats up until $y \approx 1$, and vice versa. Since the y-parameter also determines the slope of the X-ray spectrum, the model yields a fixed spectral slope, which is in the range of the observed values. This is the reason for the success of the HM model. Whereas in HM the depth of the layer has to be assumed, we are now in a position to compute it, by requiring the stopping depth τ_s to be consistent with the resulting electron temperature. A simple estimate is obtained by setting

$y = 1$, or $\theta = 1/(4\tau_s)$, and inserting into (2). This yields, assuming again that the protons are near their virial temperature:

$$\tau_s^{5/2} = \frac{m_p}{8\sqrt{6}\ln \varLambda m_e}\left(\frac{r_g}{r}\right)^{1/2},\qquad (4)$$

or

$$\tau_s \approx 1.3\left(\frac{7r_g}{r}\right)^{1/5},\qquad (5)$$

and

$$kT_e \approx m_e c^2/(4\tau_s) \approx 60\left(\frac{r}{7r_g}\right)^{1/5}\quad \text{keV}.\qquad (6)$$

We conclude that proton illumination yields optical depths and temperatures in the right range, with only a weak dependence on the assumed distance from the black hole. Obviously, the estimate is rather crude, and more detailed calculations of the energy transfer from the protons to the electrons, as well as the Comptonization process are needed to test the model. In the following we make a first step in this direction, by means of a radiative transfer calculation.

2.2   Model problem

The aim of the calculation reported below is to compute the electron temperature as a function of depth, together with the emergent photon spectrum from a layer heated by protons who deposit their energy according to 2. We do this in two steps. First, we assume the heating rate to be distributed uniformly over a layer with an assumed depth τ_s. In this layer, we solve the radiative transfer equation together with the electron temperature $T_e(\tau)$, such that the heating is in balance with cooling by comptonization of the soft photons. The loop is then closed by comparing the electron temperature found with the assumed stopping length 2.

The assumptions and simplifications that go into the model are as follows. The heating by the protons is approximated as uniform between $\tau = 0$ (the surface of the disk) and τ_s. In this layer, the only photon process is electron scattering (Comptonization). Below this, we assume that electron scattering continues to be the dominant process down to some depth τ_b. At τ_b, the downward photons are assumed to be absorbed and their energy reradiated upward as a black body spectrum. Thus, the gradual thermalization with depth by free-free processes is simplified by a step at depth τ_b. The value of τ_s is determined from the proton velocity $\beta = v/c$ and the mean electron temperature in the heated layer. Obvious improvements are possuible on these simplifications, by explicitly taking into account photon production/destruction process, an by a more accurate treatment of the energy loss of the ions as they penetrate into the disk.

The radiative transport part of the problem is simplified by reducing the angular dependences to a one-stream model: only vertically upward and downward moving photons are considered, and the electron distribution is similarly reduced from a 3-D to a one-dimensional Maxwellian distribution. For the scattering cross section and the Maxwellian the proper relativistic expressions are used. This simplification is made for programming convenience only: leaving out the full angular dependences leads to very simple expressions. Discretization by a reasonable number of angles would still yield a very modest problem in terms of computing time, and is an obvious next step to improve the calculations.

In the following an outline is given of the method used, a full description will be given elsewhere.

2.3   Numerical method

The transport equation is of the form (e.g. Rybicki and Lightman, 1976):

$$\frac{dn}{dz} = \int d^3 p \int d\Omega' \frac{d\sigma}{d\Omega}[f_e(p')n(\omega')(1 + n(\omega)) - f_e(p)n(\omega)(1 + n(\omega'))], \quad (7)$$

where $n(\omega)$ is the photon occupation number, f_e the electron momentum distribution, $\mathbf{p}$ (respectively $\mathbf{p}'$) the electron momentum, and $\boldsymbol{\omega}$ ($\boldsymbol{\omega}'$) the photon momentum vectors before (after) scattering. The dependences of $\mathbf{p}'$ and $\boldsymbol{\omega}'$) on $(\mathbf{p}, \boldsymbol{\omega}, \boldsymbol{\Omega}')$ follow from the collision kinetics.

The equation is discretized in N_ω logarithmically spaced photon-energy bins. As photon energy scale we use $\omega = h\nu/m_e c^2$. As depth scale we use the Thompson optical depth τ_T. If n_i^+ and n_i^- are the occupation numbers of the upward and downward moving photons in bin i, the result is the set of $2N_\omega$ equations

$$\mp \frac{d}{d\tau}n_i^{\pm} = n_i^{\pm}\sum_j[(B_{ji} - B_{ij})n_j^{\mp} - B_{ij}] + \sum_j B_{ji}n_j^{\mp}, \quad (8)$$

where $B_{ij}(T_e)$ is the scattering cross section integrated over the appropriate frequency and electron momenta, for scattering from bin i into bin j, in units of the Thompson cross section σ_T. The boundary conditions are

$$n_i^- = 0 \qquad (\tau = 0), \quad (9)$$

$$n_i^+ = n_{BB}(\omega_i) = 1/[1 - \exp(\omega/\theta_b)] \qquad (\tau = \tau_b). \quad (10)$$

Here n_{BB} is the black-body occupation number at temperature θ_b, which follows from the condition that the net energy flux at depth τ_b vanishes:

$$\int \omega^3[n^-(\tau_b) - n_{BB}(\omega, \theta_b)] = 0. \quad (11)$$

The condition of energy balance between the assumed heating rate $h(\tau)$ and the Compton cooling by the soft photons is

$$\frac{dF}{dz} = h, \quad (12)$$

where the energy flux F is given by

$$F = \sum_i \omega^3 (n_i^+ - n_i^-).$$ (13)

The equations (8) are discretized in optical depth by centered first order differences. The resulting set of nonlinear algebraic equations is solved by an iterative process. It turned out that full simultaneous linearization of the transfer equation, the boundary conditions and the energy equation had very poor convergence properties. Instead, an iteration was done in which only the transfer equation and the upper boundary condition were linearized, while the lower bc and the energy equation were dealt with by a modified succesive-substitution process after each iteration of the transfer equation. Convergence, however, was still problematic for large optical depths and for cases where the assumed photon energy range extended too far beyond the cutoff energy. For the cases reported here, where the optical depth is not too large, on the order of 30–100 iterations were required for an accuracy of 10^{-4} in luminosity.

3   Results

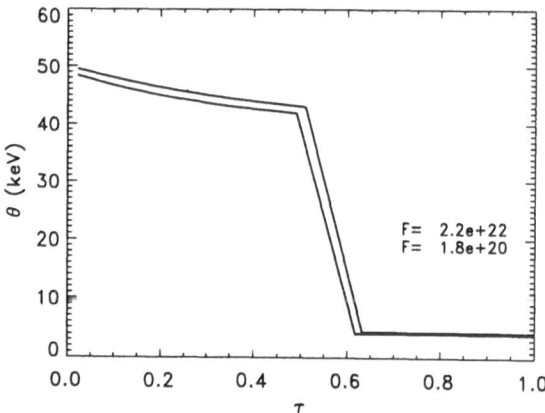

Fig. 1. Temperature as a function of optical depth in an electron scattering layer heated by virialized protons incident with energy flux F (erg/cm^2/s), for $r/r_g = 7$.

The parameters of the problem are the total energy flux F (per unit surface area of the disk), the velocity of the incident protons, and the optical depth τ_b of the thermalizing lower boundary. Assuming the protons to be thermal and virialized,

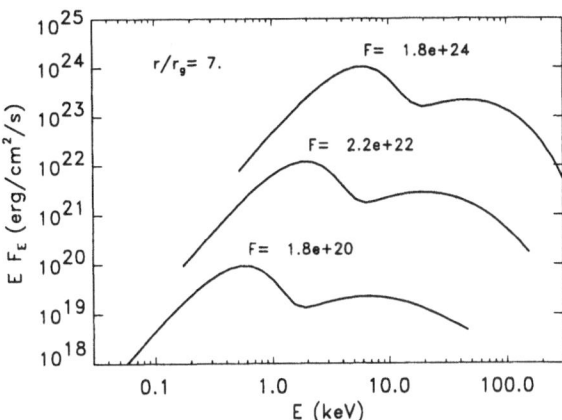

Fig. 2. Emergent spectrum of a layer heated by protons at $r = 7r_g$ for three values of the energy flux.

their mean vertical velocity component β_z as a function of the distance r from the compact object is

$$\beta_z = \left(\frac{r_g}{6r}\right)^{1/2} \tag{14}$$

The electron temperature as a function of depth is shown in figure 1 for a few values of F for $r/r_g = 7$. These fluxes correspond to effective temperatures $\theta_{\text{eff}} = (F/\sigma)^{1/4}/(m_ec^2)$ of $3\,10^{-4}$, 10^{-3} and $3\,10^{-3}$, respectively. The emergent spectrum for these energy fluxes is shown in figure 3, for $r/r_g = 7$. The dependence of the spectrum on r/r_g at a fixed F is shown in figure 2. The penetration depths are shown in figure 4. All of these cases were computed for $\tau_b = 1$.

The similarity of the spectra, apart from shifts in amplitude and photon energy, is remarkable. The temperature of the heated layer increases only very weakly with increasing energy flux. The cutoff energy decreases somewhat with distance from the hole, but again this is dependence is rather weak. On account of modest optical depth of the layer, the spectrum shows a prominent contribution from unscattered soft photons from the reprocessing depth τ_b. This peak are smeared out somewhat when the spectra are convolved over distance from the hole, at a given accretion rate (not shown here).

4 Discussion

With an admittedly somewhat simplified radiative transfer model I have shown that heating of a cool disk by protons from an ion supported advection torus reproduces the main features of the hard X-ray spectra of accreting black holes. Like the Maraschi and Haardt model (and for the same reason), it yields approximately the right spectral slope, but in addition, it also predicts approximately

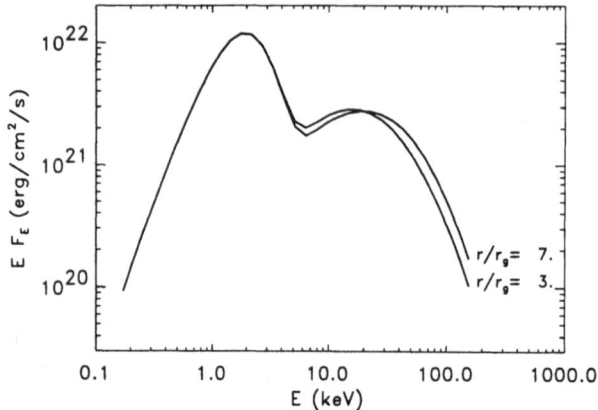

Fig. 3. Dependence of the spectrum on distance from the compact object, for a fixed energy flux.

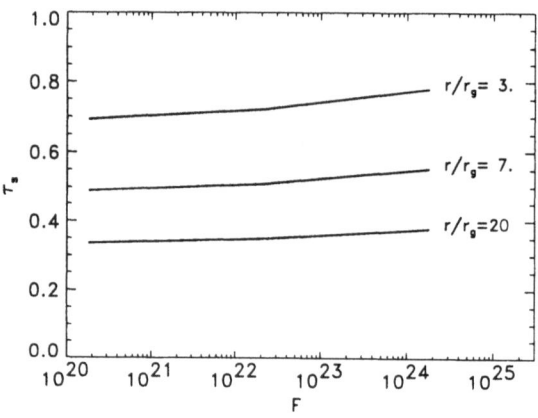

Fig. 4. Thomson depth of the heated layer as a function of energy flux and distance from the compact object.

the right optical depth of the comptonizing layer and the cutoff energy of the spectrum.

The temperature of the heated layer is insensitive to the energy flux, and stays around 40–60 keV. Instead of getting hotter at high energy flux, the incident energy is spent in upscattering a larger number of soft photons. As in the Maraschi and Haardt model, the reason for this lies in the energy balance condition. In order for the incident energy to be radiated as Comptonized flux, the Compton y-parameter has to be of of the order unity, and the temperature of the layer adjusts accordingly. What is new in the results presented here is that the optical depth of the Comptonizing layer also comes out naturally in the right range, due to the physics of Coulomb interaction between the virialized protons with the electrons in the cool disk. This process also is fairly insensitive to the proton temperature (within the relevant range), so that the emergent spectrum is only a weak function of distance from the accreting object (figure 3).

The agreement of the result with one of the most puzzling features of the hard X-ray spectra of accreting compact objects, viz. the uniformity of the spectral shape, makes it likely that ion illumination plays a major role in the physics of these objects.

4.1 Lithium

An independent observational indication for ion illumination is the observation of high Li abundances in the secondary stars of X-ray binaries (Martín et al. 1992, 1994a, 1995). The energy of virialized protons hitting the cool disk at $r = 7r_g$ (where the gravitational energy release peaks for accretion onto a hole) is around 50 MeV, just in the range where Li production by spallation of CNO elements becomes efficient. Since the observed secondaries are of spectral types known to destroy Li on a rather short time scale, a significant continual source of Li is needed. As shown by Martín et al. (1994) the energetics of the accretion process is enough to explain the observed amount of Li on the secondary, if a fraction 10^{-3} of the Li produced in the disk finds its way to the secondary (in the form of a disk wind, for example). Another consequence of ion illumination would be Li and Be production by He ions from the virialized flow reacting with He ions in the disk. These reactions peak around 50 Mev/nucleon, and are accompanied with emission of $\gamma-$lines at 431 and 478 keV. It is possible that the $\gamma-$lines observed sometimes around this energy (Gilfanov et al. 1991, Sunyaev et al. 1992) are another signature of ion illumination (Martín et al. 1994a,b).

Acknowledgement: This work was done in the context of Human Capital and Mobility network 'Accretion onto compact objects', CHRX-CT93-0329.

References

Fabian A.C., Rees M.J., 1995, MNRAS 277, 55
Gilfanov M., et al., 1991, SvA Lett. 17(12), 1059
Haardt F., Maraschi L., 1991, ApJ 413, 507
Haardt F., 1997, Mem Soc. Astron. It., in press
Maraschi L., Haardt F., 1996, in Accretion Phenomena and Related Outflows, D. Wick-
    ramasinghe, L. Ferrario & G.Bicknell (eds.), IAU Conference, in press
Liang E. P. T., 1979, ApJ 218, 247
Martín E., Rebolo R., Casares J. Charles P.A., 1992, Nature 358, 129
Martín E., Rebolo R., Casares J. Charles P.A., 1994a, ApJ 435, 791
Martín E., Spruit H.C. van Paradijs J., 1994b, A&A 291, L43
Martín E., Casares J. Charles P.A., Rebolo R., 1995, A&A 303, 785

Meyer F. Meyer-Hofmeister E., 1994, A&A 288, 175

Narayan R., Yi I., 1995, ApJ 452, 710

Narayan R., McClintock J.E., Yi I., 1996, ApJ 457, 821

Rees M.J., Begelman M.C., Blandford R.D., Phinney E.S., 1982, Nature 295, 17

Rybicki–L xx

Ryter C., Reeves H., Gradsztajn E., Audouze J., 1970, A&A 8, 389

Shapiro S.L., Lightman A.P., Eardley D.M., 1976, ApJ 204, 187

Sunyaev R.A., et al., 1992, ApJ 389, L75

Anisotropic illumination of accretion disc in Seyfert I galaxies

P.O. Petrucci, G. Henri

Laboratoire d'Astrophysique, Observatoire de Grenoble, B.P 53X, F38041 Grenoble Cedex, France

Abstract: We present a new emission model of Seyfert I galaxies where the accretion disk luminosity is entirely due to the reprocessing of hard radiation impinging on the disk. The hard radiation itself is emitted by a hot point source above the disk, that could be physically realized by a strong shock terminating an aborted jet. This hot source contains ultra-relativistic leptons scattering the disk soft photons by Inverse Compton (IC) process. Using simple formula to describe the IC process in an anisotropic photon field, we derive a self-consistent solution in the Newtonian geometry, where the angular distribution of soft and hard radiation, and the radial profile of the disk effective temperature are determined in a univocal way. This offers an alternative picture to the standard accretion disk emission law, reproducing individual spectra and predicting new scaling laws that fit better the observed statistical properties. General relativistic calculations are also carried out. It appears that differences with the Newtonian case are weak, unless the hot source is very close to the black hole.

1 Introduction

It is widely believed that the high energy emission of radio-quiet AGNs is produced by Comptonization of soft photons by high energy electrons or pairs. Two classes of models have been proposed so far: Comptonization by a thermal optically thick plasma, resulting in a lot of scattering events associated with small energy changes, or Inverse Compton (IC) process by one or few scattering events from a highly relativistic, non thermal particles distribution, which can result from a pair cascade.

Detailed observations in the X-ray range by the Ginga satellite have shown that a simple power law is unable to fit the X-ray spectrum of Seyfert galaxies. Rather, the spectra are better reproduced by a complex superposition of a primary power law, with an index $\alpha \simeq 0.9 - 1.0$, a reflected component from a cold thick gas, a fluorescent Fe Kα line and an absorption edge by a warm absorber (Pounds et al. Pounds (1990); Nandra & Pounds 1994). The second and third components could be produced by the reflection of primary hard radiation on an accretion disk surrounding the putative massive black hole powering the AGN (Lightman & White 1988; George & Fabian 1991; Matt, Perola & Piro 1991). This has led to consider various geometries where the hot source is located above the disk and reilluminates it, producing the observed reflection features. The hot source can be a non-thermal plasma (Zdziarski et al. 1990), or a thermal hot corona covering the disk (Haardt & Maraschi 1991, 1993; Field & Rogers 1993).

In another context, some observational facts have motivated the development of so-called reillumination models, where high energy radiation reflected on a cold surface (presumably again the surface of an accretion disk), produces a fair fraction of thermal UV-optical radiation. Firstly, long term observations have shown that for some Seyfert galaxies, such as NGC 4151 (Perola et al. Perola (1986)) and NGC 5548 (Clavel et al. Clavel (1992)), UV and optical luminosities were varying simultaneously, and correlated with X-ray variability on time scales of months, whereas the rapid, short-scale X-ray variability was not seen in optical-UV range. This is in contradiction with the predictions of a standard, Shakura-Sunyaev (SS) accretion disk model (Shakura & Sunyaev 1973), where any perturbation causing optical variability should cross the disk at most at the sound velocity, producing a much larger lag between optical and UV than what is actually observed. Rather, these observations support the idea that optical-UV radiation is largely produced by reprocessing of X-rays emitted by a small hot source, the UV and optical radiation being emitted at larger distances. The main problem is that the apparent X-ray luminosity is usually much lower than the optical-UV continuum contained in the Blue Bump, whereas one would expect about the same intensity in both components if half of the primary hard radiation is emitted directly towards the observer and the other half is reprocessed by the disk.

In many cases also, the equivalent width of the Fe $K\alpha$ line requires more impinging radiation than what is actually observed if explained by the reflection model (Weaver et al. 1995). As an explanation, Ghisellini et al. (1991), have proposed that the anisotropy of soft radiation could lead to an anisotropic IC emission, with much more radiation being scattered backward than forward. Due to the complexity of their calculation, they have restricted themselves to the emission by a hemispheric bowl (equivalent to an infinite plane), that could model a flared accretion disk with a constant temperature. The thermal disk-corona model faces the same kind of difficulties, for it predicts nearly the same luminosity in X-ray and UV ranges. Furthermore, it is difficult to explain very rapid X-ray variability if the corona covers a large part of the disk.

Although the spectral break observed by OSSE around 100 keV seems to favor thermal models and disprove the simplest pair cascade models, such a break could also be obtained by a relativistic particles distribution with an appropriate upper energy cut-off, such can be provided for example by pair reacceleration to avoid pair run-away (Done et al. 1990, Henri & Pelletier 1993).

The subject of this paper is to present a new model develop by Henri & Petrucci (submitted) where, on the one hand, the primary power-law emission is associated with the emission from a hot point source of relativistic leptons (the hot source) located above the accretion disk, and, on the other hand, the UV-optical component (the Blue Bump) is associated with the reprocessed radiation on the disk, together with a Compton backscattered component in X-rays (not taken into account in the present model). The crucial hypothesis is to suppose that the emission of the disk is entirely due to the reprocessing of hard radiation coming from the hot source, which is at turn produced by IC process on the soft photon emitted by the disk. In this case, a unique angular distribution of hard radiation and a unique (properly scaled) disk temperature radial profile are predicted. We derive evenly new scaling laws for luminosity and central temperature as a function of the mass. We show that the predictions of the model are sensitively different from the standard ones, and that they could better explain the observations. We have treated both Newtonian and Kerr metrics cases. The organization of these notes is as follow. We first present the main characteristics of the model in section 2 as well as the general equations governing the radiative balance between the hot source and the accretion disk in an axisymmetric gravitational field. Then, in section 3, we give the most interesting results obtained with this model before concluding in section 4.

2   The model

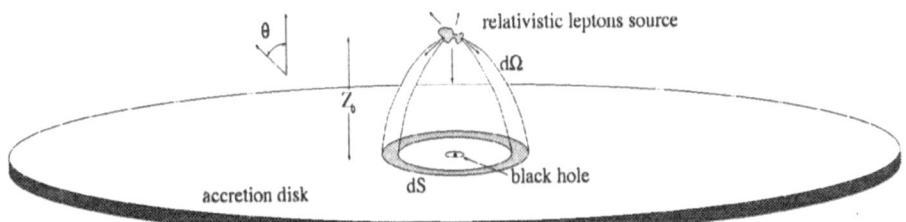

Fig. 1. The general picture of the model in Kerr metrics. We have also drawn the trajectory of a beam of photons emitted by the hot source in a solid angle $d\Omega$ and absorbed by a surface ring dS on the disk. In a Newtonian metrics, the photons trajectories are obviously straight lines. The scale is not respected.

2.1   Assumptions of the model

As we have previously explained, we consider a self-consistent model where Inverse Compton process takes place on soft photons from the accretion disk, which are themselves emitted as thermal radiation due to the heating of the disk by hard radiation. The disk is modelized by an infinite slab radiating isotropically like a black-body at the same equilibrium temperature (cf. Fig. 1). The high energy source is assumed to be an optically thin plasma of highly relativistic leptons, at rest at a given distance Z_0 above the disk, on its rotation axis. Its size is small enough to be considered as a point source. Although the complete study this source is out of the scope of this paper (and will be the subject of a future work), we have in mind that it can be physically realized by a strong shock terminating an aborted jet. We have considered both Newtonian and Kerr geometries, taking into account, in the later case, the curvature of photon geodesics as well as gravitational and Doppler shifts.

2.2   The main formulae

In this section, I only treat, briefly, the Newtonian case. Let us consider a relativistic charged particle, characterized by its the Lorentz factor $\gamma = (1-\beta^2)^{-1/2}$, and the soft photon field, characterized by the intensity distribution $I_\nu(\mathbf{k})$. We assume that the Thomson approximation is valid; in this limit, the rate of energy emitted by the particle by inverse Compton process is:

$$dP = \sigma_T \gamma^2 \int I_\nu(\mathbf{k})(1 - \beta \mathbf{k_0}.\mathbf{k})^2 d\Omega d\nu \tag{1}$$

$\mathbf{k}$ and $\mathbf{k_0}$ are respectively the unit vectors along the photon and the particle velocity. Thus, one deduced with the hypothesis of an isotropic distribution of high energy particles, the plasma emissivity of the hot source (Henri & Petrucci submitted):

$$\frac{dP}{d\Omega} \propto [(3J - K) - 4H\mu + (3K - J)\mu^2] \tag{2}$$

where the 3 Eddington parameters, J, H and K are defined by:

$$J = \frac{1}{2} \int I_\nu(\mathbf{k})d\mu d\nu \tag{3}$$

$$H = \frac{1}{2} \int I_\nu(\mathbf{k})\mu d\mu d\nu \tag{4}$$

$$K = \frac{1}{2} \int I_\nu(\mathbf{k})\mu^2 d\mu d\nu \tag{5}$$

and $\mu = \cos\theta$ is the cosine of the impinging angle of radiation (cf Fig. 1). Under the hypothesis that the disk reprocesses the whole radiation impinging on it, we equalize the power absorbed and emitted by a surface element dS of the disk at a distance r of the black hole (cf Fig. 1):

$$F(r)dS = \frac{dP}{d\Omega}d\Omega = -\mu^3 \frac{dP}{d\Omega}\frac{dS}{z_0^2} \tag{6}$$

$$= \pi I(r)dS \tag{7}$$

where Z_0 is the height of the hot source above the disk (cf Fig. 1). In equation (7), one supposes that the disk radiates like a black body. Combining Eqs. (2), (6) and (7), one can express the specific intensity $I(r)$, emitted by the disk at radius r, as a function of the Eddington parameters. Finally, using Eqs. (3)-(5), one obtains a linear system of equations between J, H and K. By setting its determinant to zero, we finally find universal solutions for the hot source and disk emissivity laws (Eqs. (2) and (7)). The relativistic case of a Kerr metrics has also been solved (Petrucci & Henri submitted). In this case, the overall spectra depend on a, the angular momentum by unit mass of the black hole, and on the ratio Z_0/M of the height of the hot source on the black hole mass. The whole computations in the Newtonian and Kerr geometry can be find, respectively, in (Henri & Petrucci submitted) and (Petrucci & Henri submitted).

3 Results

3.1 Angular distribution of the hot source

It appears from Eq. (2) that the anisotropy of the soft photon field about level with the hot source, leads to an anisotropic Inverse Compton process, with much more radiation being scattered backward than forward. Such an anisotropic re-illumination could naturally explain the apparent X-ray luminosity, usually much

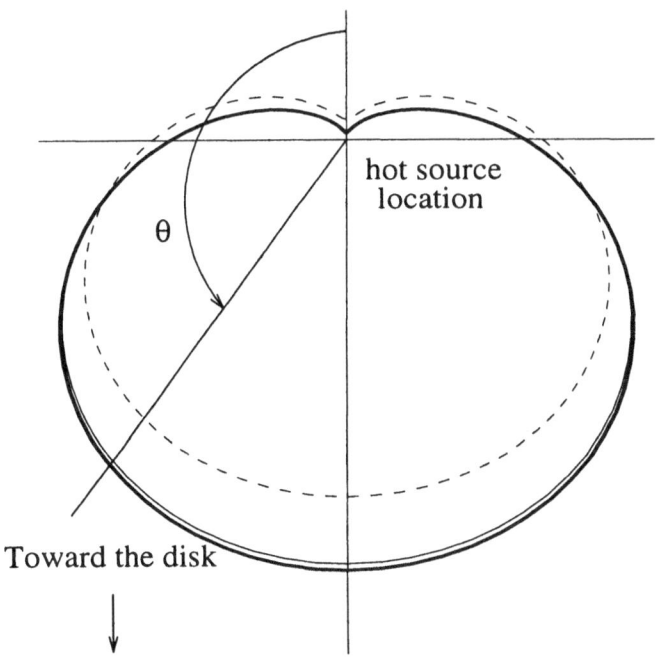

Fig. 2. Polar plots of $\dfrac{dP}{d\Omega}$ for $Z_0/M = 100$ (solid line) and $Z_0/M = 10$ (dashed line) in Kerr metrics with $a = 0.998$. The bold line corresponds to the Newtonian metrics

lower than the optical-UV continuum emitted in the blue bump. It can also explain the equivalent width observed for the iron line, which requires more impinging radiation than what is actually observed. We plot in Figure 2 the angular distribution of the power emitted by the hot source in Newtonian metrics and for different values of the source height in Kerr metrics. It appears that the closer the source to the black hole is, the less anisotropic the photon field is. This is principally due to the curvature of geodesics making the photons emitted near the black hole arrive at larger angle than in the Newtonian case.

3.2 Disk temperature profile

The radiative balance between the hot source and the disk allows to compute the temperature profile on the disk surface. It is, in fact, markedly different from "standard accretion disk model" as shown in Figure 3. Indeed, even if at large distances, all models give the same asymptotic behavior $T \propto R^{-3/4}$, in the inner part of the disk it keeps increasing in "standard model" whereas, in our model, for $R \leq Z_0$, the temperature saturates around a characteristic value T_c.

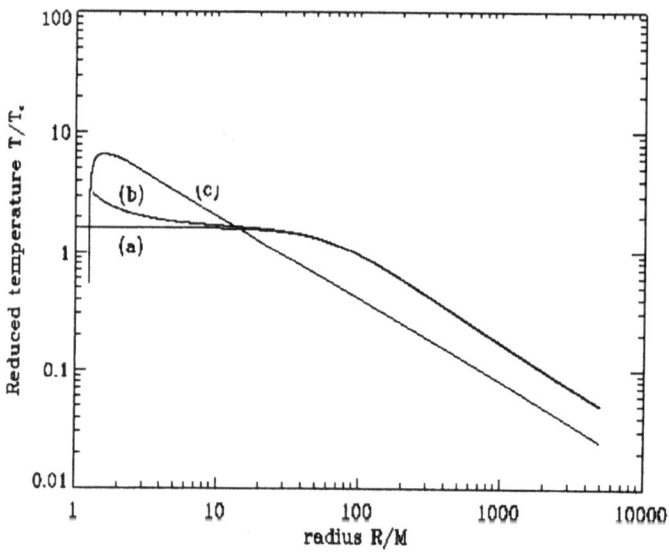

Fig. 3. Effective temperature versus r for $Z_0/M = 70$.
a) Our model in Newtonian metrics
b) Our model in Kerr metrics with $a = 0.998$
c) The standard accretion disk in Kerr metrics with $a = 0.998$

Indeed, the power radiated by the disk is essentially controlled by the angular distribution of the hot source $\dfrac{dP}{d\Omega}$ (cf Eq. 6) which is approximatively constant for $R \leq Z_0$ (i.e. $\theta \simeq \pi/4$, cf. Fig. 2). The differences between Newtonian and

Kerr metrics comes only from gravitational and Doppler shifts, which are only appreciable for $R \leq 5M$. Thus, unless Z_0 is itself small enough, these shifts concern only a small fraction of the emitting area at $T = T_c$, and modified hardly the UV to X-ray spectrum.

3.3   The overall spectra

The overall UV to X-ray spectra can be deduced from this model. The bulk of the energy coming from the disk is emitted on the blue and the ultraviolet, giving the well-known "blue-bump" observed in most quasars and many AGNs. On the other hand, the high energy spectrum depends directly on the relativistic particle distribution adopted. As reminded in the introduction, the high energy emission of Seyfert galaxies can be well reproduced by a primary power-law spectrum with a spectral index $\alpha \simeq 0.9$, superimposed on more complex structures, that can be produced by the reflection of the hard X-rays on a cold surface. Noticeably, this power law is exponentially cut-off above a characteristic energy of about 100 keV, with some uncertainty its precise value. Contrarily to thermal models, where this cut-off is related to the temperature of the hot comptonizing plasma, we propose to interpret it as a high energy cut-off of the relativistic energy distribution. Although a detailed model of the high energy source is out of the scope of this work, one can note that a model associating pair production and pair reacceleration can provide such upper cut-off, to avoid catastrophic run-away pair production (Done et al. 1990 ; Henri & Pelletier 1991).

To account for the high energy cut-off, we will assume that the particles (electrons or positrons) distribution function has the form:

$$f(\gamma,\mu) \propto \gamma^{-s} \exp\left(-\frac{\gamma}{\gamma_0}\right), \quad \gamma_{min} < \gamma < \gamma_{max} \tag{8}$$

As we have previously said, the following spectra do not include secondary components, like the Compton reflection feature and the fluorescent Fe Kα line that are observed in many Seyferts. Work is currently in progress to take these features into account. The observations seem to indicate a particle spectral index $s \simeq 3$, and a high energy cut-off $\gamma_0 \simeq 100$. We have kept these values for all simulations and an extremal value for the angular momentum by unit mass $a = 0.998$.

Influence of the inclination angle   One can see on Figure 4 Newtonian and Kerr maximal spectra for different inclination angles, ranging from $\theta = 0°$ to $60°$, for $Z_0/M = 10$. For all inclination angles, the Kerr spectra are always weaker in UV and brighter in X-ray than the Newtonian ones. However, the difference tends to be less visible for the highest inclination angles. These results can be easily explained: in the X-ray range, as shown in Fig. 2, it is due to the decreasing of the relative difference of the angular distribution between Newtonian and Kerr metrics. In the UV band, the relativistic effects (the gravitational shift and the

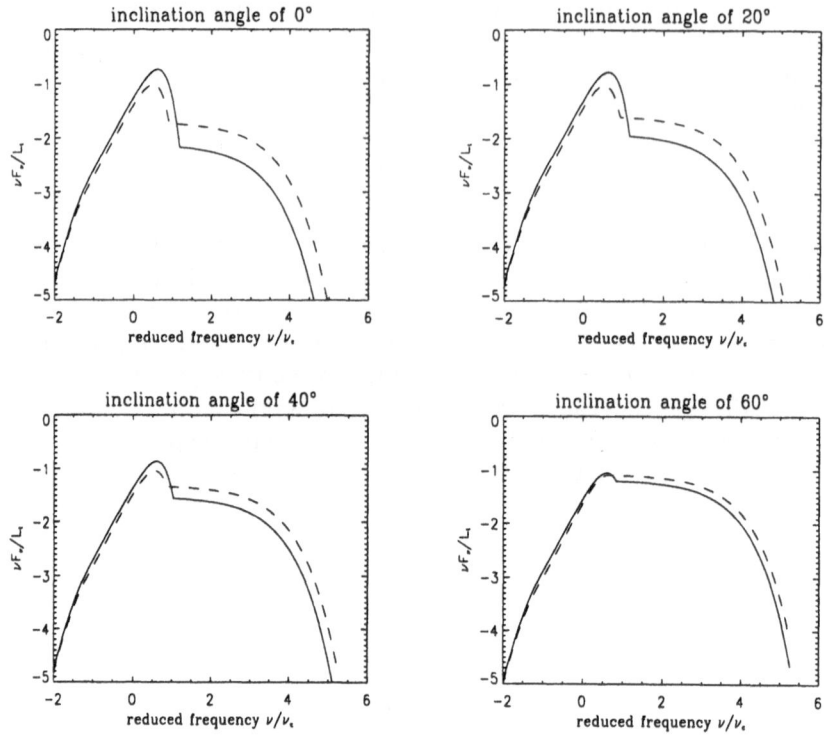

Fig. 4. Differential power spectrum for different inclination angle, in the Newtonian (solid lines) and the Kerr maximal (dashed lines) cases for $Z_0/M = 10$. We use reduced coordinates.

Doppler transverse effect) produce a net redshift in the face-on case ($\theta = 0°$) compared to the Newtonian case. For higher inclination angle, the redshifted radiation is compensated by the blueshifted one, coming from the part of the disk moving toward the observer. These effects are much less pronounced for high Z_0/M values because the emission area is much larger, and thus is less affected by relativistic corrections. Higher inclination ($\theta > 60°$) angles will probably lead to strong absorption through the external parts of the disk, presumably a molecular torus: in the unification scheme, they would correspond to Seyfert 2 galaxies.

Influence of the hot source height Figure 5 shows the overall spectrum, for different values of Z_0/M for $\theta = 0°$. The relativistic effects become important for values of Z_0/M smaller than about 50. They produce a variation of intensity lowering the blue-bump and increasing the hard X-ray emission. The change in the UV range is due to the transverse Doppler effect between the rotating disk

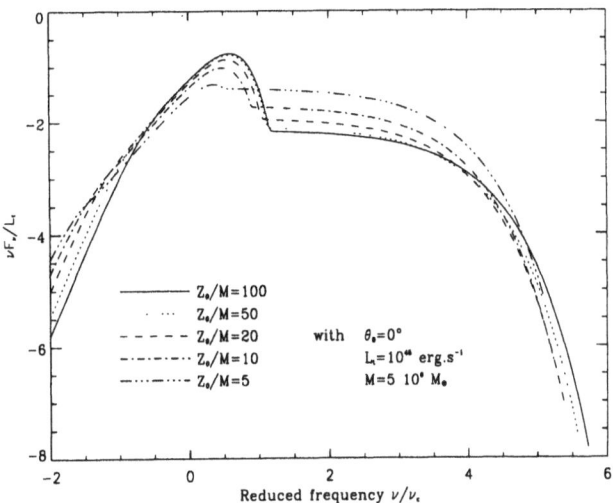

Fig. 5. Differential power spectrum for different values of Z_0/M for the Kerr maximal case. We use reduced coordinates.

and the observer, producing a net red-shift. In the X-ray range, the variation is due to the high energy dependence on Z_0/M (cf Fig. 2). The observed X/UV ratio can then be strongly altered by these effects. Quantitatively, the luminosity ratio between the maximum of the blue-bump and the X-ray plateau goes from $\simeq 30$ in the Newtonian case, to $\simeq 1.5$ for $Z_0/M = 5$.

3.4   Scaling laws

As is well known, a usual assumption for the mass-luminosity ratio of AGN is that the bolometric luminosity is limited by the radiation pressure to the Eddington luminosity:

$$L_E = \frac{4\pi G M m_p c}{\sigma_T}. \tag{9}$$

This relationship predicts a linear correlation between mass and luminosity for Eddington accreting black-holes, $L \propto M$; the corresponding Eddington temperature is given by

$$T_E = \left(\frac{L_t}{4\pi r_g^2}\right)^{1/4} \propto M^{-1/4}. \tag{10}$$

where r_g is the Schwarzschild radius. We show here that the present model predicts different scaling laws, if one adds some supplementary assumptions on the physics of the hot source. At first sight, all equations of the models are linear with respect to the global luminosity, so no particular relationship is predicted between the luminosity and other parameters like M or Z_0. However things are different if one considers the microphysics and a more realistic geometry of the hot source. First, let us consider again the upper energy cut-off of the IC spectrum discussed in section 3.3. Observations seem to show that all Seyfert galaxies have the cut-off around the same value, approximately 100 keV; however, when taking into account the reflection component, this estimate could be somewhat higher, up to 400 keV (Zdziarski et al. 1995). This is sufficiently close to pair production threshold to make plausible the idea that this cut-off is in some way fixed by a regulation mechanism to avoid run-away pair production. In this case, the cut-off energy is a physical quantity, determined by microphysics rather that macrophysical quantities. The maximal energy of photons produced by IC process is of the order

$$\gamma_0^2 h\nu_c \approx constant. \tag{11}$$

Now it is obvious that a point source is a convenient, but unrealistic approximation of the real geometry of the source, since it has a zero cross section and a infinite Thomson opacity. Rather, for a given number N of scattering particles, one gets a minimal size R_{min} for the source being optically thin; for a homogeneous sphere, it requires $R_{min} \geq \left(\frac{3N\sigma_T}{4\pi}\right)^{1/2}$. This hot source will sustain a solid angle $\Omega \simeq \pi(R_{min}/Z_0)^2$. This source can not intercept more luminosity from the disk than $L_{disk}\frac{\Omega}{\pi}$. By IC process, this luminosity will be boosted by a factor $\sim \gamma_0^2$ and part f of it will be reemitted toward the disk, giving again L_{disk}. So one gets

$$\frac{\Omega}{\pi}\gamma_0^2 f \approx 1 \tag{12}$$

Now it is plausible that the size of the source and its distance to the black hole are controlled by the global environment responsible for the hot source. Conceivably, it could be realized through a strong shock terminating an aborted jet. If one makes the (admittedly crude) assumption that all distances scale like the hole radius r_g, then one gets $\Omega \approx constant$. So Eq. (12) predicts $\gamma_0 \approx constant$, and Eq. (11) predicts $\nu_c \approx constant$. Finally one gets the following scaling laws:

$$T_c \propto \nu_c \propto M^0 \approx constant \tag{13}$$
$$L_t \propto M^2. \tag{14}$$

Of course, one could observe substantial variations of at least one of these quantities if Ω varies, either by a variation of Z_0 or a variation of R. Interestingly, observations seem to corroborate these behaviors: on a sample of many quasars and galaxies spanning a large range of masses, Walter & Fink (1993) found that the Blue Bump and soft X-ray excess were approximately in the same ratio, although the central masses can differ by a factor 10^4. This is very hard to explain in the frame of conventional accretion models. In another study, using a specific model to describe the width of Broad Lines from the emission by the external part of the disk, Collin-Souffrin & Joly (1991) have deduced the inclination angles and central masses of a sample of Seyfert 1 galaxies and quasars. They found a correlation between mass and luminosity under the form $L \propto M^\beta$, with $\beta = 1.8 \pm 0.6$. Although these results have been obtained on a limited sample, they are compatible with the previous results and clearly different from those predicted by the standard accretion disk models. A rather intriguing consequence is that there is a maximal mass above which the accretion becomes impossible by such a mechanism, where the luminosity predicted by Eq. (14) gets higher than the Eddington luminosity.

4   CONCLUSIONS

We have shown that a model based on reillumination of a disk by an anisotropic IC source could lead to a self-consistent picture where the angular distribution of high energy radiation and the radial temperature distribution of the disk are mutually linked and both determined in a single way. The model offers a simple explanation for the correlated long term variability of X and UV radiation, the short term variability of X-rays non correlated with UV variations, and the apparent X/UV deficit that seems contradictory with simple reillumination models. In its simplest form, it predicts a unique shape of disk spectrum and a X/UV ratio depending only on the inclination angle. The predicted values are in good agreement with observations. A precise comparison with real spectra should also include other components, such as a reflection component and a fluorescent Fe Kα line. This is deferred to a future work. Finally, a more complete work has to be done to explain the exact mechanism of emission of the hot source, supposed to be realized by a strong shock terminating an aborted jet.

References

Clavel, J., Nandra, K., Makino, F., Pounds, K. A., Reichert, G. A., Urry, C. M., Wamsteker, W., Peracaula-Bosch, M., Stewart, 1992, ApJ, 393, 113
Collin-Souffrin S., Joly M., 1991, Disks and Broad Line Regions. In: Duschl W.J., Wagner S.J. (eds) Physics of Active Galactic Nuclei. Springer-Verlag, p. 195
Done C., Ghisellini G., Fabian A. C., 1990, MNRAS 245, 1
Field G.B., Rogers R.D., ApJ 403, 94
George I.M., Fabian A.C., 1991, MNRAS 249, 352
Ghisellini G., George I.M., Fabian A.C., Done C., 1991, MNRAS 248, 14
Haardt F., Maraschi L., 1991, ApJ 380, L51
Haardt F., Maraschi L., 1993, ApJ 413, 507
Henri G., Pelletier G., 1991, ApJ 383, L7
Henri G., Pelletier G., Roland J., 1993, ApJ 404, L41
Henri, G., Petrucci, P.O., A&A submitted
Lightman A.P., White T.R., 1988, ApJ 335, 57
Matt G., Perola G.C., Piro L., 1991, A&A 247, 27

Nandra K., Pounds K. A., 1994, MNRAS 268, 405

Perola G.C., et al., 1986, ApJ 306, 508.

Petrucci, P.O., Henri, H., A&A submitted

Pounds, K. A., Nandra, K., Stewart, G. C., George, I. M., Fabian, 1990, Nature, 344, 132

Shakura N.I., Sunyaev R.A., 1973, A&A 24, 337

Walter R., Fink, H.H., 1993, A&A 274, 105

Weaver K.A., Arnaud K.A., Mushotzky R.F., 1995, ApJ 447, 121

Zdziarski A.A., Ghisellini G., George I.M., Svensson R., Fabian A.C., Done C., 1990, ApJ 363, L1

Zdziarski A. A., Johnson W.N., Done C., Smith D., McNaron-Brown K., 1995, ApJ 438, L63

Disc instabilities and binary evolution

A.R. King

Astronomy Group, University of Leicester, Leicester LE1 7RH, U.K.

Abstract: Disc instabilities are a plausible cause of soft X–ray transient outbursts, provided that one allows for the effect of irradiation of the disc by the central accreting source. The mass transfer rates required for transient behaviour are then so low that they place powerful contraints on the evolution of the host binary. In particular, short–period systems containing neutron stars must have companions which were significantly nuclear–evolved before mass transfer began, and millisecond pulsars in long–period binaries with white dwarf companions must decend from transients.

1 Introduction

Soft X–ray transients (SXTs) are a subset of the low–mass X–ray binaries (LMXBs) which brighten dramatically for a few months at irregular intervals of years. These timescales are very different from those of dwarf novae, and until recently made it hard to see how the disc instability model could provide a common explanation. However van Paradijs (1996) has pointed out that at least the basic condition for the instability, namely the presence of hydrogen ionization zones within the disc, can be understood provided that one takes into account the fact that the central accretor (neutron star or black hole) in an LMXB will irradiate the disc surfaces, raising their temperature from the usual effective temperature (e.g. Frank, King & Raine, 1992)

$$T_{\rm eff}^4 = \frac{3GM\dot{M}}{8\pi\sigma R^3},$$ (1)

where $M, \dot{M}, \sigma, R$ are the mass of the accreting star, the local accretion rate, the Stefan–Boltzmann constant and the disc radius respectively. Irradiation would instead give the disc a temperature

$$T_{\rm irr}^4 = \frac{\eta\dot{M}c^2(1-\beta)}{4\pi\sigma R^2}\frac{H}{R}\left(\frac{{\rm d}\ln H}{{\rm d}\ln R} - 1\right).$$ (2)

Here η is the efficiency of rest–mass conversion into luminosity, β the disc surface albedo, and $H(R)$ the local disc scale height. Since the last two factors on the rhs vary little with R, one sees that $T_{\rm irr} \propto R^{-1/2}$ as compared with $T_{\rm eff} \propto R^{-3/4}$. Dividing the equations and using typical values, including $\eta c^2 \sim GM/R_{\rm acc}$, we find

$$\frac{T_{\rm irr}^4}{T_{\rm eff}^4} \sim 10^{-3}\frac{R}{R_{\rm acc}},$$ (3)

where R_{acc} is the radius of the accretor. We see that irradiation is unimportant in CVs ($R_{acc} \sim 10^9$ cm, $R \lesssim 10^{11}$ cm) but must be taken into account in a neutron–star or black–hole binary ($R_{acc} \sim 10^6$ cm, $R \sim 10^{11}$ cm). Thus the temperature of the outer parts of LMXB discs is likely to be closer to T_{irr} than T_{eff}. If T_{irr} exceeds the typical ionization temperature $T_H \sim 6500$ K at the outer disc edge (and thus throughout the disc structure), the characteristic disc instability will be suppressed. Clearly, since

$$T_{irr} > T_{eff}, \tag{4}$$

this means that even lower values of the accretion rate $\dot{M}$ are required if the disc is to be unstable in a soft X–ray transient rather than a dwarf nova disc of similar size. Van Paradijs (1996) shows that indeed the known low–mass X–ray binaries are correctly divided into transients and persistent sources by the simple criterion (4) for the latter. This constitutes very strong circumstantial evidence in favour of a common origin of both phenomena in terms of the disc instability picture.

2 Consequences

The criterion (4) implies that SXTs must have remarkably low mass transfer rates $\dot{M}$, i.e.

$$\dot{M} < \dot{M}_{crit} \simeq 5 \times 10^{-11} m_1^{2/3} P_3^{4/3}\ M_\odot\ \text{yr}^{-1} \simeq 8 \times 10^{-10} m_1^{2/3} P_d^{4/3}\ M_\odot\ \text{yr}^{-1}, \tag{5}$$

where m_1 is the mass of the accretor in $M_\odot$ and P_3, P_d are the binary period in units of 3 hr and days respectively. Since $\dot{M}$ is determined by the long–term evolution of the binary, the constraint $\dot{M} < \dot{M}_{crit}$ has very powerful consequences for the latter. These have been explored in a series of papers which will appear shortly, so I simply summarize the main results here.

1. For short–period ($P \lesssim 2$ d) LMXBs containing neutron stars, King, Kolb & Burderi (1996) show that $\dot{M}$ is only low enough for transient behaviour (i.e. violating (3)) if the companion star is already significantly nuclear–evolved before mass transfer (driven by orbital angular momentum loss) begins. The companion masses are thus much lower than would be expected for a main–sequence star filling the Roche lobe at the observed periods, explaining the extreme mass ratios observed in SXTS. It is unclear whether a similar conclusion holds for black–hole binaries: for black–hole masses $\gtrsim 10 M_\odot$, uncertainties in T_H and the orbital magnetic braking rate may conceivably allow transient behaviour even with unevolved companions.

2. The fact that short–period SXTs containing neutron stars are observed must (by 1.) mean that nuclear–evolved companions are not rare in LMXBs with $P \lesssim 2$ d. This in turn requires these companions to be massive enough for significant nuclear evolution in the age of the Galaxy. King & Kolb (1996) examine the constraints on the formation of neutron–star LMXBs via the usual helium–star channel (e.g. Webbink & Kalogera, 1994). They show that indeed the

companions are likely to be nuclear–evolved, with pre–contact masses $1.3M_\odot \lesssim M_{2,i} \lesssim 1.5M_\odot$, provided that a sufficiently large number of pre–LMXBS receive supernova kick velocities which are less than typical pre–SN orbital velocities (~ 100 km s^{-1}). These large initial masses also explain why the LMXB period histogram does not resemble that of CVs. The latter ensemble is dominated by systems emerging from common–envelope evolution with very small companion masses $M_{2,i} \sim 0.1M_\odot$ (e.g. King et al., 1994), raising the CV population below the well–known periods gap ($P \lesssim 2$ hr).

3. Long–period LMXBs ($P \gtrsim 2$ d) must have mass transfer rates driven by the nuclear expansion of the companion, which is a low–mass subgiant. King, Kolb, Frank & Ritter (1996) show that persistent X–ray behaviour is rare among such systems, being confined to neutron–star systems with comparatively massive ($M_2 \gtrsim 0.75M_\odot$) companions. All black–hole systems are transient, and every neutron–star system eventually becomes transient as the companion is whittled down by mass transfer. As is well known, the endpoint of this evolution is a millisecond pulsar in a long–period ($P \sim 100$ d) binary. This work show that all such systems descend from soft X–ray transients. This may well be relevant to the claimed discrepancy between the numbers of millisecond pulsars and their LMXB progenitors.

3 Conclusions

The work described above supports the view that soft X–ray transients are a manifestation of disc instabilities, suitably modified for the effects of irradiation. In its turn, the requirements of this model place extremely tight constraints on LMXB formation and evolution.

Acknowledgement: I thank Uli Kolb, Hans Ritter, Juhan Frank and Jan van Paradijs for many valuable discussions. I am grateful to the UK Particle Physics and Astronomy Research Council for support as a Senior Fellow.

References

Frank, F., King, A.R., Raine, D.J., 1992, *Accretion Power in Astrophysics*, Cambridge University Press, Cambridge

King, A.R., Kolb, U., 1996, ApJ, in press

King, A.R., Kolb, U., Burderi, L., ApJ 464 L127

King, A.R., Kolb, U., Frank, J, Ritter, H., 1996 ApJ, submitted

van Paradijs, J., 1996, ApJ 464, L139

Webbink, R.F, Kalogera, V., 1994, in *The Evolution of X–ray Binaries*, eds S.S. Holt, C.S. Day, AIP Press, New York, p 321

Dwarf nova outbursts: a unification theory

Y. Osaki

Department of Astronomy, University of Tokyo, Bunkyo-ku, Tokyo, 113

Abstract: Outburst mechanisms of dwarf novae are discussed. There is a rich variety in outburst behaviors of non-magnetic cataclysmic variable stars, starting from non-outbursting nova-like stars to various sub-classes of dwarf novae (i.e., U Gem-type, Z Cam-type, and SU UMa-type). A unification model for dwarf nova outbursts is proposed within the basic framework of the disk instability model in which two different intrinsic instabilities (i.e., the thermal instability and the tidal instability) within accretion disks play an essential role. In this model, the non-magnetic cataclysmic variables are classified in the orbital-period versus mass-transfer-rate diagram into four regions depending on different combination to these two instabilities, and their observed outburst behaviors are basically understood in a unified way on this diagram.

1 Introduction

Dwarf novae (or U Gem stars) are eruptive variable stars exhibiting repetitive outbursts with a typical amplitude of $2-5$ mag and with a typical recurrence time of order of a few weeks to months. They belong to a more general class of eruptive variables called Cataclysmic Variables (CVs). A monograph covering both observations and theories of CVs has recently been published by Warner (1995a) and those who want to know more on CVs are advised to consult with that monograph. It is now well established that cataclysmic variable stars are semi-detached close-binary systems consisting of a Roche-lobe filling cool dwarf secondary and a white dwarf primary. Mass transferred from the secondary star via the inner Lagrangian point is accreted onto the white dwarf primary. Cataclysmic variables are further classified into "magnetic CVs" and "non-magnetic CVs". In magnetic CVs, the white dwarf has strong magnetic fields and mass transferred from the secondary star is channeled onto the magnetic poles of the white dwarf along the magnetic field lines. On the other hand, the white dwarf component of non-magnetic CVs has weak or no magnetic fields and mass accretion occurs via an accretion disk. Dwarf novae belong to non-magnetic disk systems.

Cataclysmic variables may also be classified into classical novae (CN), dwarf novae (DN), and nova-like variable stars (NL) depending on their outburst amplitudes. In a classical nova, an explosion has been observed only once so far with an outburst amplitude exceeding ten mag and mass is observed to be thrown out from the system. On the other hand, the nova-like stars are those binary systems which share all the binary characteristics with CN and DN, but in which no outbursts have so far been observed. They may be understood to be those same systems as CN but observed in quiescence as it is thought that the nova outburst may occur once every $10^3 \sim 10^4$ years.

Dwarf novae are further classified into three sub-classes from outburst light curves. They are (1) the U Gem-type (or SS Cyg-type) exhibiting more or less regular quasi-periodic outbursts, (2) the Z Cam-type showing rather frequent outbursts which are occasionally interrupted by a "standstill" in which the system stays in more or less constant brightness with a level intermediate between the outburst maxima and minima, and (3) the SU UMa-type showing two distinct outbursts of a short "normal" outburst with a duration of a few days and a long "superoutburst" with a duration of about two weeks. Usually, one superoutburst is followed by several short normal outbursts before the next superoutburst, and this sequence is called "supercycle". A typical length of supercycle is about one year.

The observed orbital-period distribution of non-magnetic CVs shows a bimodal character, in which their periods are either longer than 3 hour or shorter than 2 hour and systems having periods between 2 hour and 3 hour are very rare. Thus this period range is called the "period gap". U Gem stars and Z Cam stars belong to long period systems above the gap while the SU UMa stars belong to the short period ones below the gap.

As for the cause of outbursts in CVs, it is now well established that the classical nova explosion is caused by the thermonuclear runaway of accreted hydrogen on the surface of the white dwarf. A natural question that follows is then: are dwarf nova outbursts caused by the same mechanism as the classical nova but with a small scale ? Its answer is definitely *no*. Various arguments can be made against this but the strongest evidence is the observational one, that is, the seat of the outburst is not on the white dwarf itself but the outburst is caused by a sudden brightening of the accretion disk around the white dwarf, that is, a sudden increased accretion onto the white dwarf.

In order to explain a sudden increased accretion in the dwarf-nova outburst, two models were proposed in 1970s: the mass transfer burst model and the disk instability model. A fierce contest was fought between the two models in 1970s. However, a very promising intrinsic instability in the accretion disk was discovered around 1980. The disk instability model is now widely accepted at least for the ordinary dwarf novae of U Gem-type. The main objective of this paper is to convince the readers that a wide variety in outburst behaviors of dwarf novae can be understood in a unified way within the basic framework of the disk instability model. Before presenting this unification theory, we give a brief historical review of the development of the disk instability model.

2   A brief history of the development of the disk instability model

In early 1970s, Warner and Nather (1971) and Smak (1971) have found that it is the accretion disk itself that brightens during a dwarf nova outburst. Since the ultimate energy source of the accretion disk is a release of gravitational potential energy, this indicates that accretion onto the white dwarf occurs *intermittently* in dwarf novae. The question is now: why does accretion occur intermittently ?

To explain this, two models were proposed in 1970s. One model is the so-called mass-transfer burst model proposed by Bath (1973). In this model, the mass transfer from the secondary star to the accretion disk is thought to be unstable because of an intrinsic instability in the envelope of the secondary star and mass transfer rate from the secondary star is not stable but sometime increases dramatically. Mass accretion onto the white dwarf via accretion disk consequently increases that explains an outburst of dwarf nova.

Another model is the disk instability model (DI model) first proposed by Osaki (1974). He showed that many observed features of dwarf novae could be explained if one assumed that mass transferred from the secondary star during quiescence was not accreted onto the white dwarf but instead stored in the outer part of the accretion disk. If the mass thus stored exceeds some critical amount, some kind of instability (unknown at that time) was postulated to set in and mass accumulated by then would suddenly be accreted onto the white dwarf, explaining an outburst of the dwarf nova. This was the first proposal of the so-called disk instability model. These two models contested fiercely in 1970s. However, a physical mechanism responsible for the disk instability was discovered around 1980.

Hoshi (1979) was the first in realizing that the accretion-disk theory allows double solutions in outer parts of accretion disks in CVs; one solution is a low viscosity optically thin cool one (where hydrogen is neutral) and the other is a high-viscosity hot one (where hydrogen is ionized) . Although his basic idea was correct, his solution unfortunately did not allow a full limit cycle instability.

The full limit cycle instability was first discovered by Meyer and Meyer-Hofmeister (1981), who obtained the now well known S-shaped thermal equilibrium curve (see, Figure 1) by solving the vertical structure of accretion disks by taking into account of all details of opacities and equation of state and convective transport of energy.

They demonstrated that the dwarf nova outburst could be explained by the thermal limit cycle instability based on this S-shaped thermal equilibrium curve. That is, mass is stored in the disk during quiescence in a low-viscosity cold state. When mass is sufficiently accumulated in the disk to reach the turning point of the S-curve, the thermal instability sets in, which brings the disk to a high-viscosity hot state, now dumping matter onto the white dwarf, corresponding to an outburst state. The outburst ends when mass is drained in the disk and the disk drops back to a cold state once again.

The thermal limit-cycle instability for dwarf nova outbursts was then studied by five independent groups; they were Meyer and Meyer-Hofmeister in Germany, Smak in Poland, Cannizzo and Wheeler's group in USA, Faulkner, Papaloizou, and Lin in England, and Mineshige and Osaki in Japan. They have shown that the dwarf nova outburst can basically been understood by the disk instability model based on the thermal limit cycle instability. The further development of

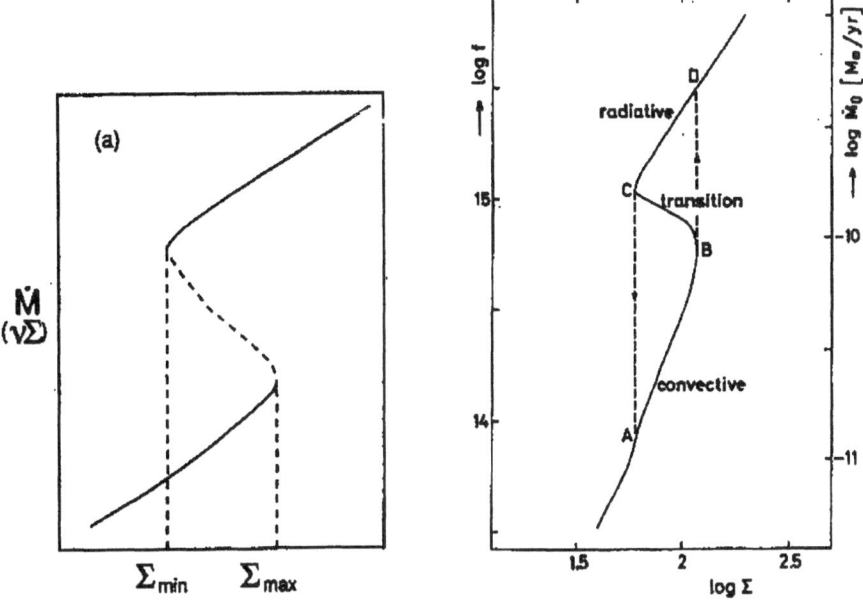

Fig. 1. The S-curve explaining the thermal limit cycle oscillation for dwarf nova outbursts. The left panel shows a schematic S-curve between the surface density Σ and viscosity required for the limit cycle oscillation. The right panel shows the original S-curve obtained by Meyer and Meyer-Hofmeister (1981) who integrated the vertical structure of the accretion disks.

the thermal limit cycle instability for dwarf nova outbursts was reviewed by Cannizzo (1993).

Another development of the theory of dwarf nova outburst comes from observational recognition of a very curious phenomenon called the superhump in SU UMa stars. Vogt (1974) and Warner (1975) independently discovered in the December 1972 superoutburst of VW Hyi, a proto-type of SU UMa stars, that a prominent periodic hump appeared at about the superoutburst maximum and was visible almost to the end of the superoutburst. The most curious of all is that its period is longer by a few percent than the orbital period. The periodic hump was then called the superhump because it appeared only during a super-outburst. Since then, superoutbursts in various SU UMa stars were observed and superhumps described above appeared in almost all cases. The superhump phenomenon now becomes the defining characteristics of superoutbursts and thus of SU UMa stars.

As for the explanation for the superhump phenomenon, various models were proposed in the past but the only model that has survived is the eccentric disk model. The eccentric disk model in its crude form was first suggested by Vogt

(1982) to explain the γ-velocity variations during superoutbursts of SU UMa stars. It was extended by Osaki (1985) who demonstrated that the superhump phenomenon could well be explained at least phenomenologically by the precessing eccentric model. However, the basic question why an accretion disk becomes eccentric in the first place remained unanswered.

A theoretical breakthrough in this problem was first provided by Whitehurst (1988) who discovered a new instability in accretion disk now called tidal instability or tidally driven eccentric instability. He found in his two-dimensional hydrodynamic simulations that an accretion disk evolved into eccentric form and its eccentric pattern slowly rotated in the inertial frame of reference. The orbiting secondary star then gives rise to tidal stressing periodically on the eccentric disk, which explains the superhump light maximum. Whitehurst (1988) has argued that this tidal instability occurs only in the binary system having a low mass secondary with the mass ratio $q = M_2/M_1 < 0.25$, which is consistent with the known fact of short orbital periods of SU UMa stars.

The tidal instability in accretion disks was further studied and confirmed by Hirose and Osaki (1990) and Lubow (1991). The tidal instability is found to occur only when the outer edge of an accretion disk reaches the so-called 3:1 resonance radius (called the eccentric Lindblad resonance) and this condition requires a large accretion disk in units of the binary separation and it can be met only in binary systems having a low mass secondary star, which correspond to those binary systems below the well-known CV period gap.

Osaki (1989) proposed that the superoutburst phenomenon of SU UMa stars may be explained by the basic framework of the disk instability model in which the two intrinsic instabilities (i.e., the thermal and the tidal instabilities) are properly combined, and this model for SU UMa stars is now called the "thermal-tidal instability" model.

3   Unification theory of dwarf nova outbursts

An intriguing possibility has now opened in that almost all variety of outburst light curves of dwarf novae may be explained by a single paradigm: the DI model. In this unification model, different outbursting behaviors among non magnetic cataclysmic variable systems are basically classified by two-parameters characterizing accretion disks in these systems; that is, orbital period of the system and mass transfer rate from the secondary star. For a given orbital period the mass transfer rate from the secondary determines the thermal stability nature of accretion disks (Smak 1983) in a sense that CV systems with high mass transfer rate yield hot "stable" disks corresponding to nova-like systems while those with mass transfer rate below the critical one give rise to thermally unstable disks, producing dwarf nova outbursts.

The critical mass-transfer rate is given in the disk-instability model by (see, Smak 1983)

$$\dot{M}_{\text{crit}} = \frac{8\pi}{3}\sigma T_{\text{eff,crit}}^4 \frac{R_{\text{d}}^3}{GM_1},\tag{1}$$

where σ and G are the Stefan-Boltzmann constant and the gravitational constant, respectively, R_d is the disk radius and $T_{eff,crit}$ is the critical effective temperature of an accretion disk below which no hot state exists. Here, we use $\log T_{eff,crit} = 3.9 - 0.1 \log R_{d,10}$ where $R_{d,10} = R_d/10^{10}$cm. We then find

$$\dot{M}_{crit} \simeq 2.7 \times 10^{17} \text{g s}^{-1}(P_{orb}/4 \text{ hr})^{1.7}, \qquad (2)$$

where we have assumed that $R_d \simeq 0.35A$ and $M_1 + M_2 \simeq 1M_\odot$ and A is the binary separation.

On the other hand, the orbital period of binary determines whether the tidal instability is possible or not. As mentioned before, the period gap gives approximately a dividing line between the tidally stable systems (above the gap) and tidally unstable ones (below the gap).

Figure 2 illustrates different classes of non-magnetic CVs in the $(P_{orb}, \dot{M})$-diagram. Various outbursting behaviors of non-magnetic CVs are understood in a unified way (an unification model) within the basic framework of the DI model in this diagram. This diagram is divided basically into four regions by different combination of stability behaviors to these two intrinsic instabilities in the accretion disks. Corresponding to these four regions different kinds of outburst behaviors are found in non-magnetic CVs.

The accretion disks in the upper right region are stable both for the thermal instability and the tidal instability and observationally we find steady accretors called nova-like systems (NL). The ordinary dwarf novae called U Gem-type (UG) are located in the lower right region where disks are thermally unstable but tidally stable. The Z Cam stars (ZC) are located in the borderline zone between these two, which show both dwarf nova outburst (thermal limit cycle oscillation) and "standstill" (steady accretion).

Moving to the upper left region in this diagram, we find another interesting class of CVs called "permanent superhumpers" (designated here as PS), which are nova-like stars (steady accretors) but nevertheless exhibit permanent superhump phenomenon (see, e.g., Skillman and Patterson 1993). These stars are understood to be permanently stuck in superoutburst because they are thermally stable but tidally unstable. The CV systems that are located in the lower left corner are in general classified to SU UMa stars which exhibit superoutbursts and superhumps. These stars can be both thermally and tidally unstable.

In the case of SU UMa stars there is a wide range in outburst recurrence time-scales ranging from ER UMa stars with supercycle less than 50 days to classical SU UMa stars with supercycle of order of a year and to the extreme of WZ Sge stars with recurrence time as long as 30 years. Their difference is basically understood by difference in mass transfer rate in the thermal-tidal instability model as discussed in the next section.

4   Thermal-tidal instability model of SU UMa stars

The present author (Osaki 1989) has proposed a model to explain the supercycle of the SU UMa stars based on the basic framework of the disk instability model.

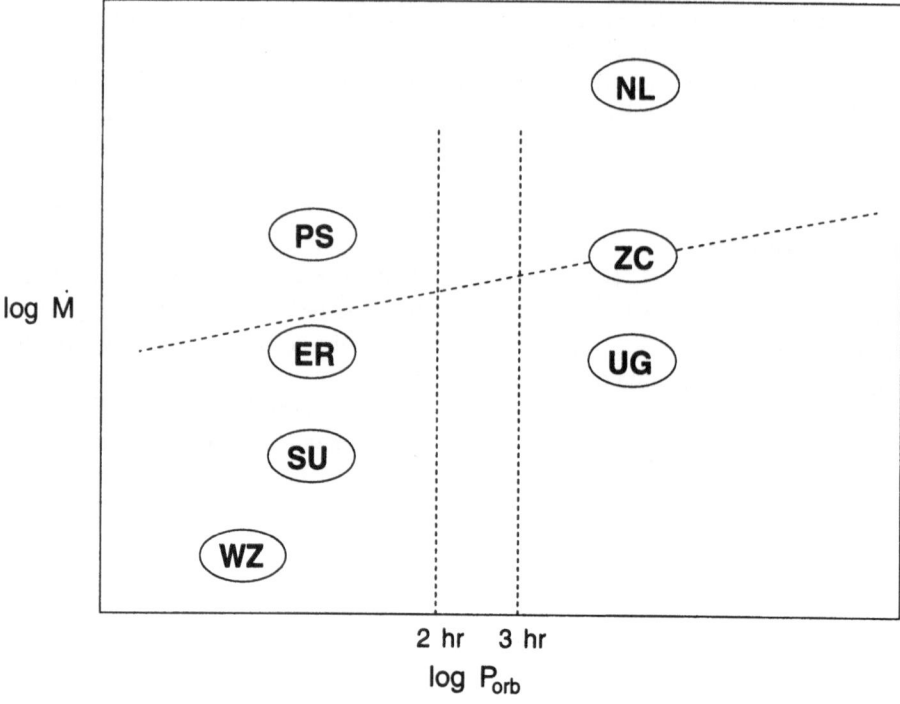

Fig. 2. $P_{orb} - \dot{M}$ diagram showing different outburst behaviors of non-magnetic CVs. The region surrounded by dotted vertical lines shows that of the CV's period gap with 2-3 hr. Dashed line shows the borderline between the thermally stable and unstable disks given by equation (3). Symbols in the figure are ; NL: nova-like stars, ZC: Z Cam stars, UG: U Gem stars, PS: "permanent superhumpers", ER: ER UMa stars, SU: SU UMa stars, and WZ: WZ Sge stars.

This model uses the two intrinsic instabilities of an accretion disk and it is thus called the thermal-tidal instability model.

In this model, both the normal outburst and superoutburst are caused by the thermal instability in the accretion disk. During the early phase of the supercycle, the disk is compact and the thermal instability produces quasi-periodic episodes of accretion, which are observed as normal outbursts but the accreted mass in each normal outburst is less than that transferred during quiescence because of inefficient tidal removal of angular momentum from the disk. Both the mass and angular momentum of the disk are gradually built up. The disk radius expands further with each successive outburst until it eventually exceeds the critical radius for the 3:1 resonance; this final normal outburst triggers the tidal instability, producing a precessing eccentric disk (observed as superhumps). The resulting outburst greatly clears the disk mass (producing "superoutburst") because of greatly enhanced tidal torques due to the eccentric disk. After the end of the superoutburst, the disk returns to the starting compact state. This is the basic idea of the thermal-tidal instability model for SU UMa stars. Figure 3 illustrates results of light-curve simulations based on a simplified treatment of the thermal-tidal model developed by Osaki (1989). The supercycle of an SU

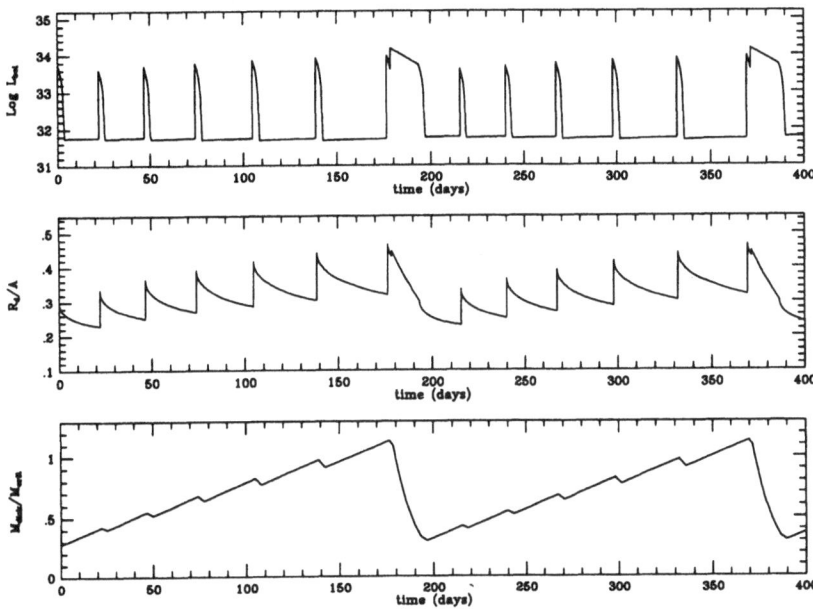

Fig. 3. A numerical simulation of supercycle of an SU UMa-type star based on the thermal-tidal instability model (Osaki 1989). ¿From top to bottom, (a) the bolometric light curve, (b) the disk radius, R_d in units of the binary separation, A, and (c) the total disk mass normalized by the critical mass above which the disk is tidally unstable.

UMa star is understood in this model as a relaxation oscillation cycle of the disk radius.

It has recently been realized that outburst parameters of SU UMa stars extend in very wide ranges; that is, they extend from the shortest extreme for the supercycle length of 20 days and the shortest recurrence time of normal outbursts of 4 days with an outburst amplitude of 3 mag for the recently discovered SU UMa star RZ LMi to the other extreme of WZ Sge which has the recurrence time of 30 years with an outburst amplitude of 8 mag. It thus encompasses a range of a factor of a thousand in time-scale.

Vogt (1993) and Warner (1995b) have noted that there exists a continuous sequence in outburst activities in SU UMa stars, in the sense that the number of normal outbursts between two consecutive superoutbursts decreases with an increase of the supercycle length and that those WZ Sge-type stars which exhibit very large amplitude outbursts with extremely long intervals seem to show only superoutbursts.

As already discussed in details in the review paper on the dwarf nova outburst which appeared in PASP (Osaki, 1996), wide ranges in outburst parameters in SU UMa stars can basically be understood as a sequence of different mass transfer rate (denoted as $\dot{M}$) in the thermal-tidal instability model. Here we do not go into any details of this model but we simply summarize this picture below.

In the thermal-tidal instability model (Osaki 1996), the basic clock of the supercycle length (denoted as T_S) in the SU UMa stars is provided by mass accumulation time for the disk to reach the 3:1 resonance radius and the supercycle length is thus inversely proportional to the mass transfer rate, that is, $T_S \propto \dot{M}^{-1}$. The classical SU UMa stars, such as VW Hyi and Z Cha having supercycle length of the order of 200-400 days can be explained in this model as systems having mass transfer rate around $0.5 \times 10^{16} \mathrm{g\ s^{-1}}$, a value expected from the standard CV evolution scenario below the period gap in which mass transfer is driven by angular momentum loss by the gravitational wave radiation. On the other hand, the recurrence time, T_N, of the normal outburst is shown to be proportional to the inverse square of mass transfer rate, i.e., $T_N \propto \dot{M}^{-2}$. Then the number, N, of the normal outbursts in a supercycle should be inversely proportional to the supercycle length, that is, $N = T_S/T_N \propto \dot{M} \propto T_S^{-1}$, which explains the observed correlations.

The WZ Sge stars could be understood as systems having extremely low mass transfer rates. However, it has been argued by various authors (see the review by Osaki, 1996) that a condition of low mass transfer rate is not sufficient but also a very low viscosity in quiescence is required to explain extremely long recurrence times of WZ Sge stars. In an extremely low mass transfer rate, the recurrence time of outbursts is determined by the viscous diffusion time in the disk, which is independent of mass transfer rate, and the thermal instability is triggered in the inner-part of accretion disk by viscous diffusion of the disk matter and this type of outbursts is called "inside-out outburst". The recurrence time in this case would then be given by the viscous diffusion time in the disk, which is thought to be of order of a year or less in the case of the standard viscosity, much shorter than the recurrence time of 30 years in the case of the WZ Sge.

There is however an alternative possibility other than that of a very low viscosity to avoid an early occurrence of inside-out outburst. Meyer and Meyer-Hofmeister (1994) suggested that the inner part of accretion disk in quiescence of dwarf novae is evacuated by a coronal siphon flow. This would then explain X-ray emissions in quiescence of dwarf novae and evaporation of matter in the inner part of accretion disks in quiescence. The WZ Sge itself in quiescence is known to be an X-ray source. Mass accretion in quiescence is too low in the standard DI model to explain observed X-ray flux in WZ Sge. This indicates that mass seems to trickle down from the disk to the central white dwarf even in quiescence by some mechanism, which fits in Meyer and Meyer-Hofmeister's picture.

The other extreme SU UMa stars are those ER UMa stars which exhibit extremely short normal and superoutburst recurrence times. These systems are also shown to be explained nicely as systems having high mass transfer rates in the thermal-tidal instability model. That is, they are the borderline systems between the classical SU UMa stars and "permanent superhumpers" and thus they may be considered in some sense to be a Z Cam counterpart below the period gap.

5   Summary

This talk may be summarized in the following one sentence:
Dwarf nova outbursts can basically be understood in a unified way within a single conceptual framework, that is, the disk instability model.

More details of this unification theory may be found in the review paper which appeared in PASP (Osaki 1996).

Acknowledgement: I would like to thank Professor Simon D. M. White, Director of Max Planck Institut für Astrophysik, for inviting me to give this talk in such an honorable occasion of Dr. Friedich Meyer's official retirement from Max Planck Institut für Astrophysik. I would like to express my sincere thanks to Drs. Friedrich Meyer and Emmi Meyer-Hofmeister for their warm hospitality during my one-year stay in the Institute in 1983-1984. I was very much benefitted from enjoyable discussions with Friedrich.

References

Bath G.T., 1973, Nature Phys. Sci. 246, 84
Cannizzo J. K., 1993, in Accretion Disks in Compact Stellar Systems, ed. J. C. Wheeler, Singapore: World Scientific Publishing, 6
Hirose M., Osaki Y., 1990, PASJ 42, 135
Hoshi R., 1979, Progr. Theor. Phys. 61, 1307
Lubow S. H., 1991, ApJ 381, 259
Meyer F., Meyer-Hofmeister E., 1981, A & Ap. 104, L10
Meyer F., Meyer-Hofmeister E., 1994, A & Ap. 288, 175
Osaki Y., 1974, PASJ 26, 429
Osaki Y., 1985, A &Ap 144, 369
Osaki Y., 1989, PASJ 41, 1005
Osaki Y., 1996, PASP 108, 39
Skillman D. R., Patterson J. 1993, ApJ 417, 298
Smak J., 1971, Acta Astronomica 21, 15
Smak J., 1983, ApJ 272, 234
Vogt N., 1974, A & Ap. 36, 369
Vogt N., 1982, ApJ 252, 653

Vogt N., 1993, in 2nd Technion-Haifa conference on Cataclysmic Variables and Related Physics, eds O. Regev, G. Shaviv, The Israel Physical Society, Jerusalem, Israel, p. 63

Whitehurst R., 1988, MNRAS 232, 35

Warner B., Nather R. E., 1971, MNRAS 152, 219

Warner B., 1975, MNRAS 170, 219

Warner B., 1995a, Cataclysmic Variable Stars, Cambridge: Cambridge University Press

Warner B., 1995b, Ap&SS 226, 187

Radiation transfer in disks of CVs

W. Hummel[1], K. Horne[2], T. Marsh[3], J.H. Wood[4]

[1] Universitätssternwarte München, Scheinerstr. 1, D–81679 München, Germany
[2] University of St. Andrews, School of Physics & Astronomy, North Haugh, St. Andrews, Fife KY16 9SS, Scotland, UK
[3] University of Southampton, Department of Physics, Highfield, Southampton SO17 1BJ, England, UK
[4] Keele University, Department of Physics, Keele ST5 5BG, England, UK

Abstract: On the basis of LTE radiative transfer calculations for axisymmetric accretion disks we fit the most prominent optical emission lines of U Gem and T Leo during quiescence. A nearly isothermal disk with a mean kinetic temperature of $\langle T \rangle = 11\,500$ K and a radial decreasing density $\sim R^{-2.3}$ can account for the observed emission line spectrum. It is shown that Stark broadening is not important in U Gem. Also a macroscopic supersonic velocity dispersion above the threshold of the spectral resolution does not improve the fit. The spectrum of T Leo can be fitted with a similar accretion disk seen at $i = 30^0$.

1 Introduction

Recently we succeeded in fitting the phase-averaged optical spectrum of U Gem in quiescence (Hummel et al. 1997) assuming a power law for the density $N(R) = N_0 R^m$ and the kinetic temperature $T(R) = T_0 R^k$. An improved fit is shown in Fig. 1, where we have taken the finite spectral resolution of the observations $R = 10\,000$ into account. Both the low excitation (Ca II) and the high excitation lines (He I) show good agreement (Fig. 1) using a nearly isothermal accretion disk.

The outstanding problem visible in the fit are the broad wings in the Balmer emission lines. They are presumably produced by

- Stark broadening
- Supersonic macroscopic velocity dispersion
- Thomson scattering
- The radial runs of the gas temperature $T(R)$ and/or the density $N(R)$ do not follow power-laws.

We check the impact of the first two mechanisms on the emission line formation in U Gem in quiescence.

2 Stark broadening

Stark broadening in dwarf novae accretion disks was first considered by Williams & Shipman (1980). The mean surface density $\langle \Sigma \rangle \simeq 0.02$ gcm^{-2} of our model

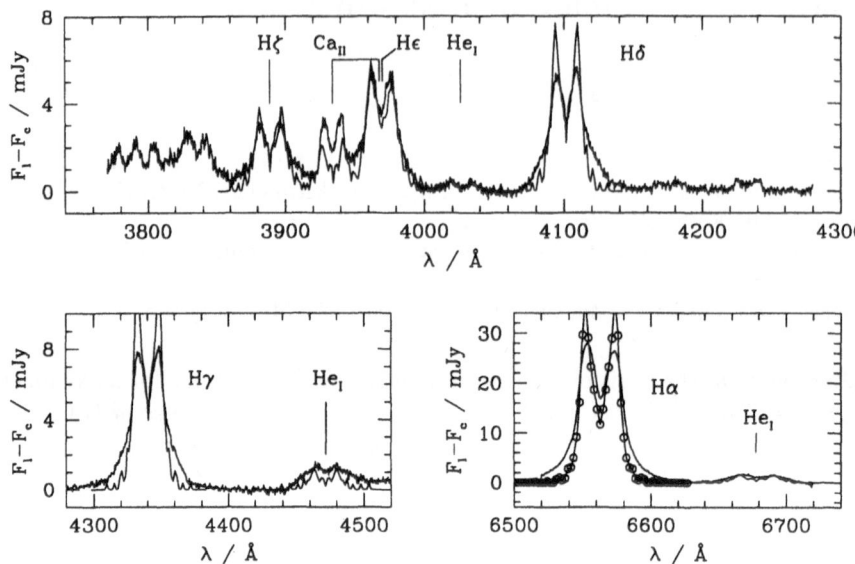

Fig. 1. Quiescent spectrum of U Gem fitted by a model accretion disk ($\chi^2 = 10$) with $N = 1.16 \times 10^{15} R^{-2.31} \mathrm{cm}^{-3}, T = 14\,000 R^{-0.07}$ K $D = 72$ pc, and microscopic turbulence $V_T = C_s$ (C_s = sound speed). The mean gas temperature is $\langle T \rangle = 11\,500$ K, the total mass of the disk is $M_d = 7.8 \times 10^{18}$ g.

disk is about a factor 100 below the critical Σ for which Stark broadening becomes important (Lin et al. 1988). In order to test this suggestion quantitatively we also include Stark broadening in the radiative line transfer calculations.

Hydrogen Stark profiles have been tabulated by Vidal et al. (1978) and extended to higher densities by Schöning & Butler (priv. comm.). Since we know from section 1 that the accretion disk is nearly isothermal, we neglect the small dependence of the Stark profiles on the temperature. The density dependence of the profiles is taken into account. For He I and Ca II we use the HWHM Lorentz widths of Dimitrijević & Sahal-Bréchot (1990) and Griem (1964) respectively. The HWHM Lorentz widths have been scaled linearly with the local electron density and interpolated to the mean gas temperature of the disk. We then folded the Lorentz profile with the Doppler profile for thermal motions and microscopic turbulence.

As for our initial model without Stark broadening (Fig. 1), we carried out multi-parameter fitting using the temperature and the density as free parameters. The comparison with the observations is shown in Fig. 2. Clearly, Stark broadening has no effect on the emission line shapes (here at $i = 70^0$), in particular on the broad Balmer line wings, and does not play a role for the line formation in U Gem during quiescence.

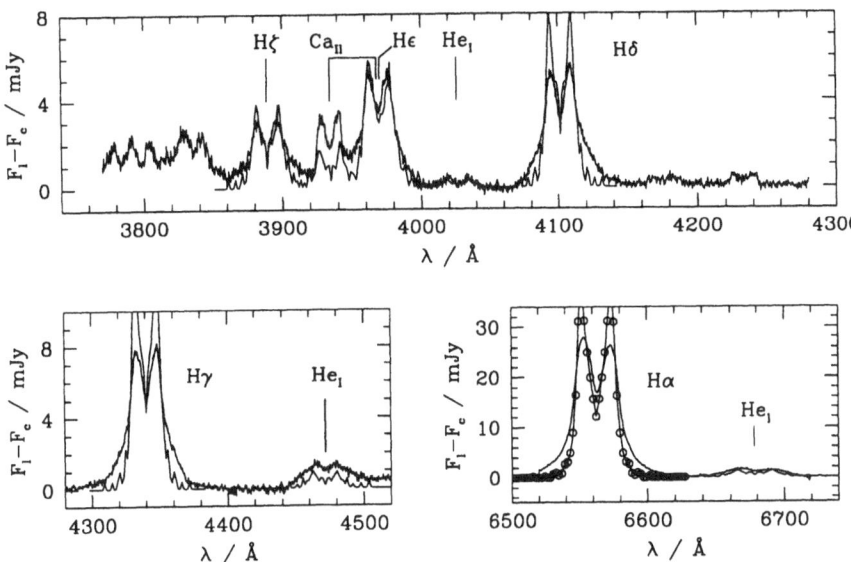

Fig. 2. Best fit model spectrum including Stark broadening. Best fit parameters are: $N = 7.7 \times 10^{14} R^{-2.12}$ cm^{-3}, $T = 14\,400 R^{-0.07}$ K with $\chi^2 = 12.3$ at $D = 70$ pc.

3   Supersonic Velocity Dispersion

In addition to the micro-turbulence V_T we allow for a macroscopic, isotropic and supersonic velocity dispersion V_M as a further possible source of the broad Balmer line wings. In practice we handle macro-turbulence by convolving the resulting line profile with a Gaussian of width V_M. Since the finite spectral resolution is taken into account in the same way we consider the total broadening V_B to be a composition of macro-turbulence and spectral resolution

$$V_B = \sqrt{\Delta V_R^2 + V_M^2} \geq 18 \text{ kms}^{-1} \tag{1}$$

and as an additional free parameter. No better χ^2 is obtained for $V_M > 0$. Our initial parameter set is the best fit. We conclude that there is no evidence for macro-turbulence above the spectral resolution threshold of $\Delta V_R = 18$ kms^{-1}.

4   T Leo

Having studied U Gem we apply our radiative transfer code to T Leo, another dwarf nova of type SU UMa (Slovac et al. 1987) with a low inclination. The optical spectra were obtained during quiescence in the same observing run as U Gem. System parameters are taken from Shafter & Szkody (1984). We use a fixed inclination of $i = 30°$ and an initial distance of $D = 80$ pc. Since the peak separations of the emission lines are poorly resolved (Fig. 3) we use the same disk radius of $R_d = 50R_{wd}$ as for U Gem. We use the best-fit values of U Gem as initial parameters for the density and the temperature. ¿From our experience

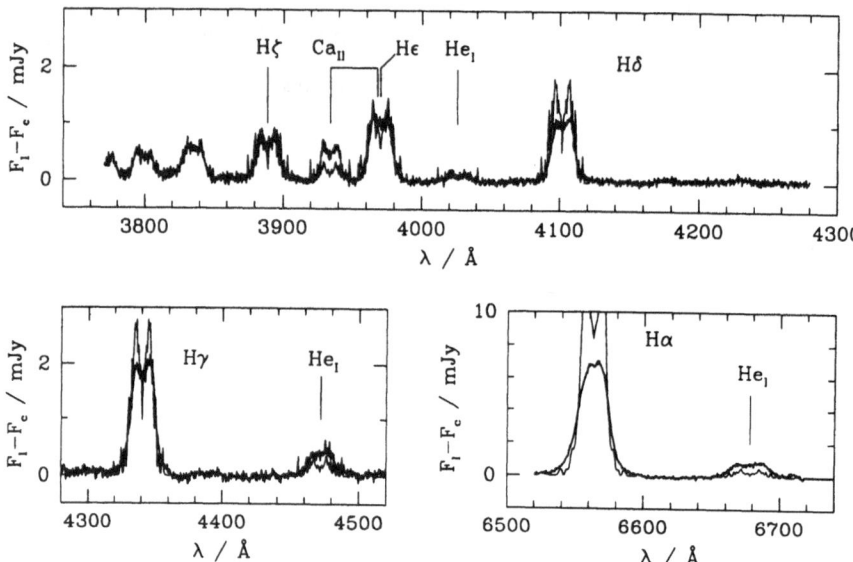

Fig. 3. Continuum subtracted optical spectrum of T Leo during quiescence and the best-fit calculated spectrum for $T - 14\,400R^{-0.07}$ K, $N = 7.9 \times 10^{14}R^{-2.6}$ cm^{-3} and $\chi = 16$. The mean gas temperature is $\langle T \rangle = 11\,500$ K and the total disk mass is $M_d = 1.4 \times 10^{18}$ g.

with U Gem we neglect macroscopic turbulence. In order to be on the safe side we include Stark broadening because of the low inclination.

The best-fit continuum subtracted spectrum of T Leo during quiescence is shown in Fig. 3. The best fit values are $T = 14\,400R^{-0.07}$ K, $N = 7.9 \times 10^{14}R^{-2.6}$ cm^{-3} and $D = 78$ pc with $\chi^2 = 16$.

The ionisation structure of T Leo shows hydrogen to be completely ionised, He II and He I are equally populated and Ca III exceeding Ca II. Since the disk is

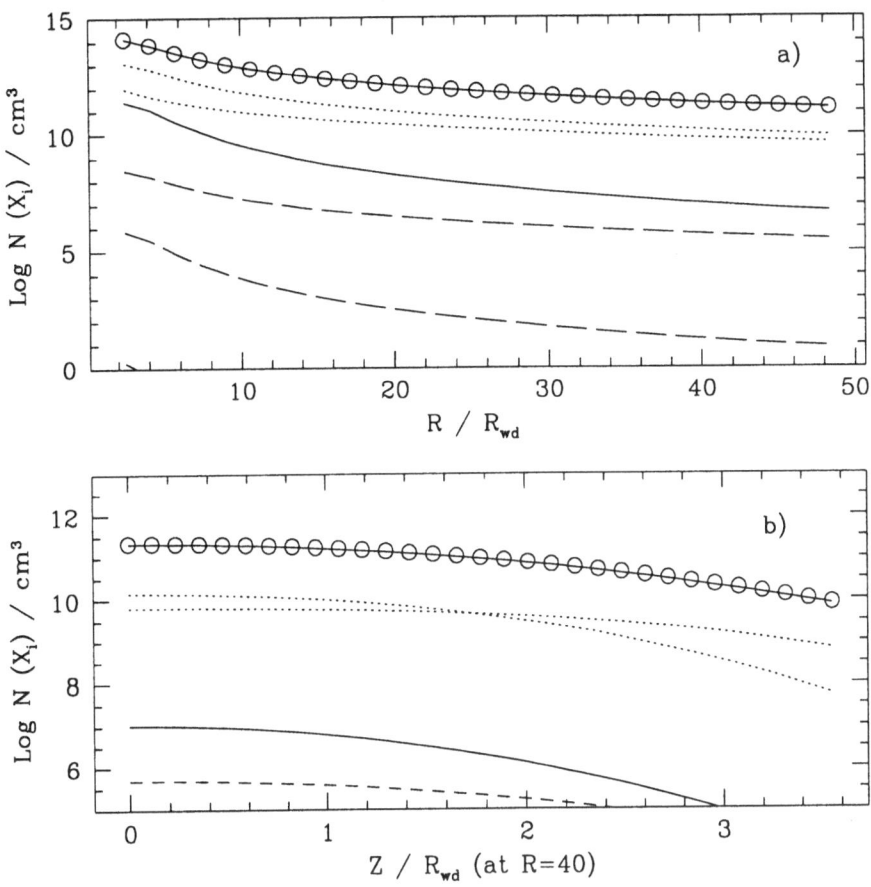

Fig. 4. Radial (a) and vertical (b) distribution of the most abundant ionisation stages in the model disk for T Leo. ¿From top to bottom: H II (—), electron density N_e (o), He I ($\cdots$), He II ($\cdots$), H I (—), Ca III (- - -), Ca II (- - -)

almost isothermal, the density distribution controls the ionisation degree in the disk. Lower densities at higher latitudes increase the ionisation (Fig. 4 b).

5   Discussion

We have shown, that Stark broadening and macroscopic turbulence cannot explain the broad wings of the Balmer emission lines in U Gem. Since our relatively hot model disk is completely ionised and the $H\alpha$ intensity is much larger than the local continuum, we expect Thomson scattering as a promising source for the wing extensions.

For T Leo the wings of the Balmer emission lines are much less developed and are better reproduced by the model. Again, as for U Gem the best fit is a compromise to account for both Ca II and He I. The T Leo model predicts Hα to be much larger than observed. A reason for this could be the following: The low inclination causes a larger optical depth in the lines due to the smaller kinematical broadening. If the calculated continuum is underestimated, Hα will saturate at stronger intensities in the model. In order to test for this we will consider the influence of electron scattering and absorption due to H^- as additional continuum opacity sources. Another possibility for the overestimated Hα emission could be the breakdown of the LTE-assumption at higher latitudes in the disk.

6   Summary

Our tests have shown that Stark broadening is insufficient to broaden the emission lines (at $i = 70^0$) in U Gem during quiescence. A supersonic macro-turbulence does not improve the fit for this object. We found, that the disk structure of U Gem and T Leo during quiescence are essential identical.

Acknowledgment: WH thanks the *Deutsche Forschungsgmeinschaft* (KU 474/22-1) for funding. We would like to thank K. Butler for some routines from his *DETAIL*-code and T. Schöning for providing his extended Stark broadening tables for hydrogen in digital form.

References

Dimitrijević, M.S., Sahal-Bréchot, S., 1990, A&ASS **82**, 519
Griem, H.R., 1964, *Plasma Spectroscopy*, McGraw-Hill, New York
Hummel, W., Horne, K., Marsh, T., Wood, J.H., 1996, *Emission Lines of Dwarf Novae Accretion disks*, in IAU Coll. **163**, on *Accretion Phenomena and Associated Outflows*, in press
Lin, D.N., Williams, R.E., Stover, R.J., 1988, ApJ **327**, 234
Shafter, A.W., Szkody, P., 1984, ApJ **276**, 305
Slovac, M.H., Nelson, M.J., Shafter, A.W., 1987, IAU Circular **4314**
Vidal, C.R., Cooper, J., Smith, E.W., 1978, ApJSS **214**, 37
Williams, G.A., Shipman, H.L., 1980, ApJ **326**, 738

Modelling magnetised accretion discs

A. Brandenburg, C. Campbell

Department of Mathematics, University of Newcastle upon Tyne NE1 7RU, UK

Abstract Some recent results are reviewed that lead us now to believe that accretion discs are basically always magnetised. The main components are Balbus-Hawley and Parker instabilities on the one hand and a dynamo process on the other. A mechanical model for the Balbus-Hawley instability is presented and analysed quantitatively. Three-dimensional simulations are discussed, especially the resulting magnetic field structure. Possibilities of reproducing the field by an $\alpha\Omega$ dynamo are investigated, especially its symmetry with respect to the midplane.

1 Introduction

Until quite recently the origin of turbulence in accretion discs was considered to be rather obscure (see, for example, the excellent textbook by Frank et al 1992, Sect. 4.7). However, this seems to have changed considerably over the past few years. There is now strong numerical evidence that turbulence may be generated by the Balbus-Hawley instability (e.g. Hawley et al 1995, Matsumoto & Tajima 1995, Brandenburg et al 1995). Moreover, three-dimensional simulations show that even in the absence of an external magnetic field there will be self-excited turbulence, because the flow can regenerate the magnetic field by dynamo action (Brandenburg et al 1995, Hawley et al 1996, Stone et al 1996). We refer to this mechanism as dynamo-generated turbulence.

There remain several outstanding problems. Firstly, we need to gain a deeper understanding of how the dynamo process works. Is it some kind of an $\alpha\Omega$ dynamo, or is it something completely different? Secondly, what can we learn from local models of dynamo-generated turbulence if the real accretion disc is global? We begin by discussing briefly how keplerian shear flows become unstable in the presence of some coupling. We then consider the structure of large scale magnetic fields that might be generated in an accretion disc. Finally, we discuss the problem of global models of magnetised accretion discs.

2 A mechanical model of the instability

In order to understand the nature of the Balbus-Hawley (1991, 1992) instability it is useful to consider mechanical models displaying similar behaviour as a magnetised fluid in keplerian motion. An imposed uniform magnetic field, for example, holds the fluid particles in place like beads on an elastic string. In both cases there is a restoring force (with a given spring constant in the mechanical model, and the magnetic tension force in the hydrodynamical model). The

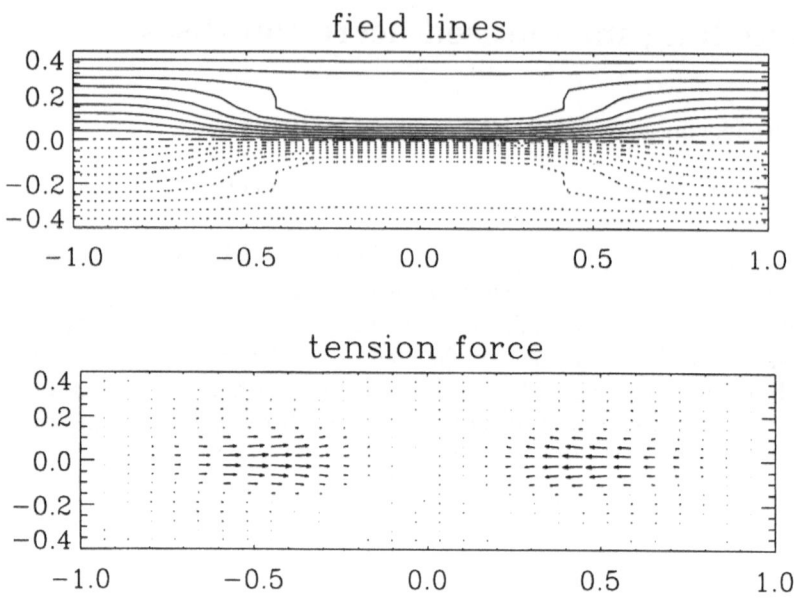

Fig. 1. Field lines representing a magnetic flux tube (upper panel). The Lorentz force $(\nabla \times \boldsymbol{B}) \times \boldsymbol{B}$ has no component in the direction of $\boldsymbol{B}$. However, the Lorentz force has two contributions, a magnetic pressure gradient, $-\frac{1}{2}\nabla(B^2)$, which is readily balanced by the gas pressure gradient, and the tension force, $(\boldsymbol{B} \cdot \nabla)\boldsymbol{B}$, which is plotted in the lower panel. The tension force tends to contract the tube. (In the thin flux tube approximation this contraction corresponds to a pressure gradient along the tube.)

analogy works also for nonuniform (turbulent) fields which typically consist of many flux tubes. Consider now a model with only two beads connected by a rubber band. This rubber band symbolises the restoring force experienced by a magnetic flux tube; see figure 1. The main difference is now that we can consider localized flux tubes and do not need to invoke a uniform large scale field. This seems more appropriate for characterizing turbulent magnetic fields as they seem to be present in the simulations. However, in this case the analogy is by no means exact, and yet it seems to capture some typical features of the dynamics of magnetic flux tubes.

Consider two particles in a keplerian orbit. The story is best conveyed with two space crafts orbiting round the earth and trying a rendezvous maneuver. *Larry November* from Sac Peak Observatory explained to us his memories of *Gemini 3 launched March 23, 1965 with Gus Grissom and John Young. The mission tried to dock with an Agena spacecraft launched separately and put into a similar orbit. The astronauts tried maneuvering by "eye" and were unable to close their gap with Agena because of the oddities of orbital dynamics. They found that they could not get closer by "speeding up". Increasing their orbital velocity only put them in a higher orbit which caused them to lag further beyond the Agena. The unsuccessful mission demonstrated that maneuvering spacecraft was unintuitive and could only be successful by use of computers. Subsequent missions used onboard computers to estimate crossing orbits given the relative locations of the spacecrafts. That software solution completely solved the diffi-cultly and permitted entirely successful docking with the lunar excursion module with the mother Apollo craft used in all of the lunar landings. Of course soft-ware solutions could be accurate within millimeters and provide corrections that minimized expended fuel.* It is surprising to hear that the astronauts really fell into this trap, but Larry also said *I do believe that after they realized the effect they did try breaking. Unfortunately, however, the effect is difficult to gauge and I do not think they were ever closer than about 100 m, and only managed to get hopelessly separated as they tried different things.* One should notice that the mission was allocated only 3 orbits, so this was probably not meant to be a particularly serious attempt!

Anyway, the main lesson is this: in order to go faster in a keplerian orbit one has to break, and vice versa. However, rubber bands behave more straightfor-wardly. They exert a restoring force when starting to pull. Therefore, objects in a keplerian orbit that are connected in some way always go unstable. In a sense this mechanism is reminiscent of *tidal disruption* of celestial bodies pass-ing nearby a black hole (see Novikov et al 1992). Here the restoring force is the self-gravity of the passing body. In fact, the criterion for disruption is similar to the stability criterion of Balbus-Hawley (see below).

In our simple model the positions of two coupled particles, $r_1(t)$ and $r_2(t)$, are governed by the equation

$$\ddot{r}_i = -\frac{GM}{r^2}\hat{r} - f(r_i - r_j), \tag{1}$$

where G is the gravitational constant, M the mass of the central object, and $f(x)$ is the restoring force in the direction $\hat{x} = x/|x|$. The restoring force is assumed to be proportional to $|x| - d_0$, where

$$d_0 = |r_2(0) - r_1(0)| \tag{2}$$

is the initial separation between the two particles. Thus, we put

$$f(x) = \begin{cases} K_0\hat{x}(|x| - d_0) & \text{when} \quad |x| > d_0, \\ 0 & \text{otherwise.} \end{cases} \tag{3}$$

The cutoff for $|x| < d_0$ was introduced to account for the fact that no force should be exerted when the rubber band is not tight, i.e. if the particles are too close together. The rubber band has a spring constant K_0 per unit mass, which is measured in units of $\Omega^2 = GM/r_0^3$, where Ω is the keplerian angular velocity at the initial radius r_0 of the particles.

In figure 2 we plot the positions of a pair of particles at different times for two different values of K_0. In the first case $(K_0 = \Omega^2)$ the spring constant is weak enough so that the pair of particles becomes (tidally!) disrupted. However, in the second case $(K_0 = 10\Omega^2)$ the coupling is strong enough so that the pair of particles always stays together. In addition to the counterclockwise orbital motion the two particles rotate about each other also in the counterclockwise direction.

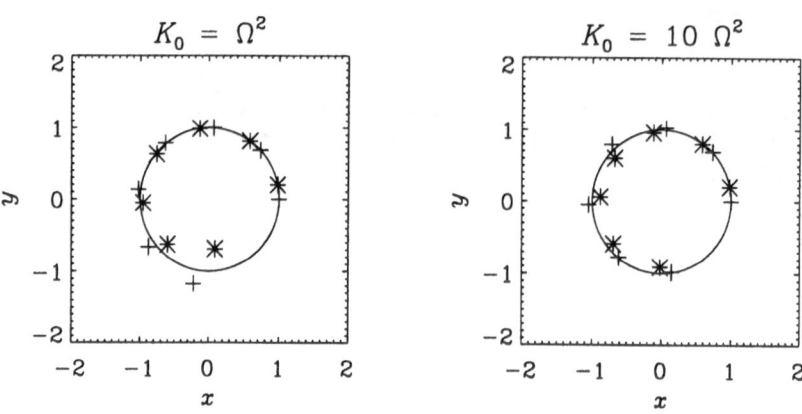

Fig. 2. The positions of a pair of particles at different times for two different values of K_0 (left panel: $K_0 = \Omega^2$, right panel $K_0 = 10\Omega^2$). Initially the leading particle (indicated by a star) moves inwards and the following particle (indicated by a plus) moves outwards.

This result can be understood by means of linear stability analysis. We assume $r_i = r_0 + \delta r$, where r_0 describes the motion of the center of mass with $\ddot{r}_0 = -\Omega^2 \delta r_0$. Linearising the gravitational acceleration yields

$$\frac{GM}{r^2}\hat{r} = \frac{GM}{(r_0 + \delta r)^3}(r_0 + \delta r) = \Omega^2 r_0 \left(1 - 3\frac{\delta r \cdot r_0}{r_0^2}\right) + \Omega^2 \delta r. \qquad (4)$$

The linearised equation of motion is then

$$\delta\ddot{r} = 3\Omega^2 \frac{\delta r \cdot r_0}{r_0^2} - \Omega^2 \delta r + K\delta r. \qquad (5)$$

To get the dispersion relation we assume $\delta r \propto e^{-i\omega t}$ and obtain after taking the inner product with r_0

$$\omega^2 = -2\Omega^2 + K. \qquad (6)$$

The system is unstable when $\omega^2 < 0$, i.e.

$$K < 2\Omega^2 \quad \text{(instability)}. \qquad (7)$$

We verified this also numerically. This dispersion relation is similar to the case of the Balbus-Hawley instability, where the criterion is:

$$\omega_A^2 < 2q\Omega^2 \quad \text{(instability)} \tag{8}$$

(Balbus & Hawley 1992), where $\omega_A = i\mathbf{k} \cdot \mathbf{B}/(4\pi\rho)^{1/2}$ is the Alfvén frequency for modes with wave vector $\mathbf{k}$ and $q \equiv -d\ln\Omega/d\ln r = 3/2$ for keplerian rotation. The criteria (7) and (8) are similar to the criterion of tidal disruption of celestrial bodies passing nearby a black hole

$$\omega_*^2 < \alpha^3\Omega^2 \quad \text{(instability)} \tag{9}$$

(Novikov et al 1992), where $\omega_* = GM_*/R_*^3$ is approximately the eigenfrequency of a star of mass M_* and radius R_* passing nearby the black hole. The coefficient α is of order unity. [Novikov et al (1992) quote the value 1.69 for an incompressible stellar model, see Kosovichev & Novikov (1991), but α is a little less for a polytropic stellar model, Luminet & Carter (1986)]. Common to all three cases is the fact that a harmonic oscillator with eigenfrequency ω can become overstable when placed in a keplerian orbit (or parabolic orbit in the case of tidal disruption) provided ω^2 is less than $\Omega^2 = GM/r_0^3$.

3   Dynamo generated turbulence

From the illustrative experiments above we have seen that both weak and strong coupling can lead to instability. A proper stability analysis of fluid in keplerian motion threaded by a magnetic field (of arbitrary orientation) confirm that there is indeed an instability (Balbus & Hawley 1991, 1992). The difference is that in the stability analysis there is a uniform imposed magnetic field, whereas the mechanical experiment presented here corresponds to an isolated magnetic flux tube connecting two points.

There are numerous subsequent investigations that extent the local analysis to global geometry (e.g. Curry et al 1994, Terquem & Papaloizou 1996, Ogilvie & Pringle 1996, Kitchatinov & Rüdiger 1997). We mentioned in the introduction that there are now also numerical simulations that show that the flow generated by the instability leads to turbulence, and that this turbulence in turn is capable of amplifying and sustaining the magnetic field by dynamo action. We briefly summarize here some of the basic results of such simulations.

The most significant result of such simulations is simply the fact that turbulence is self-sustained. An important quantitative outcome of the simulations concerns the strengths of the Maxwell and Reynolds stresses that would lead to mass accretion and angular momentum transport. Their magnitude is conveniently given in terms of the mean angular momentum gradient and a turbulent viscosity which, in turn, is specified in terms of natural units (sound speed c_s and disc scale height H) times a dimensionless factor, α_{SS}. Here the subscript SS refers to Shakura & Sunyaev (1973), who introduced this concept in the context of accretion discs. Thus, the stress is expressed as

$$\langle m_r' u_\phi' - B_r' B_\phi'/4\pi \rangle = -\langle \rho \rangle \nu_t r \frac{\partial \Omega}{\partial r}, \quad \text{where} \quad \nu_t = \alpha_{SS} c_s H. \qquad (10)$$

where $m_r = \rho u_r$ is the mass flux in the radial direction and primes indicate fluctuations about the mean. The value of α_{SS} fluctuates in time, because the system is turbulent and because the magnetic field varies strongly on long time scales. In figure 3 we show the evolution of the magnetic and kinetic contributions to α_{SS}. Comparison with the magnetic energy in the system shows that peaks in α_{SS} coincide with peaks in the magnetic energy. On average the value of α_{SS} is of the order of 0.01 (Brandenburg et al 1995, 1996a, Hawley et al 1996, Stone et al 1996). We should recall that some authors use a slightly different definition of α_{SS}, where α_{SS} could be larger by a factor $3/2$ times $\sqrt{2}$; (see the review by Brandenburg et al 1996b). Magnetohydrodynamic models of accretion discs (Campbell 1992, Campbell & Caunt 1996) show that such values of α_{SS} are sufficient to lead to dynamically important large scale magnetic fields. The resulting $\langle B_r' B_\phi' \rangle$ stresses can play a major part in the radial advection of angular momentum necessary to drive the disc inflow. Note also that, although the field is oscillatory, $\langle B_r' B_\phi' \rangle$ is always negative.

The next important result of the simulations is that there could be long term variability of the dynamo activity, which is associated with a variable large scale field. In the cartesian models investigated by Brandenburg et al (1996a) the activity varies approximately cyclically with an average period of about 30 orbits. This varying large scale magnetic field, especially the toroidal component $\langle B_\phi \rangle$, strongly affects the value of α_{SS} in a systematic manner which can be described by a parabolic fit of the form

$$\alpha_{SS} \approx \alpha_{SS}^{(0)} + \alpha_{SS}^{(B)} \langle B_\phi \rangle^2 / B_{eq}^2, \qquad (11)$$

where $B_{eq} = \langle 4\pi \rho c_s^2 \rangle^{1/2}$ is the equipartition value with respect to the *thermal* energy density, and $c_s = \Omega H/\sqrt{2}$ is the isothermal sound speed. (Both c_s and H depend on the disc temperature, which may change due to heating.) The most important contribution to (11) comes from the second term. In this second term the coefficient is $\alpha_{SS}^{(B)} \approx 0.5$.

Another remarkable result concerns the dependence of the value of α_{SS} on the rotation law. Abramowicz et al (1996) used the simulations for different values of $q \equiv -d\ln \Omega/d\ln r$. The value $q = 3/2$ is for keplerian rotation. Although the main parameter varied was q, Abramowicz et al expressed it in coordinate independent form using the magnitudes of the shear and vorticity tensors, σ and ω, respectively. They found that α_{SS} varies approximately linearly with the ratio σ/ω. Abramowicz et al (1996) suggest that this result could be used in more general circumstances as well. This would be especially important if one were to produce a global accretion disc model. By this we mean a model, as opposed to a full three-dimensional turbulence simulation. The advantage of such a model

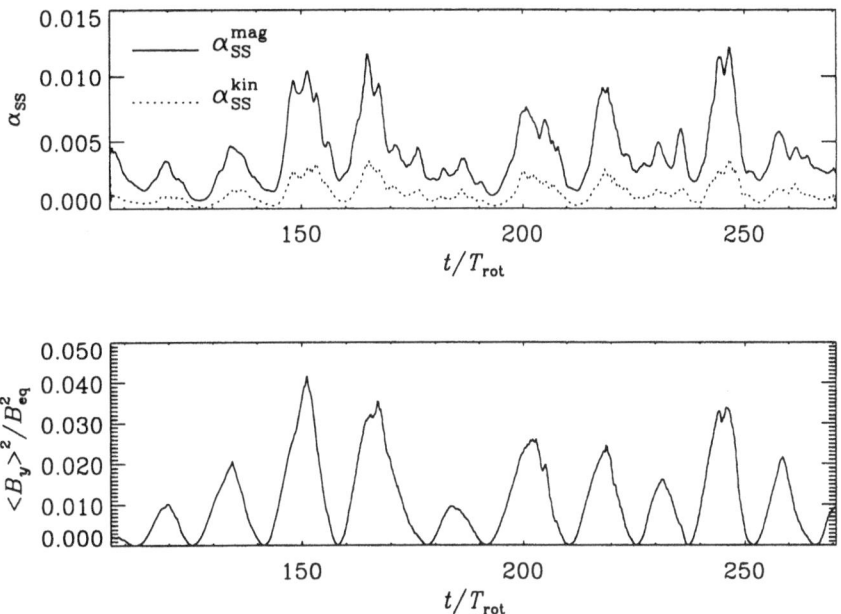

Fig. 3. Evolution of the magnetic and kinetic contributions to α_{SS}. Comparison with the magnetic energy of the mean field shows that peaks in α_{SS} coincide with peaks in the magnetic energy, which is plotted here in the form $\langle B_\phi \rangle^2 / B_{eq}^2$, where $B_{eq} = \langle 4\pi\rho c_s^2 \rangle^{1/2}$ is the equipartition value with respect to the *thermal* energy density, and $c_s = \Omega H/\sqrt{2}$ is the isothermal sound speed. The data are from Model O of the three-dimensional simulation of Brandenburg et al (1996a).

would be that it is easier to produce, and that it can be more easily applied to different circumstances. Before we consider this in more detail we need to discuss another complication that is related to the parameterization (10).

When we estimated the value of α_{SS} from eq. (10) we assumed that $\langle \rho \rangle$ was the *volume* averaged density. This seemed sensible, because the stresses do not strongly vary with height. However, it appears questionable whether a vertically averaged density would be a sensible description under more general circumstances, where the disc could be thick, for example. In that case it would seem natural to adopt an average that depends on height. However, this is not consistent with the numerical models, because then eq. (10) no longer represents a good description of the simulation's results, unless we relax the assumption that α_{SS} is independent of height.

We now allow α_{SS} to be height dependent. However, instead of assuming some unknown profile function we assume that this dependence is already captured by the dependence (11). Originally eq. (11) was obtained by considering

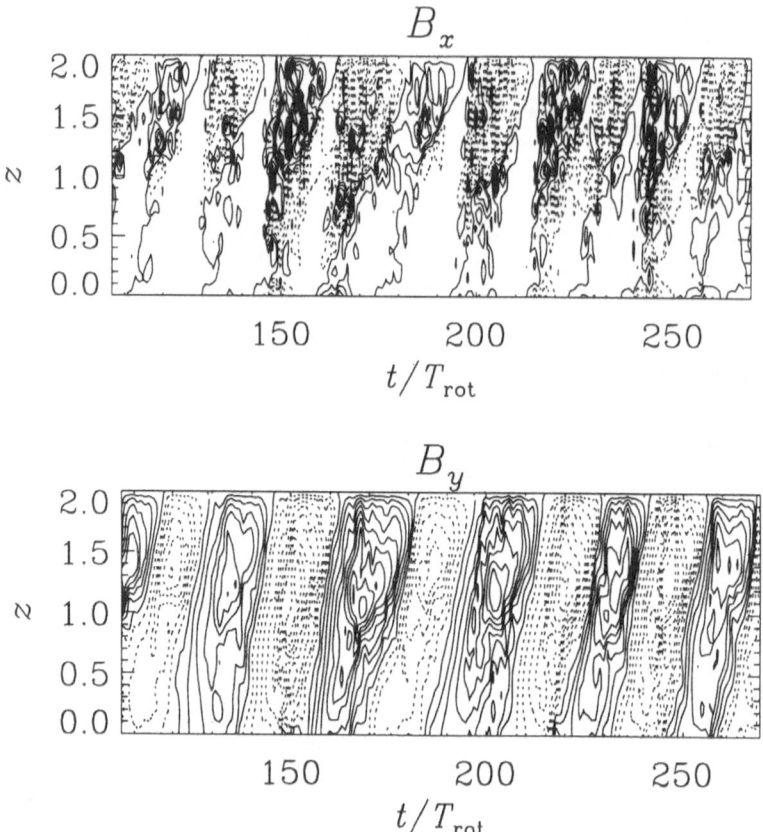

Fig. 4. Spatio-temporal pattern of the radial and toroidal components of the large scale field, $\langle B_x \rangle$ and $\langle B_y \rangle$, respectively, as a function of height and time. The data are from Model O of the three-dimensional simulation of Brandenburg et al (1996a).

volume averaged values of the stress and the density at different times during the magnetic cycle. We now assume that this relation is also valid at each height. Putting (11) into (10), and neglecting the $\alpha_{SS}^{(0)}$ term, we find that the stress varies like

$$\text{stress} = -\frac{\alpha_{SS}^{(B)}\langle B_\phi \rangle^2}{4\pi c_s^2}c_s Hr\frac{\partial\Omega}{\partial r} = -\frac{\sqrt{2}q}{4\pi}\alpha_{SS}^{(B)}\langle B_\phi \rangle^2, \qquad (12)$$

which is now independent of $\langle \rho \rangle$. However, we have to know how $\langle B_\phi \rangle$ varies with height. So, if eq. (12) were to be used in an accretion disc model one would need a sensible prediction of the variation of the mean magnetic energy density with height. Unlike the local cartesian simulations, where the magnetic energy density varied only little with height, in a truly global model the energy density ought to decrease as one goes sufficiently far away from the disc midplane. Thus, one is then seriously forced to consider accretion disc models that included the magnetic field evolution in a self-consistent manner. Let us now discuss how one can actually model the evolution of the mean magnetic field without invoking a full-blown numerical turbulence simulation. In order to appreciate the systematic behaviour of the large scale magnetic field we plot in figure 4 the spatio-temporal pattern of the radial and toroidal components of the large scale field, $\langle B_r \rangle$ and $\langle B_\phi \rangle$, respectively.

The large scale field, especially the toroidal magnetic field component, shows remarkable spatio-temporal coherence. The field varies not only cyclically, but it also migrates away from the midplane. The traditional approach to understand such organised behaviour is to adopt the mean-field approach, where the original induction equation is averaged and turbulent transport coefficients are introduced that describe the evolution of the nonlinear term, $\mathcal{E} \equiv \langle u' \times B' \rangle$ in terms of the mean field itself. In principle such parameterizations can be fairly complicated, but more importantly, they are typically extremely uncertain. We may therefore use the simulations to estimate the "transport coefficients" assuming a relation $\mathcal{E} = \mathcal{E}(\langle B \rangle)$. Such a relation should contain terms that are capable of yielding dynamo action. So, in its crudest approximation it should take the form

$$\langle u' \times B' \rangle = \alpha \langle B \rangle - \eta_t \nabla \times \langle B \rangle, \tag{13}$$

where α is the traditional dynamo α-effect (not to be confused with α_{SS}) and η_t is a turbulent magnetic diffusivity. The simulations are consistent with the following estimates: $\alpha \approx -0.001\Omega H$ and $\eta_t \approx 0.008\Omega H^2$, where the sign of α is negative in the upper disc plane, but positive in the lower disc plane; see Brandenburg et al (1995), Brandenburg & Donner (1996). This is quite peculiar. The opposite result is expected from conventional mean-field theory, where the α-effect is related the helicity of the turbulence. Whilst the helicity of the turbulence in our simulations does have the expected sign, the simulations indicate that it has not much to do with α.

The sign of α can be explained as a direct consequence of the Balbus-Hawley instability. Figure 2 illustrates this. Initially this instability turns a toroidal magnetic flux tube in the counterclockwise direction. This is because the following particle accelerates, so it moves outwards, whilst the leading particle brakes and moves inwards, corresponding to counterclockwise rotation. This alone would not lead to an α-effect, or to a component of $\langle u' \times B' \rangle$ in the direction of $\langle B \rangle$, because we also need a systematic orientation of the velocity. This bit is easy, however, because strong magnetic flux tubes are always susceptible to magnetic buoyancy, which will lift them vertically away from the midplane. So, motion in the direction of z together with a twist of the magnetic flux tube in the counterclockwise direction does lead to a systematic sign of the α-effect, and this sign is negative.

A negative α-effect has various implications. First of all, if the generated magnetic field is oscillatory, as in fact it is in the simulations, there will be a magnetic field migration associated with its cyclic variation. This is indeed what is observed in the simulations. The direction of this field migration is indeed consistent with the implied negative sign of α. Furthermore, a negative α could affect parity selection of the magnetic field. While for positive α the preferred parity of the magnetic field is always even, this does not need to be the case when the sign of α is reversed. We address this in the following section.

To test the hypothesis that the magnetic field evolution seen in figure 4 can be explained by a mean-field dynamo we now solve the horizontally averaged induction equation using eq. (13),

$$\frac{\partial \langle B_x \rangle}{\partial t} = -\frac{\partial}{\partial z} \alpha \langle B_y \rangle + \eta_t \frac{\partial^2 \langle B_x \rangle}{\partial z^2}, \tag{14}$$

$$\frac{\partial \langle B_y \rangle}{\partial t} = -q\Omega \langle B_x \rangle + \eta_t \frac{\partial^2 \langle B_y \rangle}{\partial z^2}, \tag{15}$$

where $q = 3/2$. Here and below x corresponds to radius and y to longitude.) Since $\alpha \ll \Omega H$ we have neglected the α-effect in the second equation (15). On the boundaries we assume

$$\frac{\partial \langle B_x \rangle}{\partial z} = \frac{\partial \langle B_y \rangle}{\partial z} = 0 \quad \text{on} \quad z = 0; \quad \langle B_x \rangle = \langle B_y \rangle = 0 \quad \text{on} \quad z = L_z. \tag{16}$$

This boundary condition was also used in the three-dimensional simulations (except in those cases where no symmetry was prescribed; see figure 6). The calculations of Brandenburg et al (1995) confirmed that α changes sign about the equator. The simplest functional form for α is therefore $\alpha = \alpha_0(z/H)$. For $\alpha_0 = -0.001\Omega H$ we reproduce the right cycle frequency, $\Omega_{\text{cyc}}/\Omega = 0.03$. In fact, one can show that $\Omega_{\text{cyc}}/\Omega = \mathcal{O}(|\alpha/\Omega H|^{1/2})$. In figure 5 we plot the resulting spatio-temporal pattern of $\langle B_x \rangle$ and $\langle B_y \rangle$. The qualitative agreement with figure 4 is quite striking.

4   The parity of large scale magnetic fields

A large number of different dynamo models has been studied over the years. However, the case of a negative α-effect has not received much attention, because it was thought to be unphysical. In the case of galactic dynamos the result is typically that for $\alpha_0 < 0$ oscillatory modes of dipole-type parity are most easily excited (Parker 1971, Stix 1975, see also Brandenburg et al 1990). This result is interesting in various respects. Firstly, a dipole-like magnetic field seems to be more favourable when modelling magnetically driven jets from accretion discs (eg Yoshizawa & Yokoi 1993 and references therein). However, if a dipole-like magnetic field is indeed easier to excite than a quadrupole-like magnetic field the question arises why the field found in the local simulations was actually still quadrupole-like. Also, although galactic dynamos are similar in geometry to accretion disc dynamos, there are marked differences concerning especially the form of the rotation curve. So we are encouraged to look now more systematically at the different field parities under those different conditions.

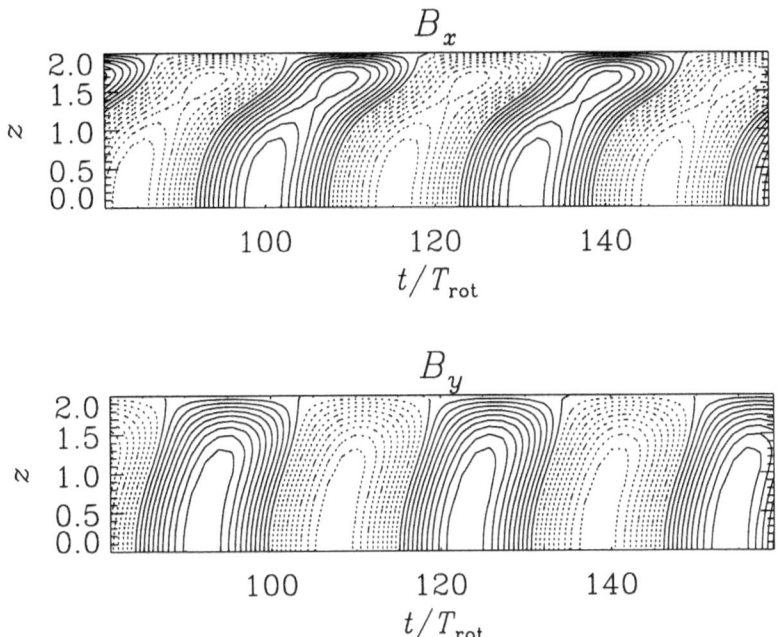

Fig. 5. Spatio-temporal pattern of $\langle B_x \rangle$ and $\langle B_y \rangle$ as a function of height and time, obtained by solving (14)-(16).

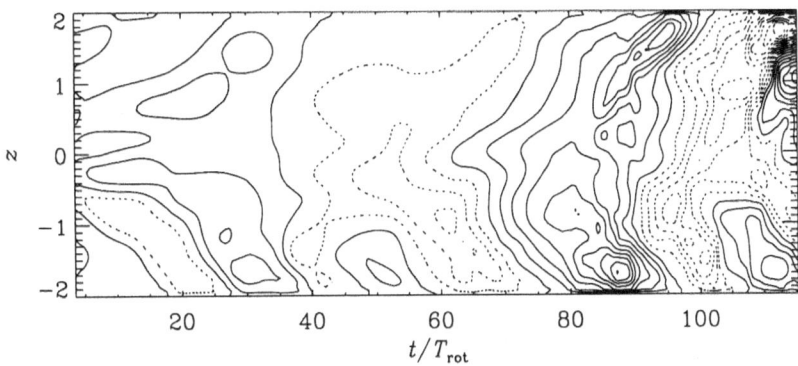

Fig. 6. Spatio-temporal pattern of $\langle B_y \rangle$ as a function of height and time for the three-dimensional simulation of Brandenburg et al (1995). No restriction to the parity is made. Note that the field is mostly symmetric about the midplane.

First we consider the magnetic field generated by a mean-field dynamo in a local box with the same boundary conditions as those used in the numerical simulations. The field structure of $\langle B_x \rangle$ and $\langle B_y \rangle$ is given in figure 7. The critical values of the dynamo number

$$D = q\frac{\alpha_0 \Omega_0 H^3}{\eta_t^2} = q\left(\frac{\alpha_0}{\Omega_0 H}\right)\left(\frac{\eta_t}{\Omega_0 H^2}\right)^{-2}, \tag{17}$$

where the dynamo is just marginally supercritically, are given in table 4. Here, $q = 3/2$ for keplerian rotation. Note that even for negative α the quadrupole-type geometry (even parity) is the most preferred one. This is a bit surprising, but it is at least not in conflict with the results of the three-dimensional simulations, which also give quadrupole-type symmetry when no symmetry restriction is imposed; see figure 6. Continuous inflow through the disc requires a quadrupole-type field structure. Unless the surroundings are very highly conducting, a dipole-type field does not lead to a magnetic torque on rings of disc material and hence cannot contribute to the radial advection of angular momentum (Campbell 1997).

However, we now need to check whether the occurrence of a quadrupole mode for negative values of α could be an artefact of the local geometry used in our model. Therefore we now consider briefly a global $\alpha\Omega$ dynamo model with a disc-like distribution of α and η_t. In this model we used the following profiles for α and Ω

$$\alpha = \alpha_0 \frac{z}{H} \exp\left\{\frac{1}{2}\left[1 - \left(\frac{z}{H}\right)^2\right]\right\}, \tag{18}$$

$$\Omega = \Omega_0 \left[1 + \left(\frac{r}{r_0}\right)^{\frac{3}{2}n}\right]^{\frac{1}{n}}, \tag{19}$$

with $n = 10$. Again, the result is surprising. The critical solutions are plotted in figure 7 and the critical values of the dynamo number D are given in table 4, where we also compare with results obtained by other authors.

The table shows that there is not a unique result regarding the parity of the easiest excited mode for negative values of α. The disks of Stepinski & Levy (1988, SL88) are relatively thick and represent only this innermost parts of the disc. The general behaviour of those models is similar to dynamos in spherical geometry. The models of Stepinski & Levy (1990, SL90) are thinner, and here the parity depends on whether the fields are confined to the disc (SL90-1) or whether the disc is surrounded by a vacuum (SL90-2), permitting the field to extend into the corona. In Torkelsson & Brandenburg (1994a, TB94) the parity depends on whether or not there is a cavity in the middle of the disc. In those models a cavity was introduced to model an inner boundary of the disc at the innermost stable orbit around black holes. If such a cavity is present (TB94-4c, ie Model c in their Table 4), then A0 parity is more preferred.

To shed some light on the occurrence of the different modes we now compute the full spectrum of growth rates shown as a function of dynamo number $D =$

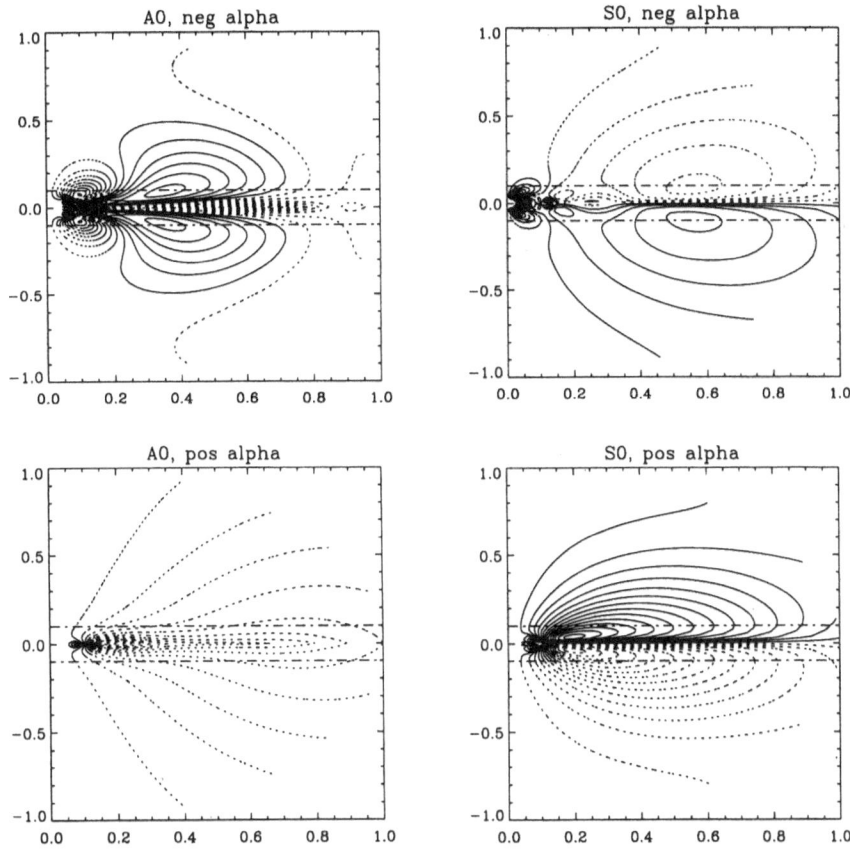

Fig. 7. Poloidal magnetic field (dotted lines indicate opposite field orientation) for an $\alpha\Omega$ dynamo in disc-like geometry using the profiles given by (18) and (19).

$q\alpha_0\Omega_0 H^3/\eta_t^2$ by solving an eigenvalue problem; see figure 8. Again we use $\alpha = \alpha_0(z/H)$. Note that for either sign of D the growth rates of quadrupole-like (symmetric or S0) solutions are largest. Those modes are oscillatory, except for a certain interval $0 < D \lesssim 400$ (positive alpha, but negative shear), where this branch splits into two non-oscillatory branches. Similar behavior is seen for the next easily excited mode of A0 type (antisymmetric), but here the two non-oscillatory solutions have merged into a single oscillatory one before its growth rate becomes positive. This illustrates where the sensitivity of either oscillatory or non-oscillatory behaviour comes from.

In conclusion, we must regard the parity of the dynamo as uncertain: it depends on geometrical aspects which could decide upon whether or not two non-

Table 1. Critical values of the dynamo number for different models (see text for explanations). Numbers in bold face indicate the most easily excited mode. A0 and S0 refer to antisymmetric (dipole-type) and symmetric (quadrupole-type) modes.

	neg alpha		pos alpha	
	A0	S0	A0	S0
cartesian	−1130 osc	−400 osc	+630 osc	+13 st
Figure 7	−130 osc	−80 osc	+160 osc	+35 osc
TB94-4O	−576 osc	−512 osc	+352 osc	+40 st
TB94-4c	−1080 osc	−1120 osc	+792 osc	+77 st
SL88-3	−138 osc	−169 osc	+201 osc	+187 osc
SL90-1	−9 st	−45 osc	+40 osc	+60 st
SL90-2	−68 osc	−45 st	+48 osc	+9 st

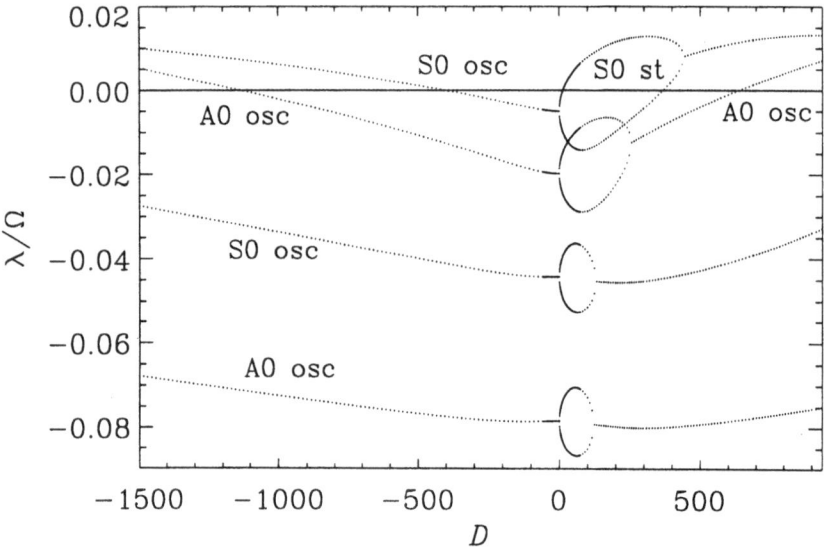

Fig. 8. Growth rates λ (in units of Ω) as a function of the dynamo number D for a dynamo in cartesian geometry using equations (14) and (15) and $\alpha = \alpha_0 z/H$ in $0 < z < H$.

oscillatory branches have merged into a single oscillatory one. Also, in the highly nonlinear regime things can change again, as was demonstrated by Torkelsson & Brandenburg (1994b), who presented a survey of models for both signs of α, different nonlinearities (buoyancy and α-quenching), and a large range of different dynamo numbers.

5   Towards a magnetised standard disc model

In the forthcoming years it will be important to design a new standard accretion disc model that includes those new effects turbulence simulations have revealed recently. Ideally, one would like to have a magnetic version of the famous Shakura-Sunyaev solution. The model by Campbell (1992) is such an example. In this section we briefly address a few issues where some adjustments to this theory could be made and where conceptual differences should need some further clarification.

An important property of the dynamo is the value of the magnetic field at which the dynamo saturates. This value is determined by the dominant feedback magnetism. In the model of Campbell (1992) it was assumed that magnetic buoyancy limits the magnetic field strength. Another possibility is α-quenching, which seems to be important in the three-dimensional simulations (Brandenburg & Donner 1996). A more urgent concern is related to the sign of the dynamo α. The simulations suggest that α is negative. This is rather surprising and could not have been predicted. This appears to be directly related to the dynamics of the Balbus-Hawley instability. So we do have some understanding of this surprising result and are tempted to include it into actual models of magnetised accretion discs. One immediate consequence would be that the magnetic field can no longer assumed to be steady. A variable level of the magnetic field could lead to variability of the temperature in the disc, which could be of interest in connection with the outbursts of cataclysmic variables (CVs). Current models describing outbursts of CVs invoke a dependence of α_{SS} on H/R (e.g. Cannizzo et al 1988), which does not seem to be supported by the simulations (Vishniac & Brandenburg 1997). Furthermore, standard outburst models require α_{SS} to be around 0.1 during outbursts. Such a large value could only be achieved in the presence of a sufficiently strong magnetic field. This seems unsatisfactory, because the origin of such a field needs to be explained. More importantly, it would then be difficult to explain the absence of strong fields during quiescent phases. Therefore a self-consistent model for CV outbursts seems to be highly desirable.

6   Conclusions

Significant progress has been made in understanding the origin of viscosity and the nature of turbulence in accretion discs. Three-dimensional simulations in local geometry can be used to address those questions. Nevertheless, it is important to consider those simulations as preliminary. Global simulations using more realistic, open, boundary conditions and including the effects of curvature are necessary. Finally, proper account of radiative transfer should be made before we can begin to address questions of observational significance, such as the CV outbursts.

References

Abramowicz, M. A., Brandenburg, A., & Lasota, J.-P. 1996 *Monthly Notices Roy. Astron. Soc.* **281**, L21-L24.

Balbus, S. A. & Hawley, J. F. 1991 *Astrophys. J.* **376**, 214-222.

Balbus, S. A. & Hawley, J. F. 1992 *Astrophys. J.* **400**, 610-621.

Brandenburg, A. & Donner, K. J. 1996 *Monthly Notices Roy. Astron. Soc.* (submitted).

Brandenburg, A., Tuominen, I., Krause, F. 1990 *Geophys. Astrophys. Fluid Dyn.* **50**, 95-112.

Brandenburg, A., Nordlund, Å., Stein, R. F., Torkelsson, U. 1995 *Astrophys. J.* **446**, 741-754.

Brandenburg, A., Nordlund, Å., Stein, R. F., Torkelsson, U. 1996a *Astrophys. J. Letters* **458**, L45-L48.

Brandenburg, A., Nordlund, Å., Stein, R. F., Torkelsson, U. 1996b Dynamo generated turbulence in disks: value and variability of alpha. In *Physics of Accretion Disks* (ed. S. Kato et al) pp.285–290. Gordon and Breach Science Publishers.

Campbell, C. G. 1992 *Geophys. Astrophys. Fluid Dyn.* **63**, 197-213.

Campbell, C. G. 1997 *Magnetohydrodynamics in Binary Stars.* Kluwer Academic Publishers, Dordrecht.

Campbell, C. G. & Caunt, S. E. 1996 *Mon. Not. R. Astr. Soc.*, submitted.

Cannizzo, J. K., Shafter, A. W., Wheeler, J. C. 1988 *Astrophys. J.* **333**, 227-235.

Curry, C., Pudritz, R. E., & Sutherland, P. 1994 *Astrophys. J.* **434**, 206-220.

Frank, J., King, A. R., & Raine, D. J. 1992 *Accretion power in astrophysics.* Cambridge: Cambridge Univ. Press.

Hawley, J. F., Gammie, C. F., & Balbus, S. A. 1995 *Astrophys. J.* **440**, 742-763.

Hawley, J. F., Gammie, C. F., & Balbus, S. A. 1996 *Astrophys. J.* **464**, 690-703.

Kitchatinov, L. L. & Rüdiger, G. 1997 *Monthly Notices Roy. Astron. Soc.* (in press).

Kosovichev, A. G. & Novikov, I. D. 1991 Preprint IOA, Cambridge.

Luminet, J.-P. & Carter, B. 1986 *Astrophys. J. Suppl.* **61**, 219-248.

Matsumoto, R. & Tajima, T. 1995 *Astrophys. J.* **445**, 767-779.

Novikov, I. D., Pethick, C. J. & Polnarev, A. G. 1992 *Monthly Notices Roy. Astron. Soc.* **255**, 276-284.

Ogilvie, G. I. & Pringle, J. E. 1996 *Monthly Notices Roy. Astron. Soc.* **279**, 152-164.

Parker, E. N. 1971 *Astrophys. J.* **163**, 255-278.

Shakura, N. I., & Sunyaev, R. A. 1973 *Astron. Astrophys.* **24**, 337-355.

Stepinski, T. F., Levy, E. H. 1988 *Astrophys. J.* **331**, 416-434.

Stepinski, T. F., Levy, E. H. 1990 *Astrophys. J.* **362**, 318-332.

Stix, M. 1975 *Astron. Astrophys.* **42**, 85-89. The galactic dynamo Erratum: A&A 68,459

Stone, J. M., Hawley, J. F., Gammie, C. F., & Balbus, S. A. 1996 *Astrophys. J.* **463**, 656-671.

Terquem, C. & Papaloizou, J. C. B. 1996 *Monthly Notices Roy. Astron. Soc.* **279**, 767-784.

Torkelsson, U. & Brandenburg, A. 1994a *Astron. Astrophys.* **283**, 677-691.

Torkelsson, U. & Brandenburg, A. 1994b *Astron. Astrophys.* **292**, 341-349.

Vishniac, E. T. & Brandenburg, A. 1997 *Astrophys. J.* **475**, -.

Yoshizawa, A., Yokoi, N. 1993 *Astrophys. J.* **407**, 540-548.

Causal viscosity
in accretion disc boundary layers

W. Kley[1], J.C.B. Papaloizou[2]

[1] Max-Planck-Society, Research Unit Gravitational Theory, Universität Jena, Max-Wien-Platz 1, D-07743 Jena, Germany
[2] Astronomy Unit, School of Mathematical Sciences, Queen Mary & Westfield College, Mile End Road, London E1 4NS, UK

Abstract: The structure of the boundary layer region between the disc and a comparatively slowly rotating star is studied using a causal prescription for viscosity. The vertically integrated viscous stress relaxes towards its equilibrium value on a relaxation timescale τ, which naturally yields a finite speed of propagation for viscous information. For a standard α prescription with α in the range $0.1 - 0.01$, and ratio of viscous speed to sound speed in the range $0.02 - 0.5$, details in the boundary layer are strongly affected by the causality constraint. We study both steady state polytropic models and time dependent models, taking into account energy dissipation and transport. Steady state solutions are always subviscous with a variety of Ω profiles which may exhibit near discontinuities. For $\alpha = 0.01$ and small viscous speeds, the boundary layer adjusted to a near steady state. A long wavelength oscillation generated by viscous overstability could be seen at times near the outer boundary. Being confined there, the boundary layer remained almost stationary. However, for $\alpha = 0.1$ and large viscous speeds, short wavelength disturbances were seen throughout which could significantly affect the power output in the boundary layer. This could be potentially important in producing time dependent behaviour in accreting systems such as CVs and protostars.

1 Introduction

The boundary layer region between a star and accretion disc is of fundamental importance for non-magnetic accreting systems. This is because up to half the total accretion energy may be liberated over a relatively small scale in this region. (Lynden-Bell and Pringle 1974, Pringle 1977). Consequently, the angular velocity changes rapidly from a near Keplerian value to a smaller value associated with the accreting star on a scale length that is expected to be comparable to the pressure scale height of the slowly rotating star.

In a thin Keplerian disc, the inflow velocity is generally highly subsonic. However, in the boundary layer where the gradients increase the radial infall velocity may become large, reaching supersonic values, if an unmodified viscosity prescription appropriate to the outer disc is used (see Papaloizou and Stanley 1986; Kley 1989, Popham and Narayan 1992). In this case, it has been argued

(Pringle 1977) that the star would lose causal connection with the outer parts of the disc so that information about the inner boundary conditions could not be communicated outward. In order to prevent such a situation, the viscosity prescription should be modified so as to prevent unphysical communication of information. Various approaches that limit the viscosity in the vicinity of the star (thus reducing the radial inflow velocity) have been suggested (Papaloizou & Stanley 1986, Popham & Narayan 1992, Narayan 1992). Here we adopt an approach frequently used in non-equilibrium thermodynamics (eg. Jou, Casa-Vasquez and Lebon 1993), and we assume that the viscous stress components relax towards their equilibrium values on a characteristic relaxation timescale τ. This leads naturally to a set of basic equations incorporating a finite propagation speed for viscous information given by $c_v = \sqrt{\nu/\tau}$, where ν is the usual kinematic viscosity.

We use these to investigate the structure of the boundary layer region between the disc and a comparatively slowly rotating star by studying vertically averaged one dimensional models, as many of their properties are expected to be manifested in the more general two dimensional case. We begin with a study of steady state polytropic disc models and then go on to study time dependent models in which energy dissipation and heat transport are taken into account using, for illustrative purposes, parameters appropriate to protostellar discs.

2   Equations

In an accretion disc the vertical thickness H is usually assumed to be small in comparison to the distance r from the centre, i.e. $H/r \ll 1$. This is naturally expected when the material is in a state of near Keplerian rotation. Then one can vertically integrate the hydrodynamical equations and work only with vertically averaged state variables. Under the additional assumption of axial symmetry, the vertically integrated equations of motion in cylindrical coordinates (r, φ, z) read:

$$\frac{\partial \Sigma}{\partial t} + \frac{1}{r}\frac{\partial}{\partial r}(ru\Sigma) = 0, \tag{1}$$

$$\frac{\partial(\Sigma u)}{\partial t} + \frac{1}{r}\frac{\partial}{\partial r}(ru\Sigma u) = r\Sigma\Omega^2 - \frac{\partial P}{\partial r} - \Sigma\frac{GM_*}{r^2} + f_\nu \tag{2}$$

$$\frac{\partial(\Sigma r^2\Omega)}{\partial t} + \frac{1}{r}\frac{\partial}{\partial r}(r\Sigma r^2\Omega u) = \frac{1}{r}\frac{\partial}{\partial r}\left(r^2 T_{r\varphi}\right) \tag{3}$$

$$\frac{\partial(\Sigma\epsilon)}{\partial t} + \frac{1}{r}\frac{\partial}{\partial r}(r\Sigma\epsilon u) = -\frac{P}{r}\frac{\partial}{\partial r}(rv) + D_v - \int_{-\infty}^{\infty}\nabla\cdot\mathbf{F}dz \tag{4}$$

Here Σ denotes the surface density $\Sigma = \int_{-\infty}^{\infty}\rho dz$, where ρ is the density. v is the radial velocity, Ω the angular velocity, P the vertically integrated (two-dimensional) pressure, M_* the mass of the accreting object, G the gravitational constant, f_ν the viscous force per unit area acting in the radial direction, and $T_{r\varphi}$ is the $r\varphi$ component of the vertically integrated viscous stress tensor. In the energy equation ϵ denotes the specific internal energy, $D_v \equiv T_{r\varphi}\frac{\partial \Omega}{\partial r}$ is the viscous dissipation rate per unit area , and $\mathbf{F}$ is the radiative energy flux.

2.1   Causal Viscosity

Viscous processes are of central importance in accretion discs in that they are responsible for the angular momentum transport that allows the radial inflow and accretion to occur. It is believed that processes such as MHD turbulence are likely to be responsible for the existence of the large viscosities, required to account for observed evolutionary timescales associated with accretion discs (see Papaloizou and Lin 1995, and references therein).

The essential component of the viscous stress tensor for accretion discs is the (r, φ) component. The prescription normally adopted is $T_{r\varphi} = T^0_{r\varphi}$, where $T^0_{r\varphi}$ is given by an expression in the form appropriate to a microscopic viscosity such that

$$T^0_{r\varphi} = r\Sigma\nu\frac{\partial\Omega}{\partial r} \qquad (5)$$

Here ν is the kinematic viscosity coefficient. In accretion disc theory, the α prescription of Shakura and Sunyaev (1973) is often used such that

$$\nu = \alpha c_{\rm s} H, \qquad (6)$$

Here α is a (usually constant) coefficient of proportionality describing the efficiency of the turbulent transport. In writing (5) and (6) it is envisaged that the turbulence behaves in such a way as to produce a viscosity through the action of eddies of typical size H and turnover velocity $\alpha c_{\rm s}$. Vertical hydrostatic equilibrium gives

$$H = \frac{c_{\rm s}}{\Omega_{\rm k}}, \qquad (7)$$

where $\Omega_{\rm k}$ is the Keplerian angular velocity which is given by $\Omega^2_{\rm k} = GM_*/r^3$.

The ansatz $T_{r\varphi} = T^0_{r\varphi}$ results in the transport of angular momentum through diffusion, with a diffusion coefficient $\nu \equiv \alpha c^2_{\rm s}/(\Omega_{\rm k})$. This leads formally to the possibility of instantaneous communication of disturbances in the angular momentum distribution, or an infinite speed c_v of propagation of viscous information. In the main part of the accretion disc this causes no serious problems, since (radial) velocities are very small in comparison to the sound speed. However, in the boundary layer where the incoming material hits the surface of the accreting object, the radial infall velocity may become large, reaching supersonic values $|v| > c_{\rm s}$ (see Papaloizou and Stanley 1986; Kley 1989; Popham and Narayan 1992).

To overcome this *causality* problem various rather ad-hoc approaches that limit the viscosity in the vicinity of the star (thus reducing the radial inflow velocity) have been suggested (see introduction). Here we follow a more general approach frequently used in non-equilibrium thermodynamics (Jou et al. 1993) and also in relativistic physics (Israel 1976) where the theory requires a finite speed of propagation for information related to a given physical process. One assumes that the actual turbulent stresses tend to approach the equilibrium value $T_{r\varphi}^0$ on a suitable relaxation time τ. This is described through an additional equation for the time evolution of the vertically integrated (r, φ) component of the viscous stress

$$\frac{dT_{r\varphi}}{dt} = \frac{(T_{r\varphi}^0 - T_{r\varphi})}{\tau} \tag{8}$$

Note that the total or convective time derivative is used here.

This prescription was used to model the central regions of discs around compact objects by Papaloizou and Szuszkiewicz (1994) who noted that the system of equations (1 - 4) are then hyperbolic and thus completely causal with a propagation speed for viscous information given by

$$c_v = \sqrt{\frac{\nu}{\tau}} \equiv c_s \sqrt{\frac{\alpha}{\Omega_k \tau}}. \tag{9}$$

Note that in the limiting case of $\tau \to 0$, the stress is given by its equilibrium value $T_{r\varphi}^0$. Also variation of α, which may be a function of (r, Σ, Ω), does not affect the causality properties of the equations.

Here we apply the above formalism using the (r, φ) component of the viscous stress as this is the most important for the one dimensional models we consider. However, the formalism can be applied to all the components of the tensor and be used in more general two dimensional models of the type developed by Kley (1989).

3   Steady state polytropic models

To illustrate, as well as simplify, we first use a polytropic equation of state. It is found in practice that such a treatment yields the essential behaviour of the radial and angular velocities. We adopt

$$P = K\Sigma^\gamma, \tag{10}$$

where K is the polytropic constant and γ is the adiabatic index. The local sound speed in the disc is then given by

$$c_s^2 = \frac{\partial P}{\partial \Sigma} = K\gamma \Sigma^{\gamma-1}. \tag{11}$$

To analyze time independent solutions for a polytropic equation of state we drop the time derivatives in the evolution equations. The continuity and angular momentum equations can then be integrated yielding

$$\dot{M} = 2\pi \Sigma v r \tag{12}$$

$$\dot{J} = \dot{M} r^2 \Omega - 2\pi r^2 T_{r\varphi}. \tag{13}$$

Here the constants of integration denote the inward mass flow rate $\dot{M}$ through the accretion disc and the total angular momentum flow rate $\dot{J}$; both are negative. The total angular momentum flux consists of the advective and viscous part. Using the radial component of the equation of motion (assuming $f_\nu = 0$) and the viscous relaxation equation, we obtain two ordinary differential equations for Ω and v (see also Papaloizou and Szuszkiewicz 1994):

$$\left(v^2 - c_s^2\right) \frac{r}{v} \frac{dv}{dr} = \left[r^2 \left(\Omega^2 - \Omega_k^2\right) + c_s^2\right] \tag{14}$$

$$v\tau \left(\frac{c_v^2}{v^2} - 1\right) \frac{d\Omega}{dr} = \left[\Omega - \frac{\dot{J}}{\dot{M}r^2}\left(1 - \frac{2\tau v}{r}\right)\right], \tag{15}$$

where we have also made use of the mass and angular momentum flux integrals. We note that a complicating feature of the above differential equations is that they have critical points whenever the infall velocity reaches the sonic or viscous speed respectively.

Once values of α, $\Omega_k \tau$ and H/r have been specified, the above system provides two first order ordinary differential equations for Ω and v with the additional parameter $\dot{J}$. Solutions can be found with v_* and Ω specified at the inner boundary with $\dot{J}$ being determined as an eigenvalue in order that the exterior solution matches onto a Keplerian disc. We present here results for illustrative examples with $\alpha = 0.01$, and $\Omega_k \tau$ in the range $0.1 - 25$. In all cases Ω/Ω_k was taken to be one third at the inner boundary, $\gamma = 2$, and $H/r \sim 0.05$ in the Keplerian part of the disc. Each model has a constant value of c_v/c_s. Details of the models are given in table 1. All of our calculations are such that the flow

Table 1. Parameter of the stationary polytropic and time dependent radiative models

Polytropic Models					Radiative Models			
Nr.	α	$\Omega_k \tau$	c_v/c_s	v_*	Nr.	α	$\Omega_k \tau$	c_v/c_s Remarks
1	0.01	0.1	0.316	0.1	11	0.01	1	0.10 stable, with overstab.
2	0.01	1.0	0.1	0.01	12	0.01	25	0.02 stable, overstab. damped
3	0.01	4.0	0.05	0.01	13	0.1	10	0.10 stable, $\dot{M} = 10^{-6}$
4	0.01	9.0	0.033	0.01	14	0.1	250	0.02 stable
5	0.01	25.0	0.02	0.01	15	0.1	10	0.10 unstable
					16	0.1	0.4	0.50 unstable

remains subviscous throughout. In all cases, except perhaps model 1, which has the largest value of c_v/c_s for $\alpha = 0.01$, the flow speed almost reaches the viscous speed at its maximum. In such cases the Ω profile becomes nearly discontinuous (Fig. 1a). For model 1, the profile is moderately extended approximately up to the pressure scale height in the slowly rotating star. However, in models 2-5 the profile approaches a discontinuity. The jump in angular velocity occurs when the flow speed is at a maximum and almost equal to the viscous speed

(Fig. 1b). At the discontinuity, there is a jump in the velocity gradient. The tendency to form a discontinuity is even more noticeable in models which have $\alpha = 0.1$ (not shown). We note that the radial equation of motion (14) implies

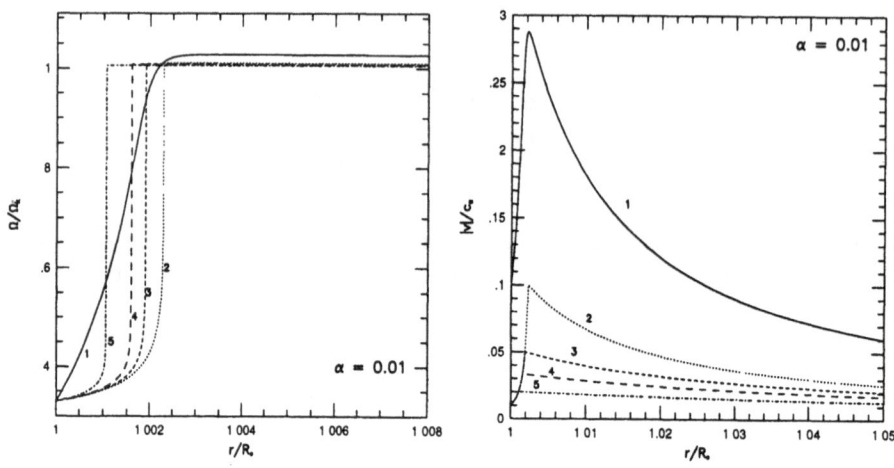

Fig. 1. Ratio of the angular velocity Ω to the Keplerian value Ω_k and radial Mach number $|v|/c_s$ versus radius for models 1 to 5.

that a discontinuity in Ω must occur at constant velocity and be accompanied by a jump in velocity gradient. The occurrence of these near discontinuities is reminiscent of the 'shear shocks' envisaged by Syer and Narayan (1993). At such locations material is instantaneously slowed down as it encounters the stellar surface. However, they occur here at the viscous speed only and do not involve a transition from super to subviscous speeds. The discontinuities tend to be approached whenever the model is strongly affected by the causality condition. The condition $R_*/H > c_{\mathrm{v}}/(\alpha c_s)$ provides a rough indication as to when this occurs in the models presented in this section.

4 Time dependent calculations

In addition to the steady state calculations described above, we also studied time dependent evolution of the flow in order to investigate any essential unsteady behaviour associated with the boundary layer. The numerical solution of equations (1) to (4) is accomplished by a finite difference method. The partial differential equations are discretized on a spatially fixed one-dimensional grid that stretches from $r = 1$ to $r = 2$. This computational domain is covered by typically 1000 grid cells. A forward time and centered space method with operator splitting and a monotonic advection scheme is used (Kley 1989).

We have studied polytropic models of the type used above in which the energy equation (4) is dropped, as well as models which include (4) with heat transport. For these cases

$$\int_{-\infty}^{\infty} \nabla \cdot \mathbf{F} dz = 2\sigma T_{\text{eff}}^4 - H/r \frac{\partial}{\partial r}(rF_r), \tag{16}$$

where T_{eff} is the effective temperature at the disc surface, σ is Stefan's constant, and F_r is the radial radiative flux. For our models we used an analytic approximation to tabulated opacities (Lin & Papaloizou 1985). The gas consists of a Hydrogen and Helium mixture where the dissociation of H_2 and the degrees of ionization of H and He are calculated by solving the Saha equation.

We adopted conditions appropriate to protostellar discs, where the protostar has $M_* = 1.0 M_\odot, R_* = 3R_\odot$, and $T_* = 4000K$. Through the surrounding disc, a mass flow rate of $\dot{M} = 10^{-7} M_\odot yr^{-1}$ is accreted (only model 13 has $\dot{M} = 10^{-6} M_\odot yr^{-1}$). At the inner boundary a fixed outwardly directed stellar flux, $F_* = \sigma T_*^4$, is assumed. The radial infall velocity at $R_{\min} = R_*$ is fixed at a given small fraction of local Keplerian velocity at R_*. We use typically $v_* = 10^{-3} v_{k*}$. The stellar angular velocity is 0.3 of the break-up velocity for the polytropic test cases, and to 0.1 for the fully radiative models.

At the outer boundary the angular velocity is Keplerian, the radial radiative flux vanishes and the radial infall velocity and the density are prescribed in such a way to ensure a given constant mass inflow rate through the system. For initial conditions we use a simple polytropic disc model with no boundary layer. The system is subsequently evolved until the region containing the boundary layer attains a quasi-steady state which was reached in most cases. Oscillations caused by viscous overstability persist typically near the outer boundary (see Kato 1978, Godon 1995).

In order to compare with the steady state calculations described above we have considered time dependent polytropic models with constant α, and $\Omega_k \tau$. There was, in general, positive agreement between the two methods. There is a tendency for the evolutionary calculations to overshoot the viscous speed somewhat, an effect which decreases with increasing spatial resolution of the calculations.

4.1 Radiative models

The parameters of the calculations with thermal effects included are listed in Table 1. Solutions appropriate to a statistically steady state are presented for models (14, 15) which both have $\dot{M} = 10^{-7} M_\odot$, and $\alpha = 0.1$. The viscous velocity $c_v = c_s \sqrt{\alpha/(\Omega_k \tau)}$ differs by a factor of 5 between the two models. Some state variables are plotted in figure 2. The structure of the v and Ω profiles is similar to that in the polytropic case, i.e. Ω displays a near discontinuity and v has a peaked maximum near the viscous speed. The rate of liberation of energy in the boundary layer is $0.5 \dot{M} R_*^2 \Delta\Omega^2$, with $\Delta\Omega$ being the jump in Ω that occurs there. Heat diffusion then occurs over a greater length scale. In the case with $\Omega\tau = 25$, the optical depth is about ten times larger than that with $\Omega\tau = 1$.

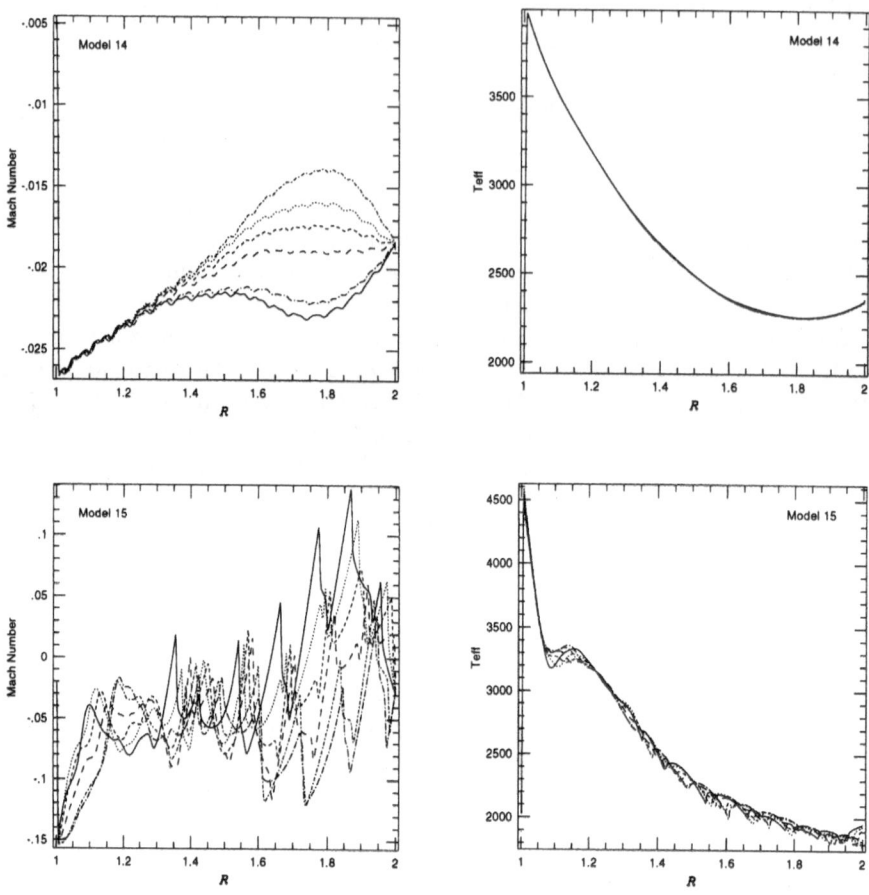

Fig. 2. The radial mach number and the effective temperature versus radius for models 14 and 15 (see table 1) at six different times.

But this only results in a 20 percent reduction in T_{eff} because the latter quantity is predominantly determined by a fixed rate of energy production. However, typically there will be an order of magnitude difference in the estimated value of $\dot{M}$ for which the boundary layer becomes optically thin. This is the main effect of increasing the relaxation time τ. Another consequence of increasing τ is damping of the present viscous overstability because of the stronger phase lag induced. In model 12 the overstability is eventually damped completely. We ran three models with $\alpha = 0.1$ which had constant values of $\Omega_k\tau$ chosen such that c_v/c_s was $0.02, 0.1$ and 0.5, respectively. The case with $c_v/c_s = 0.02$ (Model 14, figure 2) behaved in a very similar way to the cases with $\alpha = 0.01$, in that it had an almost steady and stable inner boundary layer region. However, highly

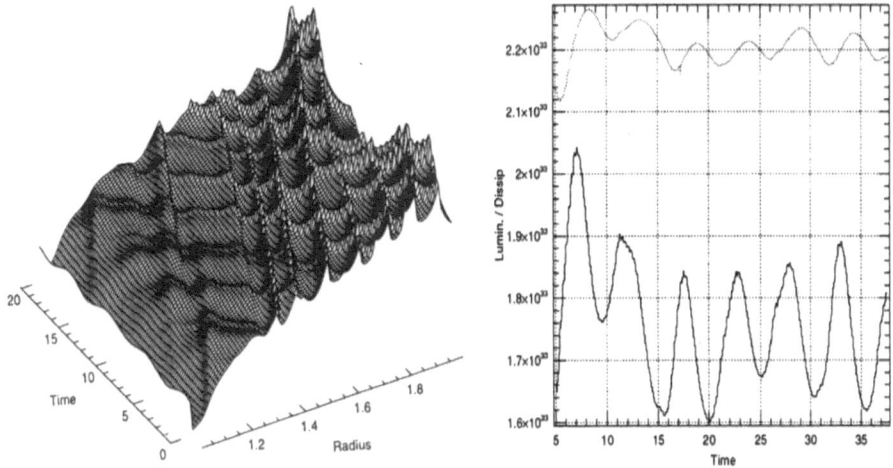

Fig. 3. Time variation of the density for model 15 (a). Luminosity (dashed line) and total dissipation (solid line) versus time (b). value Ω_k and radial Mach number $|v|/c_s$ versus radius for the same model. Time given in dimensionless units with an offset.

unstable features remained in the boundary layers of the other two models (15, 16) with $\alpha = 0.1$. These features were not seen in the other models even when they showed signs of viscous overstability. Here the characteristic wavelength is much shorter and the temperature structure and power output are significantly affected in a highly irregular manner. The instability appears more pronounced with higher c_v/c_s, here the strongest instabilities are present for $c_v/c_s = 0.5$. Then slightly supersonic speeds with shocks, as well as superviscous speeds occur. We remark that in the models with $\alpha = 0.1$, the short wavelength inward coming compressional waves are not expected to be reflected at a Lindblad resonance before they reach the boundary layer region, so they are able to affect the power output there. This would indicate that if the outer boundary condition allows, waves may exist in the boundary layer region where they may significantly affect the power output. In Figure 3b the variation of the luminosity and the total dissipation is shown over a short time interval for model 15. The luminosity varies over timescales of the order of the dynamical time (orbital Keplerian period at the stellar surface) but the variations in amplitude are less than 4%.

Note that the driving mechanism for these motions is not just simple viscous overstability which was never seen in the boundary layer region. The generation of superviscous and sometimes supersonic speeds indicates that the nature of the causal description must play a role. In figure 3a the time variation of the density is displayed in a three-dimensional space time diagram. It is clear that there are quasi-periodic wave-like perturbances moving from the inside outwards which are generated in the vicinity of the boundary layer. The perturbations interact intricately with reflected waves moving inwards (Fig. 3a).

These waves occur in a region where the central temperatures lie somewhat below $10^4 K$, where opacity rises rapidly with temperature. Hence, even though the central temperatures vary only very little, the effective temperature displays much stronger variations. The origin of the disturbances lies in an interaction of the radiative transport with the causal viscous transport. For higher $|v/c_s|$, the interaction becomes very much stronger leading to variations in luminosity of a few percent in the case of model 15. Notice that models 13 and 15 have identical parameter α, τ, and c_v/c_s, and only differ in the mass inflow rate which is ten times higher in model 13. The increased disc thickness for the higher $\dot{M}$ model leads to a larger optical depth and higher temperatures that drive the system out of the instability region.

Acknowledgement: W. K. would like thank the Astronomy unit at QMW for their kind hospitality during two visits. This work was supported by the EC grant ERB-CHRX-CT93-0329.

References

Godon P.,1995, MNRAS, 274, 61.

Israel W., 1976, Annals of Physics, 100, 310.

Jou D., Casas-Vasquez J., Lebon G., 1993, *Extended irreversible Thermodynamics*, Springer-Verlag, Berlin 1993.

Kato S., 1978, MNRAS, 185, 629.

Kley W., 1989, A&A, 208, 98.

Kley W., Papaloizou J.C.B., Lin D.N.C., 1993, ApJ, 409, 739.

Lin D.N.C., Papaloizou J.C.B., 1985, in *Protostars and Planets II*, ed. D.C. Black & M.S. Mathes (Tucson: Univ. Arizona Press), 981. MNRAS, 168, 603.

Lynden-Bell D., Pringle J. E., 1974, MNRAS, 168, 603.

Narayan R., 1992, ApJ, 394, 261.

Popham R., Narayan R., 1991, ApJ, 370, 604.

Popham R., Narayan R., 1992, ApJ, 394, 255.

Pringle J. E., 1977, MNRAS, 178, 195.

Papaloizou J.C.B., Lin D.N.C., 1995, Ann. Rev. Astron. Astrophys., 33, 505.

Papaloizou J.C.B., Stanley G.Q.G., 1986, MNRAS, 220, 593.

Papaloizou J.C.B., Szuszkiewicz E., 1994, MNRAS, 268, 29.

Shakura N.I., Sunyaev R.A., 1973, A&A, 24, 337.

Syer D., Narayan R., 1993, MNRAS, 262, 749.

The nonlinear evolution of a single mode of the magnetic shearing instability

U. Torkelsson[1], G.I. Ogilvie[1], A. Brandenburg[2], Å. Nordlund[3], R.F. Stein[4]

[1] Institute of Astronomy, Madingley Road, Cambridge CB3 0HA, United Kingdom
[2] Department of Mathematics, University of Newcastle upon Tyne, NE1 7RU, United Kingdom
[3] Theoretical Astrophysics Center, Juliane Maries vej 30, DK-2100 Copenhagen Ø, Denmark
[4] Department of Physics and Astronomy, Michigan State University, East Lansing, MI 48824, USA

Abstract We simulate in one dimension the magnetic shearing instability for a vertical magnetic field penetrating a Keplerian accretion disc. An initial equilibrium state is perturbed by adding a single eigenmode of the shearing instability and the subsequent evolution is followed into the nonlinear regime. Assuming that the perturbation is the most rapidly growing eigenmode, the linear theory remains applicable until the magnetic pressure perturbation is strong enough to induce significant deviations from the original density. If the initial perturbation is not the fastest growing mode, the faster growing modes will appear after some time.

1 Introduction

Balbus and Hawley (1991) realised that a Keplerian accretion disc penetrated by a vertical magnetic field is unstable to a shearing instability previously described by Velikhov (1959) and Chandrasekhar (1960). Three-dimensional numerical simulations (e.g. Hawley et al. 1995, 1996; Matsumoto & Tajima 1995; Brandenburg et al. 1995, 1996; Stone et al. 1996) have later demonstrated that the laminar shear flow becomes turbulent in the nonlinear domain.

Owing to the complexity of a three-dimensional simulation it is not well understood how the flow evolves from being linearly unstable to being turbulent. A related problem is to identify the saturation mechanism that sets the amplitude of the fully developed turbulence. In this paper we will try to investigate these problems in an alternative way. We will reduce the amount of data by studying the instability of a vertical magnetic field in only one dimension, and simplify the dynamics by starting from a perturbation which is already an eigenmode of the linear stability problem. Clearly we are missing much of the essential physics by imposing these restrictions, but on the other hand it is possible to understand the remaining physical effects, and by comparing our results with more complete calculations we may understand which of the missing physical effects are important.

Since the initial analysis by Balbus & Hawley (1991) a large number of papers extending the linear analysis have been published. The original work was generalised by Balbus & Hawley (1992a) who showed that the maximum growth rate of the magnetic shearing instability for any poloidal field configuration is given by Oort's A-constant. A more detailed analysis determining the axisymmetric eigenmodes for a stratified disc was later published by Gammie & Balbus (1994).

The original work concerned only the instability of an axisymmetric poloidal magnetic field, but Balbus & Hawley (1992b) extended the analysis to toroidal fields and non-axisymmetric perturbations. Several groups have later continued the analytical stability analysis of the toroidal field (e.g. Foglizzo & Tagger 1995, Ogilvie & Pringle 1996, Terquem & Papaloizou 1996, Coleman et al. 1995). An important difference in the stability properties of vertical and toroidal magnetic fields is that for the vertical magnetic field the most rapidly growing mode always has a finite wavelength, if there are any unstable modes at all, whereas for a toroidal field the most rapidly growing mode has an arbitrarily small vertical wavelength.

In this paper we will introduce briefly the linear theory for instabilities of a vertical magnetic field in an unstratified disc in Sect. 2, and present the corresponding nonlinear simulations in Sect. 3. Sect. 4 is devoted to magnetic fields in a stratified disc. Finally we summarise and discuss our results in Sect. 5.

2   The linear stability of a vertical magnetic field

The equations of ideal magnetohydrodynamics (MHD) can be written as

$$\frac{\partial \rho}{\partial t} + \nabla \cdot (\rho \mathbf{u}) = 0, \tag{1}$$

$$\frac{\partial \mathbf{u}}{\partial t} + \mathbf{u} \cdot \nabla \mathbf{u} = -\frac{1}{\rho}\nabla p + \frac{1}{\mu_0 \rho}(\nabla \times \mathbf{B}) \times \mathbf{B} + \mathbf{g}, \tag{2}$$

and

$$\frac{\partial \mathbf{B}}{\partial t} = \nabla \times (\mathbf{u} \times \mathbf{B}), \tag{3}$$

where ρ is the density, $\mathbf{u}$ the velocity, p the pressure, μ_0 the magnetic permeability, $\mathbf{B}$ the magnetic field, and $\mathbf{g}$ the gravity. Furthermore we assume an ideal equation of state and that all perturbations are adiabatic. For the rest of this paper we will assume dimensionless units such that $\mu_0 = 1$.

As our equilibrium model we choose a Keplerian accretion disc with no stratification. We transform the equations to a system rotating at the Keplerian angular velocity, Ω_0, at a reference radius R_0, and linearise the shear flow in terms of the parameter $\frac{x}{R_0}$, where x is the radial distance from R_0 (cf. Brandenburg et al. 1995). On this disc we impose a homogeneous vertical magnetic field B. We then add a perturbation of the form $\propto \exp[i(k_z z - \omega t)]$, that is the perturbations are independent of the horizontal coordinates x and y. Linearising Eqs (2) and (3) we then obtain

$$-\rho\left(i\omega\delta u_x + 2\Omega_0\delta u_y\right) = ik_z B\delta B_x, \tag{4}$$

$$\rho\left[-i\omega\delta u_y + 2\left(\Omega_0 - A\right)\delta u_x\right] = ik_z\delta B_y, \tag{5}$$

$$-\rho i\omega\delta u_z = -ik_z\delta\Pi + ik_z B\delta B_z, \tag{6}$$

$$-i\omega\delta B_x = ik_z B\delta u_x, \tag{7}$$

$$-i\omega\delta B_y = ik_z B\delta u_y - 2A\delta B_x, \tag{8}$$

and

$$-i\omega\delta B_z = 0, \tag{9}$$

where $\delta\Pi = \delta p + \mathbf{B}\cdot\delta\mathbf{B}$ is the perturbation of the total pressure, and

$$A = \frac{1}{2}\left(\Omega - \frac{d\left(R\Omega\right)}{dR}\right)_{R_0} \tag{10}$$

is Oort's constant, which is $\frac{3}{4}\Omega_0$ for Keplerian rotation. Note that Eq. (9) leads to $\delta\Pi = \delta p$. The pressure and density perturbations are related via

$$\delta p = v_s^2\delta\rho, \tag{11}$$

where v_s is the sound speed, and the density perturbation in its turn is given by

$$-i\omega\delta\rho = -i\rho k_z\delta u_z. \tag{12}$$

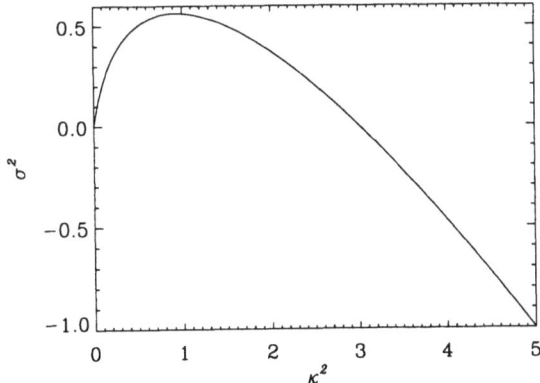

Fig. 1. The dispersion relation for an unstratified disc. κ is a normalised wave number and σ is the growth rate normalised to the Keplerian frequency

The resulting dispersion relation can be split into two parts

$$\omega^2 = v_s^2 k_z^2,\tag{13}$$

for longitudinal, 'acoustic', waves, and

$$\omega^4 - \left(2\omega_A^2 + \Omega_0^2\right)\omega^2 + \omega_A^2\left(\omega_A^2 - 3\Omega_0^2\right) = 0,\tag{14}$$

for transverse, 'magnetic' and 'inertial', perturbations. We have defined the Alfvén frequency $\omega_A = k_z v_A$ for the Alfvén velocity, $v_A = B/\sqrt{\rho}$. The magnetic mode is unstable for $0 < \omega_A^2 < 3\Omega_0^2$. We now introduce the conditions

$$\frac{\partial u_x}{\partial z} = \frac{\partial u_y}{\partial z} = u_z = 0,\tag{15}$$

and

$$B_x = B_y = \frac{\partial B_z}{\partial z} = 0\tag{16}$$

on the vertical boundaries. These boundary conditions are identical to the ones used by Brandenburg et al. (1995), and describe a stress-free surface with no flow going through it and with a vertical magnetic field. The unstable magnetic mode can be written as

$$\delta u_x = a,\tag{17}$$

$$\delta u_y = \frac{\sigma^2 + \kappa^2}{2\sigma}a,\tag{18}$$

$$\delta B_x = \frac{i\kappa}{\sigma}\rho^{1/2}a,\tag{19}$$

$$\delta B_y = \frac{i\kappa}{\sigma}\frac{\sigma^2 + \kappa^2 - 3}{2\sigma}\rho^{1/2}a,\tag{20}$$

and

$$\delta u_z = \delta B_z = \delta\rho = \delta p = 0,\tag{21}$$

where a is an arbitrary constant, and κ and σ are given by

$$\kappa = \frac{k_z v_A}{\Omega_0},\tag{22}$$

and

$$\sigma^2 = \frac{1}{2}\left(1 + 16\kappa^2\right)^{1/2} - \left(\frac{1}{2} + \kappa^2\right),\tag{23}$$

respectively (Fig 1), and $\omega = i\sigma\Omega_0$. The instability appears for $\kappa^2 < 3$, and the fastest growing mode has

$$\kappa^2 = \frac{15}{16}\tag{24}$$

so that the dispersion relation reduces to

$$\sigma^2 = \frac{9}{16}.\tag{25}$$

The maximum growth rate is thus equal to Oort's A-constant. The corresponding eigenmode can be written as

$$\delta u_x = \delta u_y = a, \tag{26}$$

and

$$\delta B_x = -\delta B_y = \frac{\sqrt{15}}{3} i \rho^{1/2} a. \tag{27}$$

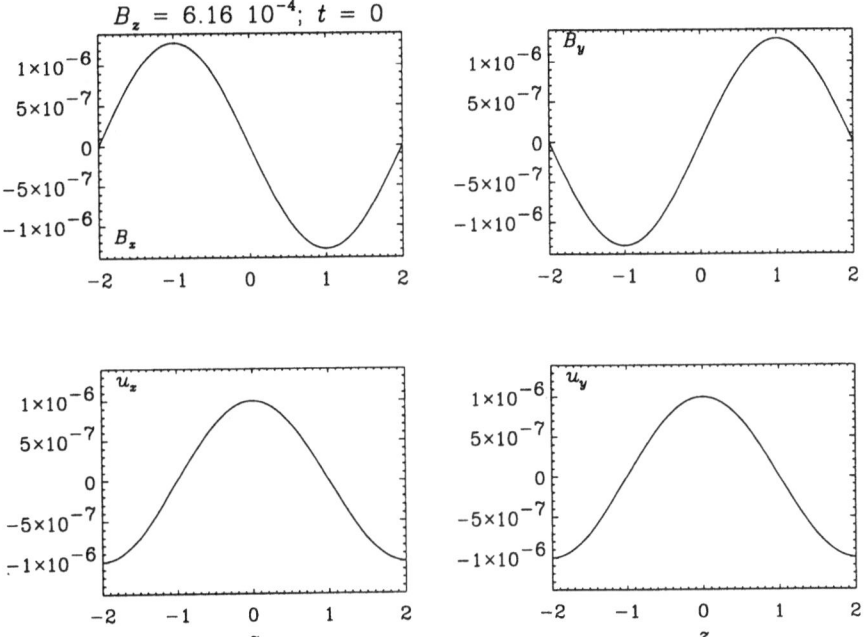

Fig. 2. The initial state of Model 1. The imposed vertical magnetic field is $6.16\,10^{-4}$, the density is 1.0 and the angular velocity is 10^{-3}. The magnetic perturbation is shown in the upper row, and the velocity perturbation in the lower row (x-components to the left and y-components to the right)

3   The evolution of a single mode in an unstratified disc

We use the numerical code of Brandenburg et al. (1995), but restrict it to one dimension, the vertical, to study the evolution of the eigenmodes described above. Our standard background model assumes an initial density $\rho = 1$, a radius of 100, so that $\Omega_0 = 10^{-3}$, and the internal energy, $e = 7.5\,10^{-7}$. The vertical

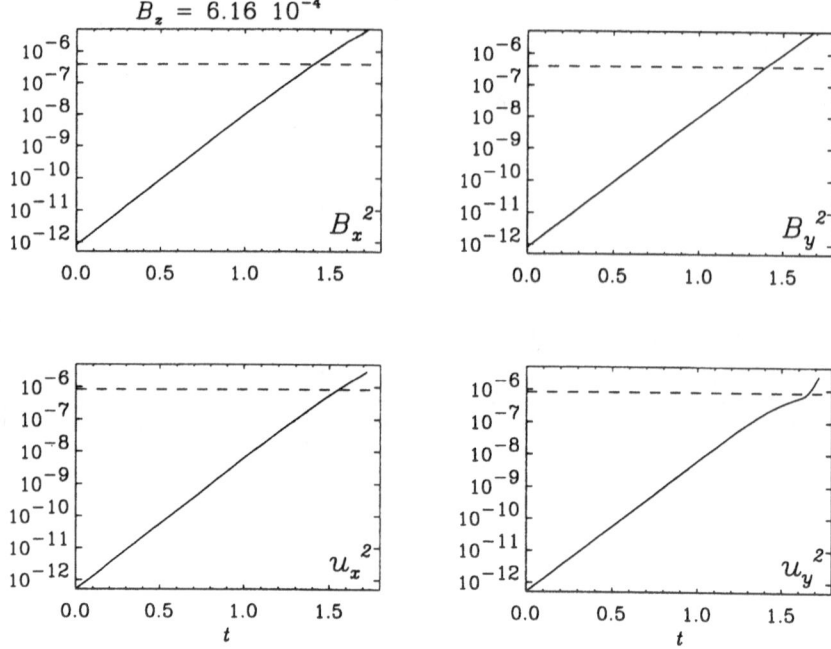

Fig. 3. The vertical averages of B_x^2 and B_y^2 (upper row), and u_x^2 and u_y^2 (lower row) as a function of time for Model 1. Time is given in orbital periods. The dashed lines denote B^2 (upper row) and c_s^2 (lower row)

Table 1. Vertical eigenmodes in unstratified discs. For each model we give the number of grid points used, N_z, the imposed vertical magnetic field, B, the wavelength of the excited mode, λ, and its growth rate, σ in units of the Keplerian angular frequency

Model	N_z	B	$\lambda = 2\pi/k_z$	σ
1	63	$6.16\,10^{-4}$	4	0.75
2	127	$1.54\,10^{-4}$	1	0.75
3	63	$1.54\,10^{-4}$	4	0.37

extent of the box is 4. The eigenmodes that we study are given in Tab. 1. λ is the wavelength of the linear eigenmode of the shearing instability. Models 1 and 2 are the fastest growing eigenmodes for their magnetic field strengths.

The initial state of Model 1 is shown in Fig. 2. The instability is growing at its linear growth rate even in what could be considered to be the nonlinear regime, as is illustrated in Fig. 3. This is a consequence of the fact that the linear

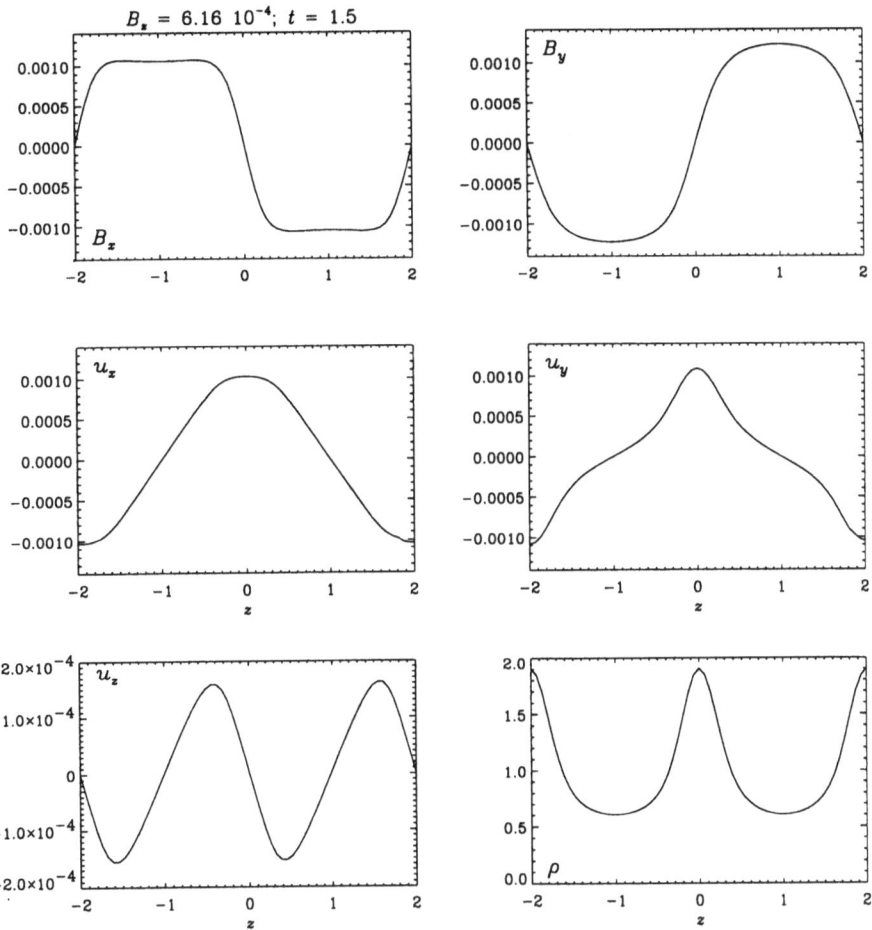

Fig. 4. The evolved state of Model 1 at 1.5 orbital periods. *Upper row:* B_x and B_y. *Middle row:* u_x and u_y. *Lower row:* u_z and ρ. Note that initially $\rho = 1$ and $v_s = 9.1\,10^{-4}$

eigenmode is an exact solution of the incompressible MHD equations (Goodman & Xu 1994). It is clear from Fig. 2 that to second order the magnetic pressure is modulated on half the wavelength of the eigenmode. The magnetic pressure gradient generates a vertical velocity, u_z, and a density fluctuation. These effects are of no consequence for the growth of the eigenmode until the density fluctuations are comparable to the background density (Fig. 4). At this late stage the mass is concentrated towards the nodes of the horizontal magnetic field as predicted by Goodman & Xu (1994) when the magnetic pressure dominates over the gas pressure.

Model 2 shows a similar pattern to Model 1, as it is the fastest growing

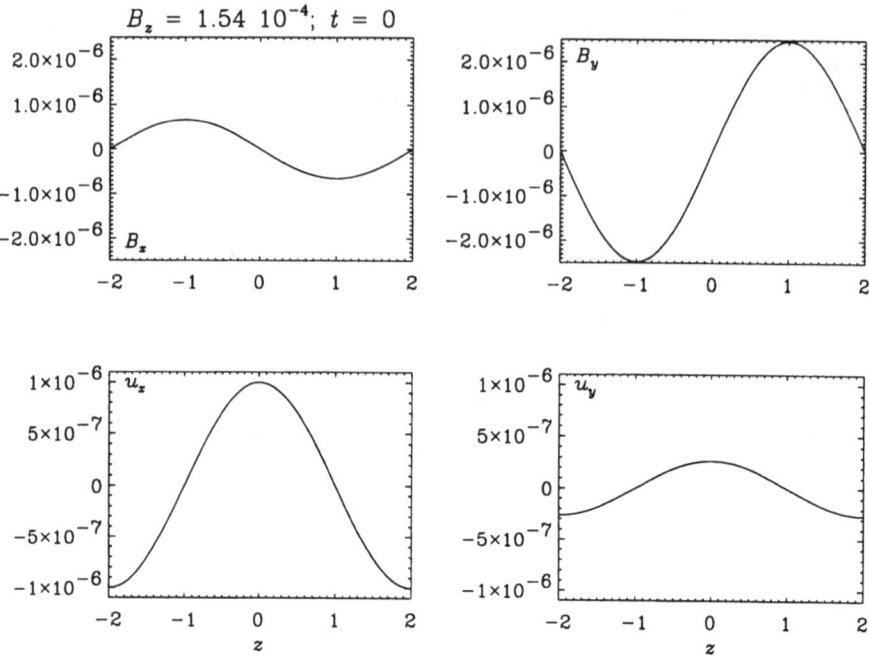

Fig. 5. The initial state of Model 3. The imposed vertical magnetic field is $1.54\,10^{-4}$. The figure is organised in the same way as Fig. 2

mode for that magnetic field strength. Model 3 has the same imposed vertical magnetic field as Model 2, but the eigenmode has a longer wavelength, and thus a lower growth rate. In general $|u_x| > |u_y|$ and $|B_x| < |B_y|$ for a mode with a wavelength longer than that of the fastest growing mode (Fig. 5), and the other way around for a mode with a too short wavelength. The nonlinear terms in the MHD equations transfer power to modes with smaller wavelengths. In Models 1 and 2 these deviations are visible only in quantities which lack a contribution from the linear mode, such as u_z, but in Model 3 the deviations appear in all quantities as they belong to modes with larger growth rates (Fig. 6).

4   A single mode in a stratified disc

We assume a vertical gravitational acceleration $g_z = -\Omega_0^2 z$ and that the disc is isothermal in the vertical direction, so that hydrostatic equilibrium gives

$$\rho = \rho_0 e^{-z^2/H^2}, \tag{28}$$

where ρ_0 is the density at the midplane of the disc, and H is the scale height. For our standard model we choose both ρ_0 and H to be unity. In this case the vertical

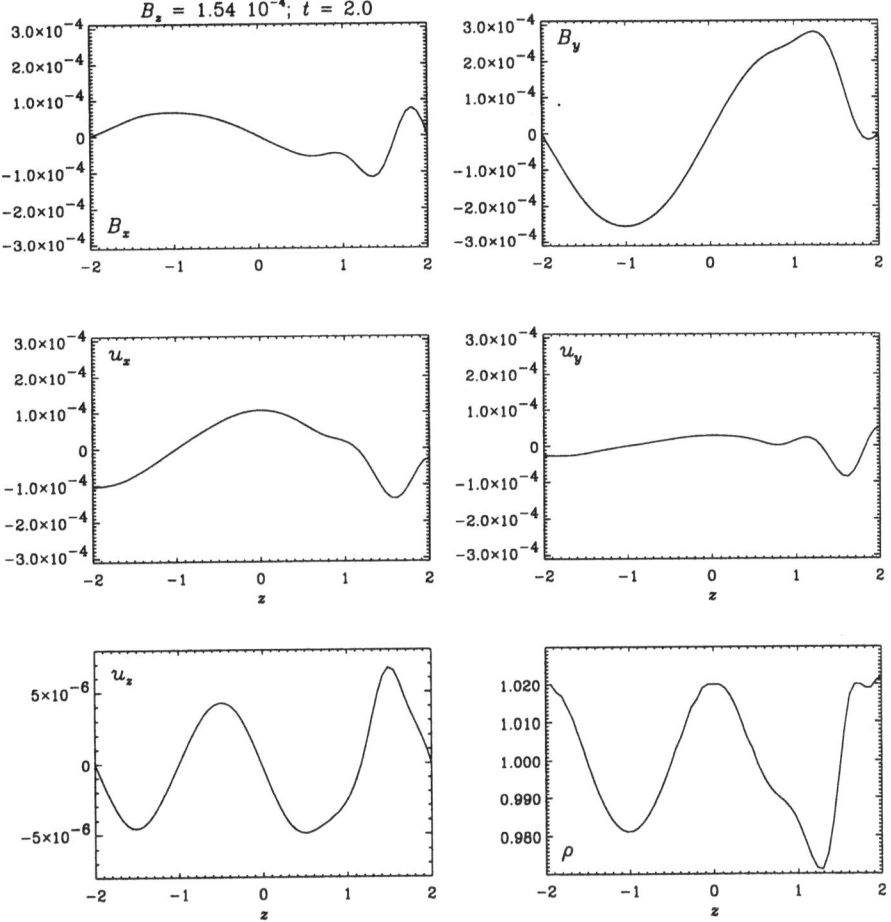

Fig. 6. The evolved state of Model 3 at 2.0 orbital periods. The figure is organised in the same way as Fig. 4

dependence is not described by $\exp(ik_z z)$. It is possible to re-write Eq. (2) as a set of second-order ordinary differential equations in the Lagrangian displacement, ξ (Gammie &Balbus 1994). Restricting ourselves as before to an imposed vertical field and modes with no dependence on the horizontal coordinates, we find

$$B^2 \frac{\mathrm{d}^2\xi_x}{\mathrm{d}z^2} = - \left(\omega^2 + 3\Omega_0^2\right) \rho\xi_x + 2i\omega\Omega_0\rho\xi_y, \qquad (29)$$

and

$$B^2\frac{\mathrm{d}^2\xi_y}{\mathrm{d}z^2} = -2i\omega\Omega_0\rho\xi_x - \omega^2\rho\xi_y, \qquad (30)$$

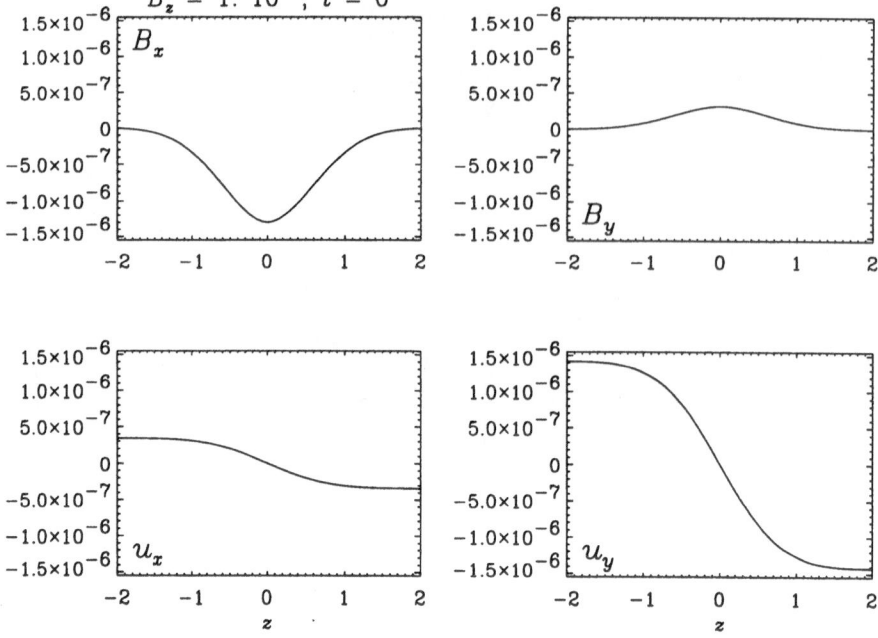

Fig. 7. The initial state of Model 4. The imposed vertical magnetic field is $1\,10^{-3}$, the density profile is a Gaussian with a maximum of 1.0, and the angular velocity is 10^{-3} (as in the unstratified models). The figure is organised in the same way as Fig. 2

Table 2. Vertical eigenmodes in isothermal, stratified discs. For each model we give the number of grid points used, N_z, the imposed vertical magnetic field, B, the number of nodes in u_x, N_{nodes}, and the growth rate, σ, in units of the Keplerian angular frequency

Model	N_z	B	N_{nodes}	σ
4	127	$1\,10^{-3}$	1	0.34
5	127	$1.54\,10^{-4}$	5	0.75
6	127	$1.54\,10^{-4}$	1	0.38
7	127	$2\,10^{-4}$	5	0.64

with the boundary conditions

$$\frac{d\xi_x}{dz} = \frac{d\xi_y}{dz} = 0 \text{ at } z = \pm 2H. \tag{31}$$

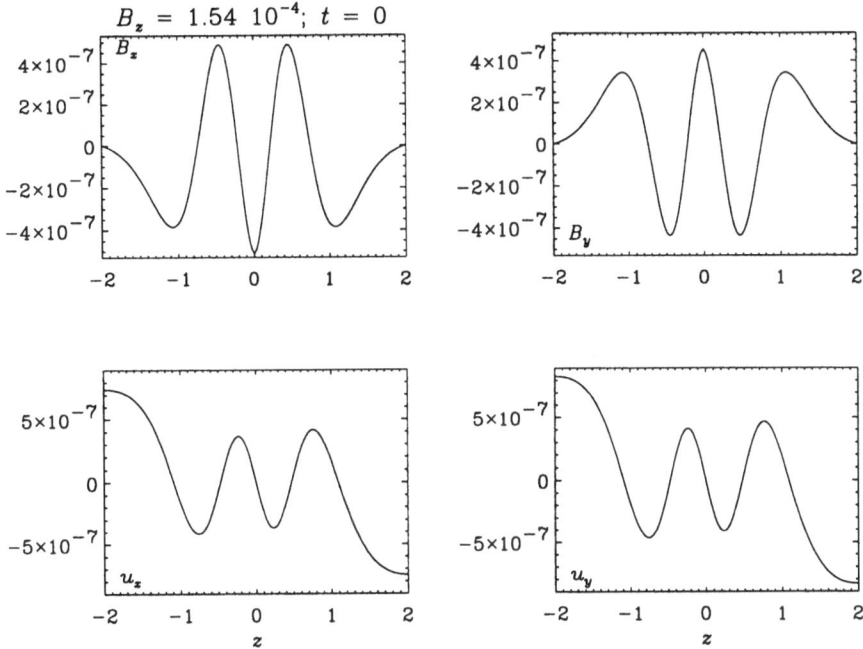

Fig. 8. The initial state of Model 5. The imposed vertical magnetic field is $1.54\,10^{-4}$. The figure is organised in the same way as Fig. 2

We calculate a set of eigenmodes using a shooting method, and normalise them such that $|\xi_x| = 1$ at $z = \pm 2H$. These eigenmodes are then orthogonal, and we can calculate $\delta\mathbf{u}$ and $\delta\mathbf{B}$ from them as

$$\delta\mathbf{u} = -i\omega\boldsymbol{\xi} + 2A\xi_x\mathbf{e_y}, \tag{32}$$

and

$$\delta\mathbf{B} = B\frac{d\xi}{dz}. \tag{33}$$

Our models are described briefly in Tab. 2. The wavelength is no longer a well-defined concept, so we prefer to identify the modes by the number of nodes in the velocity perturbation, N_{nodes}.

In Model 4 the magnetic field is so strong that the only unstable mode is the one with $N_{\mathrm{nodes}} = 1$ (Fig. 7). Furthermore the growth rate of this mode is significantly smaller than the maximal growth rate, $0.75\Omega_0$, which can only be achieved by fine-tuning the magnetic field strength to make an appropriate mode fit precisely in the disc. An imposed magnetic field of $1.54\,10^{-4}$ gives close to the maximum growth rate for a $N_{\mathrm{nodes}} = 5$ mode (Fig. 8). At this field strength there is a large range of unstable modes. An interesting alternative is therefore to excite the $N_{\mathrm{nodes}} = 1$ mode (Fig. 9). The $N_{\mathrm{nodes}} = 3$ mode is growing more

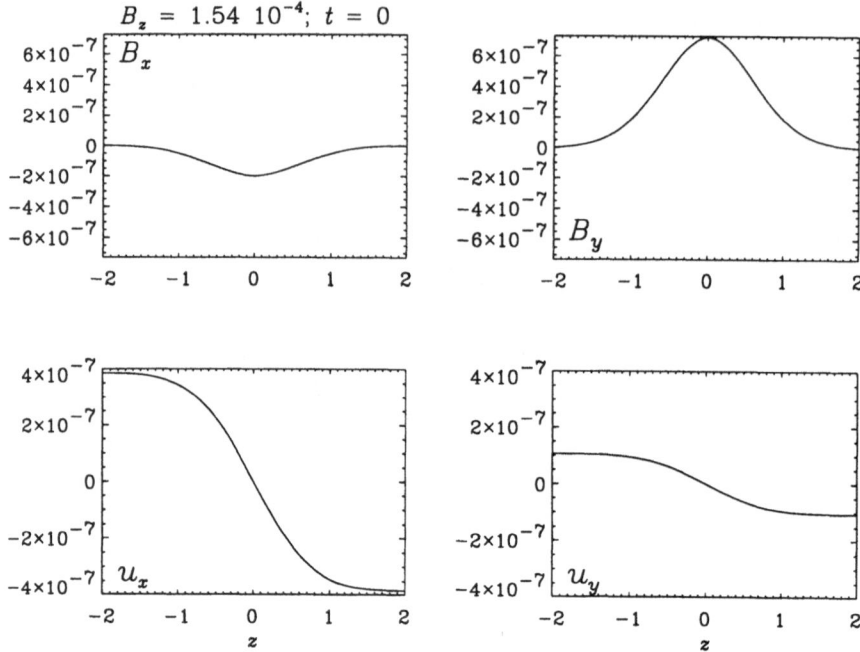

Fig. 9. The initial state of Model 6. The imposed vertical magnetic field is $1.54 \, 10^{-4}$. The figure is organised in the same way as Fig. 2

rapidly than the originally excited mode, and can be distinguished in a snapshot at 2.4 orbital periods (Fig. 10). At around this time the growth rates of, in particular, B_x and u_y increases from 0.37 to 0.67 (Fig. 11), which is close to the growth rate for the $N_{nodes} = 3$ mode, which is 0.69. At 3.0 orbital periods the $N_{nodes} = 3$ mode is dominating, and there may be a hint of an $N_{nodes} = 5$ mode too (Fig. 12). We see also that the density maxima are located at the nodes of the horizontal magnetic field as predicted by Goodman & Xu (1994).

It can be instructive here to decompose the evolved state in the linear eigen-modes. We calculate the scalar product of u_x with the corresponding velocity of the eigenmode $u_{x,i}$ as

$$a_i = \frac{\int u_x u_{x,i} \rho \mathrm{d}z}{\int u_{x,i}^2 \rho \mathrm{d}z}, \tag{34}$$

where the integrals are taken over the entire vertical extent of our box, and ρ, taken at $t = 0$ works as a weighting function. The time evolution of the amplitudes a_i are plotted in Fig 13. Initially all modes are growing exponentially with their linear growth rates, but a_5 changes sign at 2.7 orbital periods. Note that the normalisation of the eigenmodes is arbitrary, and therefore the amplitudes of the different modes should not be compared with each other. It is intriguing that a_5 changes sign at the same time as u_x changes sign on the boundaries.

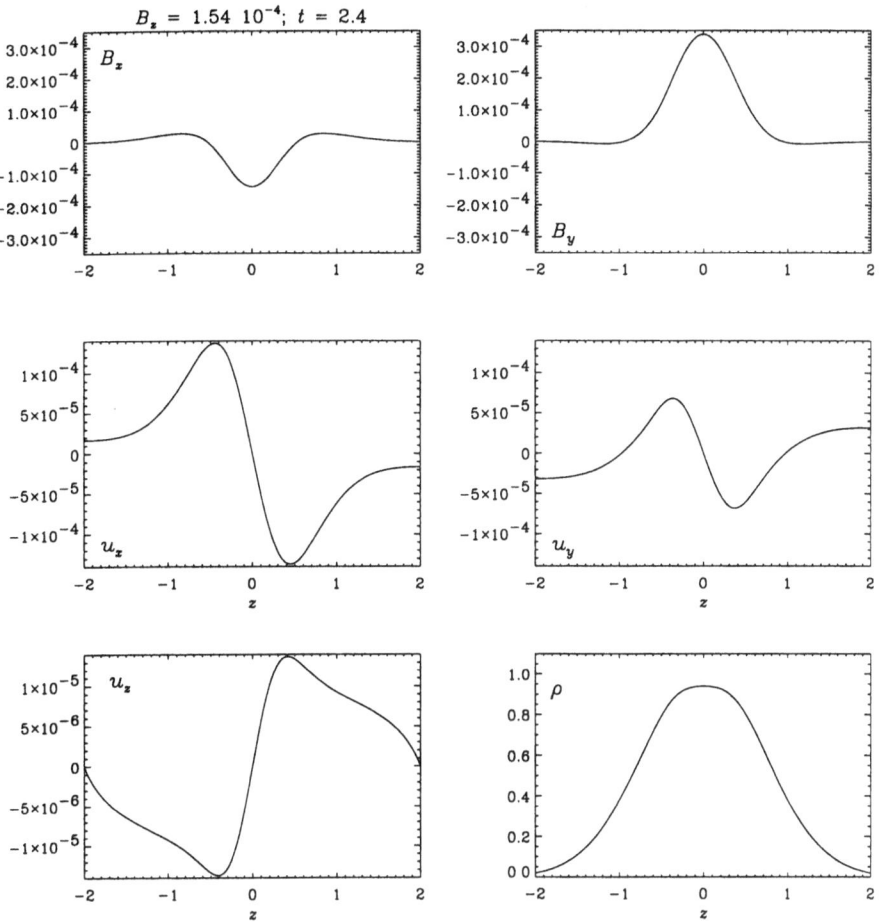

Fig. 10. The evolved state of Model 6 at 2.4 orbital periods. The figure is organised in the same way as Fig. 4

Model 7 is the opposite to Model 6. Here we start out with the $N_{\text{nodes}} = 5$ mode (Fig. 14) although the fastest growing mode is the $N_{\text{nodes}} = 3$ mode. Comparing with the evolved mode after 3.4 orbital periods (Fig. 15) one may suspect that it has obtained some power in the lower odd modes. To investigate that we do a spectral decomposition including the first three odd modes (Fig 16). ¿From the start all modes are growing exponentially at the expected growth rates, but at three orbital periods a_1 increases its growth rate to $2\Omega_0$, due to a nonlinear interaction between the modes.

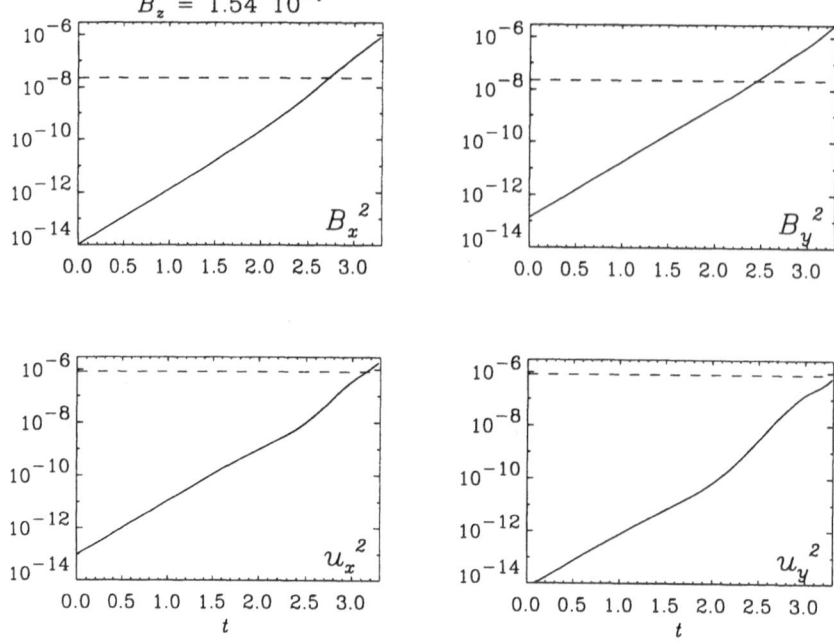

Fig. 11. The vertical averages of B_x^2 and B_y^2 (upper row), and u_x^2 and u_y^2 (lower row) as a function of time for Model 6. Time is given in orbital periods. The dashed lines denote B^2 (upper row) and c_s^2 (lower row)

5   Discussion and summary

In this paper we have been studying the nonlinear development of one-dimensional unstable modes of a vertical magnetic field in a Keplerian disc. The intention was to understand the development of turbulence in accretion discs, and in particular the saturation mechanism.

Our results show that starting from the fastest growing mode the instability grows exponentially at the linear growth rate until the magnetic pressure fluctuations are large enough to dominate over the gas pressure. This is not surprising in view of the result of Goodman & Xu (1994) that a single mode of the shearing instability is an exact solution of incompressible MHD. The parasitic instability, that was also found by Goodman & Xu (1994), is not applicable here as it needs a non-vanishing wave number in the horizontal plane. A mode different from the fastest growing mode is on the other hand unstable, as other modes with higher growth rates are generated by the nonlinear terms in the equations. These modes will eventually dominate.

Hawley & Balbus (1992) investigated the instability of a homogeneous vertical field in the two-dimensional meridional plane. An important difference is

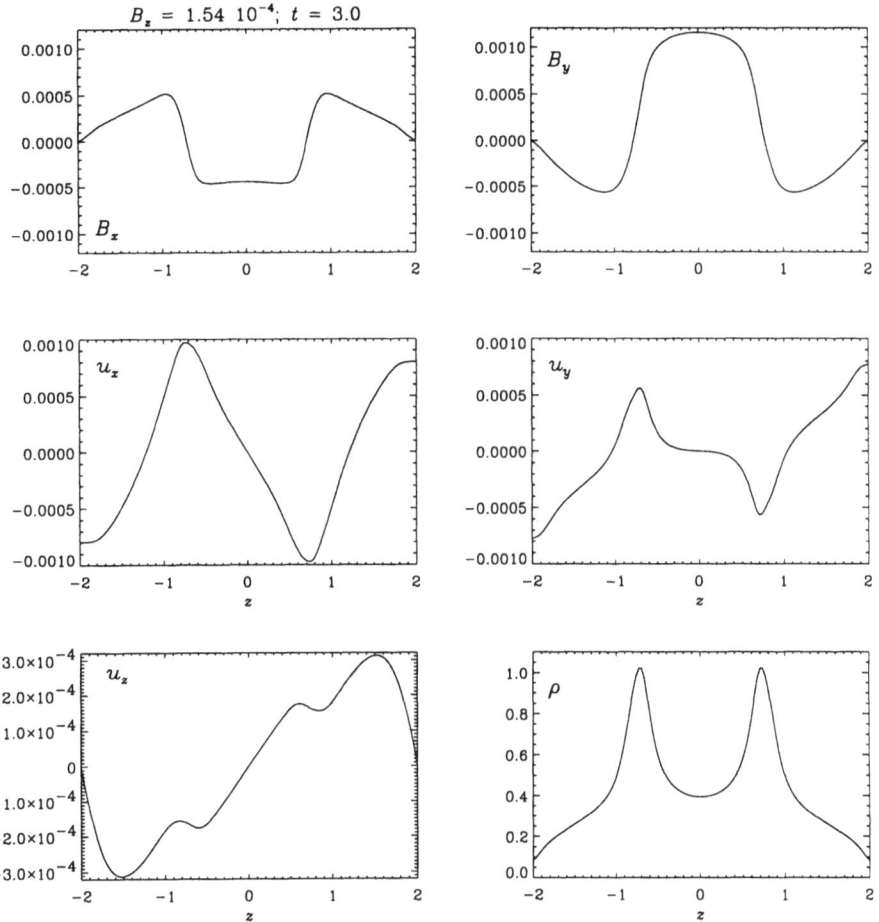

Fig. 12. The evolved state of Model 6 at 3.0 orbital periods. The figure is organised in the same way as Fig. 4

that as a perturbation they added random pressure fluctuations. As a result a large number of different modes were excited and grew with their own growth rates. By making a Fourier decomposition in the spatial coordinates they were able to determine the growth rates of each mode individually, and found that the modes initially followed the expected behaviour, but later on the growth rates decreased, and eventually the flow settled down to a so-called two-channel solution. The two-channel state appeared to be independent of the strength of the vertical field, and can thus not be interpreted as the dominating mode of the shearing instability, whose wavelength would have been a function of the magnetic field strength. As we do not find the two-channel solution in our simu-

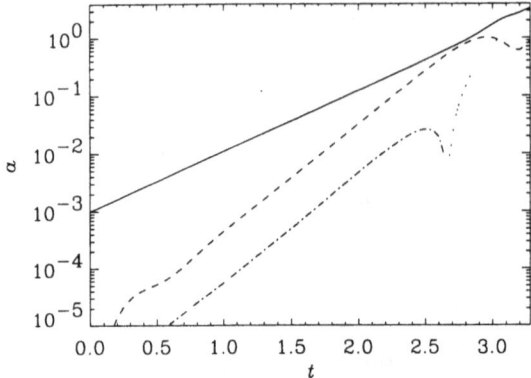

Fig. 13. The amplitudes of the first three odd eigenmodes as a function of time for Model 6. The lines represent a_1 (solid line), $-a_3$ (dashed line), a_5 (dashed-dotted line) and $-a_5$ (dotted line)

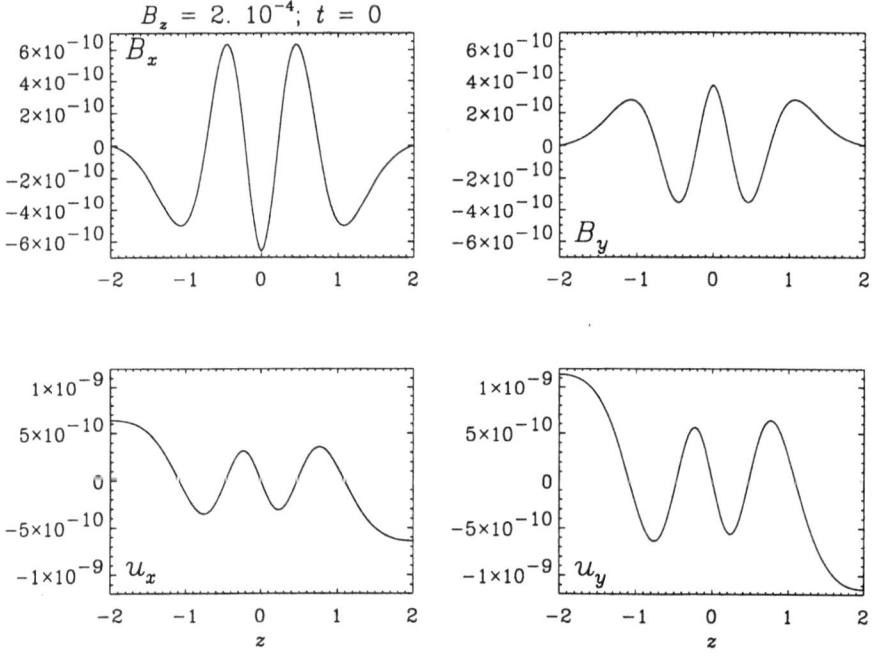

Fig. 14. The initial state of Model 7. The imposed vertical magnetic field is $2.\,10^{-4}$. The figure is organised in the same way as Fig. 2

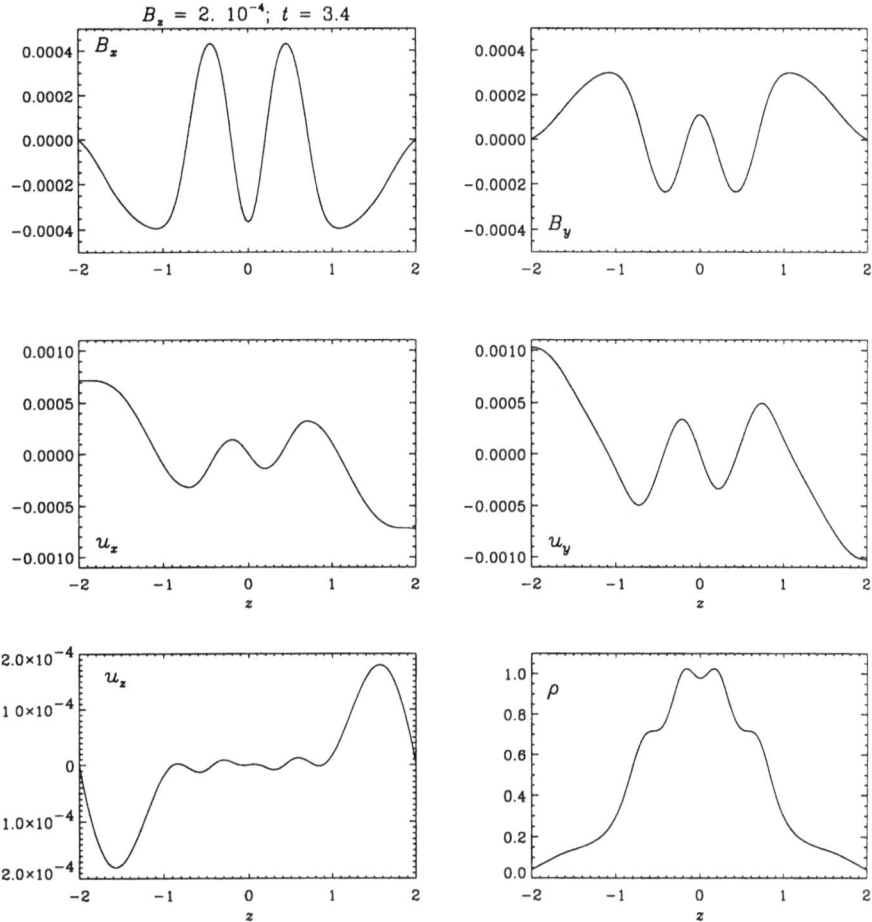

Fig. 15. The evolved state of Model 7 at 3.4 orbital periods. The figure is organised in the same way as Fig. 4

lations, we must conclude that it is a consequence of the two spatial dimensions. The two dimensions in themselves provide more freedom for dissipating magnetic fields, but equally important may be that the parasitic instability of Goodman & Xu (1994) becomes operative in two dimensions.

Both our one-dimensional results and the two-dimensional results of Hawley & Balbus (1992) are radically different from what has been found in three-dimensional simulations (e. g. Hawley et al. 1995, Brandenburg et al. 1995), as only the latter reach a turbulent state. It is of significant interest to understand the reason for these differences, as that may give us a clue to how the shearing instability leads to turbulence. The one- and two-dimensional simulations find

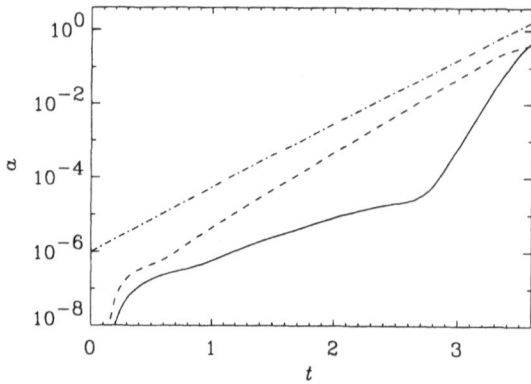

Fig. 16. The amplitudes of the first three odd eigenmodes as a function of time for Model 6. The lines represent $-a_1$ (solid line), a_3 (dashed line), a_5 (dashed-dotted line)

a preferred length scale, in our case the wavelength of the most rapidly growing mode, and in Hawley & Balbus (1992) the length scale of the two-channel solution. In the case of the shearing instability of a toroidal magnetic field (e.g. Ogilvie & Pringle 1996, Terquem & Papaloizou 1996) the growth rate increases with increasing vertical wave number so that there is not a preferred length scale. This may explain why only three-dimensional models, which do allow the toroidal field to become unstable, develop a turbulent state.

Another important question is the nature of the saturation mechanism in the turbulent models. This work shows that the one-dimensional instability does not saturate until the magnetic pressure is comparable to the gas pressure. An attractive possibility for the three-dimensional models is that the field strength is limited by the magnetic field becoming buoyantly unstable, but the turbulence seems to saturate at comparable levels independently of whether the disc is stratified or not (Torkelsson et al. 1996). A more likely alternative is that the turbulence saturates when the magnetic field has become so tangled that the rate of dissipation is comparable to the growth rate of the fundamental instabilities. This cannot happen in a system restricted to one dimension, and thus such a mechanism cannot work in the simulations described in this paper, but may work in three dimensions, where the magnetic field saturates at much lower field strengths.

In conclusion, we have studied the evolution of the shearing instability of a vertical magnetic field in a Keplerian disc in one dimension, the vertical. With these restrictions the instability does not saturate until the gas and magnetic pressures are comparable, and there is no turbulent cascade as the instability has a preferred mode with the maximum growth rate and a finite wave number.

Acknowledgement: UT is supported by an EU post-doctoral fellowship. This work was supported in part by the Danish National Research Foundation through its establishment of the Theoretical Astrophysics Center.

The stability of magnetically threaded accretion disks

R. Stehle[1,2]

[1] Max-Planck-Institut für Astrophysik, Postfach 1523, 85740 Garching, Germany
[2] Astron. Inst. 'Anton Pannekoek', Kruislaan 403, 1098SJ Amsterdam, Netherlands

Abstract: We study the non axisymmetric stability of a magnetically threaded accretion disks numerically in the (r, ϕ) plane with full hydrodynamics. The 3–D disk magnetosphere is solved in potential field approximation. We demonstrate the existence of a global instability which causes the initial surface mass– and magnetic flux density distribution to be redistributed on the rather short orbital timescale, i.e. the dynamical timescale. As long as the instability works a significant fraction of the disk mass is accreted to the central object. After the instability the surface density decreases exponentially with disk radius.

1 Introduction

It is generally thought that magnetic fields play a dominant role in the formation of protostellar objects like T Tauri stars (see Shu et al. 1993 for a general introduction). Molecular clouds are stabilized against gravitational collapse not only by rotation and thermal pressure but also by large scale magnetic fields which thread the clouds. Magnetic braking, thermal cooling or diffusive magnetic processes like ambipolar diffusion result in a supercritical cloud(fragment) due to which the total magnetic flux through the cloud and the magnetic support decrease (McKee et al. 1993). A dynamical collapse preferentially along the magnetic field lines and the rotation axis (see Tomisaka 1995) yields a massive central object (the protostar) surrounded by a rather thin accretion disk which is still threaded by the fossil interstellar magnetic field, now advected from large distances to the central star. During the collapse the magnetic field lines are compressed and the magnetic field increases significantly in strength. It is this scenario which we have in mind when we study in the following sections the stability of a magnetically threaded accretion disk which revolves around a central star. We assume that the magnetic field is advected from large distances to the central object and therefore that the total magnetic flux through the disk is non-zero. This is different to magnetic fields generated by small scale dynamo processes (e.g. Stone et al. 1996, Brandenburg et al. 1995) where the magnetic field forms closed loops on one side of the accretion disk and the total magnetic flux through the disk vanishes.

2   The basic model assumptions

In this section we briefly discuss the basic assumptions which enter in our model equations. For a full description of the model we refer to Stehle & Spruit (1997a, MNRAS in preparation).

As protoplanetary accretion disks are usually cold (i.e. $c_s \ll v_\phi$, where c_s is the sound velocity and v_ϕ the azimuthal gas velocity) the thermal pressure plays only a minor role in the time evolution and the disk can be assumed to be geometrically thin. We assume the disk magnetosphere outside the disk to be current free. Its magnetic field distribution is then given by a potential field, and electric currents $\mathbf{j}$ exist only in the disk, i.e. in cylindrical coordinates we have a current sheet $\mathbf{j}(r, \phi, z) = \mathbf{j}(r, \phi)\delta(z)$. Matching the disk magnetosphere to the magnetic flux density distribution $B_z(r, \phi)$ through the disk yields the electric currents in the disk. The disk mass is assumed to be negligible compared to the mass of the central star so that the gravitational field is that of the central point mass only. Finally we assume that the gas is sufficiently ionized everywhere that the magnetic flux is frozen into the disk gas. This determines the time evolution of $B_z(r, \phi)$.

For each timestep we calculate the 3–D magnetosphere from the $B_z(r, \phi)$ distribution in the disk. This yields the magnetic forces onto the disk, necessary to evolve the disk hydrodynamically in time. The numerical time integration is done with a two dimensional Eulerian grid with the van Leer (1977) scheme for upwind differencing (see Stone & Norman 1992, Stehle & Spruit 1997b, MNRAS in preparation for more details on the hydrodynamic part). The inner disk rim of our calculation is $r_{in} = 0.1\, r_{out}$ where r_{out} is the radius of the outer edge of the disk. The integration is advanced on a grid of 156×128 grid cells in r and ϕ respectively.

3   A model calculation

3.1   Initial conditions

An initially axisymmetric model is chosen in which the disk surface mass density $\Sigma(r)$ decreases with radius as $\Sigma \sim r^{-3/2}$ and $B_z \sim r^{-5/4}$. At the disk boundaries B_z drops exponentially to zero (Fig. 1b). The magnetic field strength g_m provides up to 32% of the support of the disk against gravitation g_g (Fig. 1d).

We compute the initial azimuthal velocity $v_\phi(r)$ from the force balance of centrifugal, gravitational and magnetic forces. The initial stationary model is perturbed with low amplitude ($\sim 10^{-7}$) point–to–point noise in v_r. We follow the time evolution numerically; as unit of time we use the Kepler period of the outer disk rim $T(r_{out})$.

3.2   The time evolution of the global instability

We find the initial axisymmetric set up to be unstable against a non axisymmetric instability. The instability starts immediately after the onset of our com-

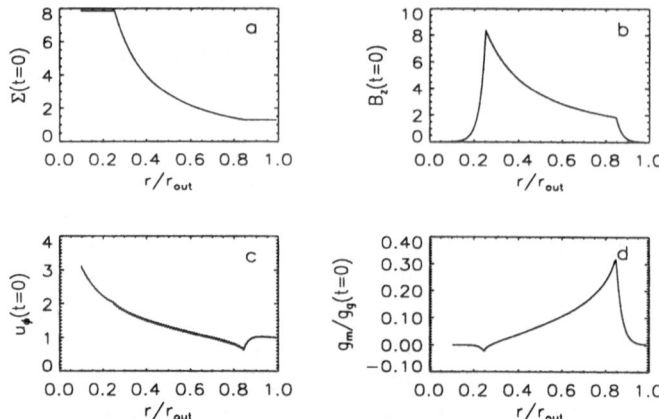

Fig. 1. The initial conditions. The radial acceleration due to the magnetic support of the disk g_m relative to the gravity g_g of the central star increases with r, with a max. of 32 %.

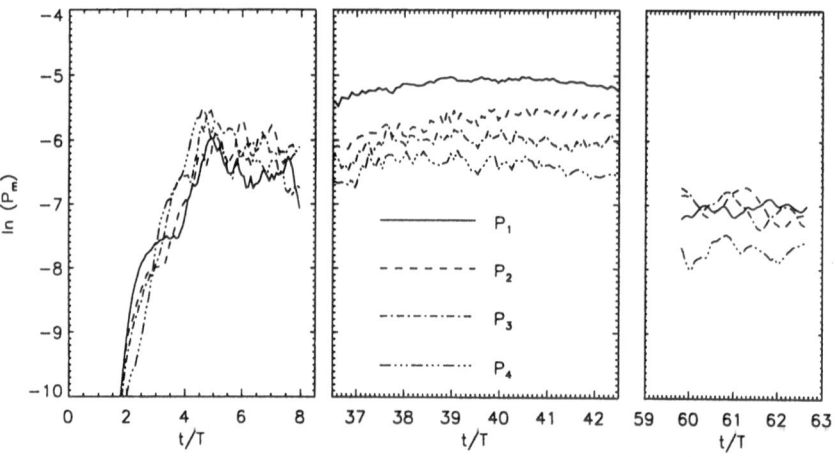

Fig. 2. The time evolution of the relative strength P_m of the Fourier modes of B_z compared to the $m = 0$ mode. Left: The rapid initial increase saturates at $\sim 2.5\,10^{-3}$. Middle: During the violent instability phase the m=1 component is by far most prominent. Right: At the end of the calculation ($t/T \sim 60$) the magnetic waves and especially P_1 are damped.

putation and is well seen in the time evolution of the relative strength P_m of the power in the Fourier components $m = 1, 2, 3, 4$ of B_z integrated over the whole disk compared to the $m = 0$ mode (Fig. 2 left). The initial perturbation increases in strength within a few orbits, i.e. on a dynamical timescale. The instability saturates when the power in the Fourier modes m are of the order of $\sim 2.5\,10^{-3}$ of that in the $m = 0$ component.

The instability starts with rather high values of $m \simeq 5...8$ (see $\Sigma(r, \phi)$ in Fig. 3a) and it is only later that smaller values of m become dominant. By the magnetic forces in this non axisymmetric field pattern mass is accreted towards the central star and piles up near the inner and outer edges of the disk. The instability acts much like "viscous spreading" and transports angular momentum outward so that mass is allowed to fall inward. At $t/T \simeq 30$ a prominent $m = 1$ spiral arm develops (Fig. 2 middle) causing mass accretion on even shorter timescales (Figs. 3 c–e). This indicates that the instability is more efficient in transporting angular momentum outward when lower Fourier modes m dominate the evolution. An explanation might be that the magnetic forces transmitted by the potential field of the magnetosphere have their largest range at the lowest values of m. At the end of this phase of the instability the $m = 1$ wave damps (Fig. 2 right, Fig. 3f) and the mass–distribution of the disk with r is completely restructured. This is shown in the next section.

3.3   The disk model at t/T=62.4

We stop our calculation at $t/T = 62.4$ after the $m = 1$ arm instability has disappeared and the waves have already somewhat decayed. At that time a significant fraction of the mass has been brought inward and $\Sigma(r)$ now decreases exponentially with disk radius (see Fig. 4, the spread in the dots at a particular radius results from the spread of the values in ϕ). The inclination of the field lines increases with r, starting normal to the disk plane at $\sim 0.2\,r_{out}$. For $r < 0.2\,r_{out}$ they first bend inward before they open up again to get closed on the other side of the disk. The poloidal structure of these magnetic field lines is ideal to confine bipolar outflows or jets (Blandford 1993, Spruit, Foglizzo & Stehle 1997) which are observed in young stellar objects. We note that at $r \simeq 0.2\,r_{out}$, where the magnetic field lines are normal to the disk plane, the magnetic forces vanish. At that point the surface density shows a sharp jump. Our computation suggests that the instability is unable to accrete the mass beyond this point.

For $r > 0.4\,r_{out}$ the inclination of the magnetic field lines to the disk plane is less than $60°$ that according to the Blandford & Payne (1982) model a disk wind can be launched along these magnetic field lines.

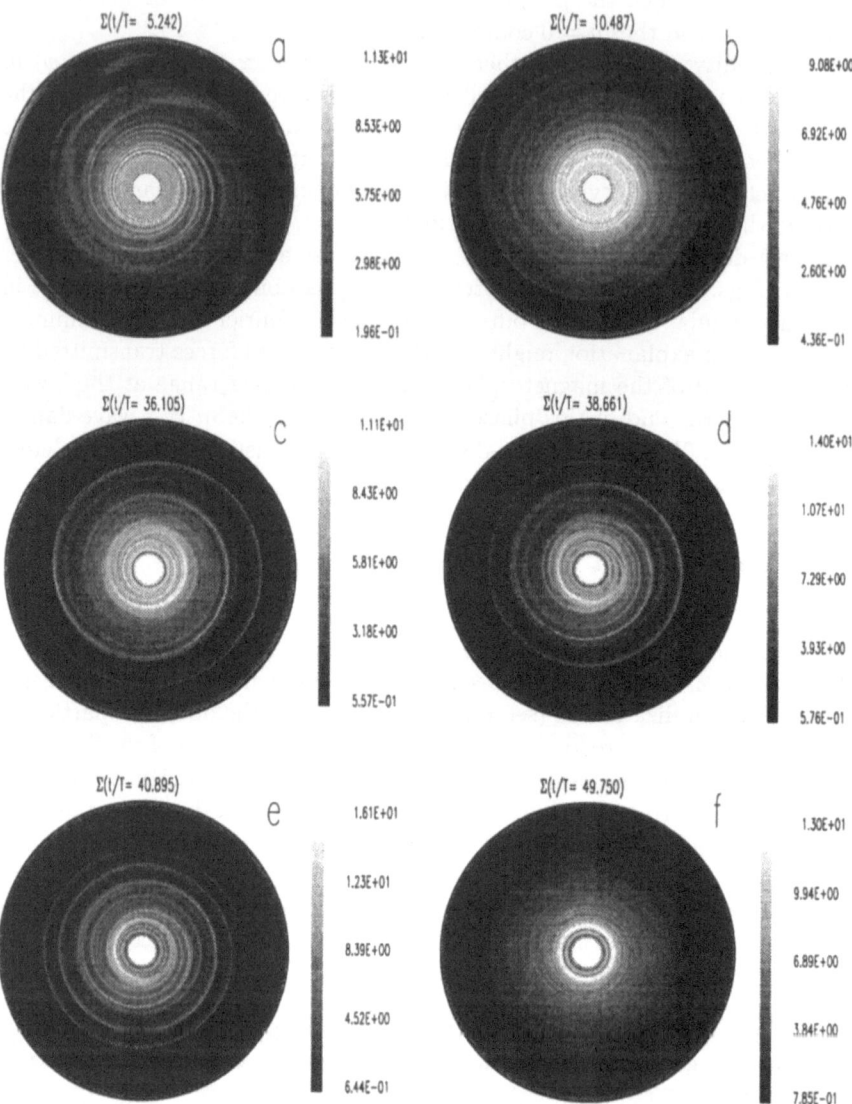

Fig. 3. The surface density distribution $\Sigma(r, \phi)$ of the disk as a function of time. The grey–scaling is done for each plot individually. We observe a prominent one arm instability which sets in at $t/T \sim 30$ and which acts like "viscous spreading". See the text for further explanations.

4   Summary

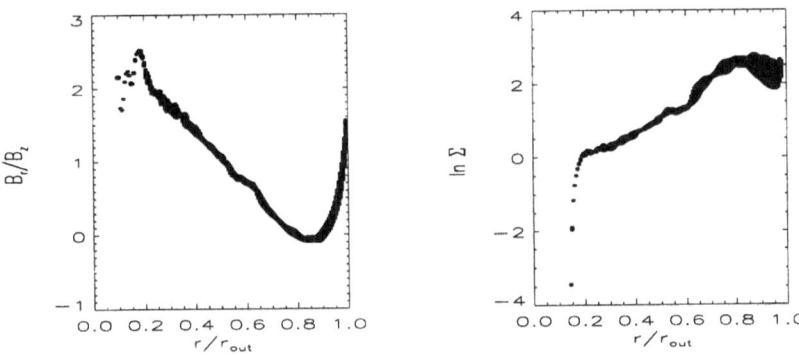

Fig. 4. The distribution of B_r/B_z and $\ln \Sigma(r)$ at $t/T = 62.4$. It is evident that the surface density decreases exponentially with disk radius and that the inclination of the magnetic field lines to the disk decreases.

With the particular calculation above we demonstrated the existence of a global instability in accretion disks which are threaded by large scale magnetic fields. The instability works on the rather short dynamical timescale and redistributes angular momentum in such a way that mass is accreted towards the central star. The mass accretion due to that instability is most efficient at that time where the m=1 component of the Fourier spectrum of $B_z(r, \phi)$ is most prominent. We observe an exponential decrease of surface mass and magnetic field distribution with r for our "final" model. For this distribution the waves are significantly damped in strength (Fig.2 right). As the mass redistribution takes place on the short dynamical timescale we expect that the surface density in magnetically threaded accretion disks falls of exponentially. We also suggest that this instability can operate in a semi periodic way if ongoing mass accretion from the surrounding molecular cloud feeds the outer part of the accretion disk until enough mass had been piled up that the instability can set in again. It has to be verified, however, if this instability can be identified with the FU Orionis outbursts, observed in protoplanetary accretion disks (Hartmann, Kenyon & Hartigan 1993).

Acknowledgement: R.S has been supported by the 'DAAD aus Mitteln des 2.ten Hochschulsonderprogramms' and partly form the research network 'accretion onto compact objects and protostars' by EC grant ERB–CHRX–CT93–0329. He wants to thank J. Papaloizou, A. King and especially H. Spruit for ongoing discussions and many useful suggestions.

References

Blandford, R.A.: 1993, in D. Burgarella, M. Livio and C. O'Dea, eds., *Astrophysical Jets*, Cambridge: Cambridge University Press, 15

Blandford, R.A., Payne, D.G.: 1982, MNRAS 176, 465

Brandenburg, A., Nordlund, A., Stein, R.F., Torkelsson, U.: 1995, ApJ, 446, 741

Hartmann, L., Kenyon, S., Hartigan, P.: 1993, in E.H. Levy and J.I. Lunine, eds., *Protostars and Planets III*, Tucson: The University of Arizona Press, 497

van Leer, B.: 1977, J. Comput. Phys. 23, 276

McKee, C.F., Zweibel, E.G., Goodman, A.A., Heiles, C.: 1993, in E.H. Levy and J.I. Lunine, eds., *Protostars and Planets III*, Tucson: The University of Arizona Press, 327

Shu, F., Najita, J., Galli, D., Ostriker, E., Lizano S.: 1993, in E.H. Levy and J.I. Lunine, eds., *Protostars and Planets III*, Tucson: The University of Arizona Press, 3

Spruit, H.C., Stehle, R., Papaloizou, J.C.B.: 1995, MNRAS 275, 1223

Spruit, H.C., Foglizzo, T., Stehle, R., 1997, MNRAS submitted

Stone, J.M., Norman, M.L.:1992, ApJSS 80, 753

Stone, J.M., Hawley, J.F., Gammie, C.F., Balbus, S.A.: 1996, ApJ 463, 656

Tomisaka, K.: 1995, ApJ 438, 226

The equilibrium and stability of a thin accretion disc containing a poloidal magnetic field

G.I. Ogilvie[1,2]

[1] Institute of Astronomy, University of Cambridge, Madingley Road, Cambridge CB3 0HA, UK
[2] Department of Applied Mathematics and Theoretical Physics, University of Cambridge, Silver Street, Cambridge CB3 9EW, UK

Abstract: The equilibrium of an accretion disc containing a magnetic field is considered using a simplified model. A perfectly conducting, non-self-gravitating fluid, in differential rotation about a massive central object, contains a purely poloidal magnetic field which bends as it passes through the disc, enforcing isorotation on magnetic surfaces. The solutions for thin discs are obtained from an asymptotic analysis. Two families of solutions are found, which involve different balances of forces in the radial and vertical directions. The stability of the equilibria to axisymmetric perturbations is determined by obtaining a dispersion relation which is local in radius but is nevertheless capable of describing global normal modes. In particular, the stability boundary for the magnetorotational instability is obtained.

1 Introduction

1.1 MHD in accretion discs

There are compelling reasons for believing that magnetic fields are essential to the physics of accretion discs. The anomalous transport of angular momentum that is required for accretion to proceed is most convincingly explained by magnetohydrodynamic (MHD) mechanisms. One possibility is that a large-scale magnetic field removes angular momentum from the disc by driving a wind that accelerates to a super-Alfvénic velocity (Blandford and Payne 1982). Such rotating MHD winds are known to collimate into jets perpendicular to the disc (Heyvaerts and Norman 1989). Alternatively, the disc may be in a state of MHD turbulence as a result of the magnetorotational instability (Velikhov 1959; Chandrasekhar 1960; Balbus and Hawley 1991), which is a robust, linear and essentially local instability that operates on any sufficiently weak magnetic field in an accretion disc. The Reynolds and Maxwell stresses associated with the resulting turbulence transport angular momentum radially outwards, as is required for accretion (Brandenburg et al. 1995, 1996; Hawley, Gammie and Balbus 1995, 1996; Stone et al. 1996).

In investigating the linear theory of the magnetorotational instability, some authors have used a local analysis, equivalent to a zeroth-order WKB approximation (Balbus and Hawley 1991), while others have represented the disc either as a stratified shearing sheet (Gammie and Balbus 1994; Foglizzo and Tagger 1995) or as a cylinder uniform along its axis (Curry, Pudritz and Sutherland 1994; Curry and Pudritz 1995; Ogilvie and Pringle 1996). Papaloizou and Szuszkiewicz (1992) demonstrated that the axisymmetric stability of a much more general equilibrium can be determined using a variational principle, but did not apply their result to any specific disc models. The primary aim of the present work is therefore to construct more realistic equilibria of MHD accretion discs and to determine their stability, although not necessarily using the variational principle.

A secondary aim is to address the connection between discs and winds or jets. The work of Blandford and Payne (1982) demonstrated that, if a Keplerian disc contains a poloidal magnetic field which is inclined at an angle greater than 30° to the vertical at the surface of the disc, then a centrifugally driven wind results. In their model, however, the disc is treated as an infinitely thin sheet without regard for its internal equilibrium. It remained to be shown, therefore, that a steady disc equilibrium exists with a magnetic field that is sufficiently inclined. Furthermore, as Shu (1991) has pointed out, the bending of poloidal field lines in the disc is associated with a radial Lorentz force, and the resulting deviation from Keplerian rotation could invalidate the argument that leads to the critical angle of 30°. These matters have been addressed by Wardle and Königl (1993), who showed that the difficulties can be resolved when the internal structure of the disc is taken into account consistently. The asymptotic analysis of thin discs described in this paper handles all these matters in a precise and unambiguous way.

Some authors have addressed the disc–wind connection using radially self-similar models (Königl 1989; Li 1995), by solving for a disc of non-zero angular thickness and connecting it to the wind solution of Blandford and Payne (1982). There remain difficulties with this approach, however. In particular, the correct matching conditions are uncertain, and, more significantly, the wind solution itself behaves unphysically at infinity. It appears that the finite size of the disc must be taken into account in order to study the behaviour of winds and jets at large distances.

1.2   A general model

As a simplified representation of an accretion disc, one may consider the following model (Papaloizou and Szuszkiewicz 1992; Ogilvie 1997a, hereafter 'Paper I'). A perfectly conducting, non-self-gravitating fluid rotates differentially in the gravitational potential of a central mass M, and contains a purely poloidal magnetic field which bends as it passes through the disc. The system is steady and axisymmetric, and is arranged such that the density, pressure and angular velocity are all symmetrical about the equatorial plane, while the magnetic field has dipolar symmetry. The accretion flow and any toroidal magnetic field are neglected at this stage.

This model may be considered as a special case of the general problem of steady, axisymmetric MHD flow studied by Mestel (1968) in the context of stellar winds. In this case there are two quantities which are constant on each magnetic surface: the angular velocity and a Bernoulli function. The principal equation governing this model is a non-linear, elliptic partial differential equation in two dimensions (a version of the Grad–Shafranov equation). It would be possible in principle to solve this numerically, but very specific choices would have to be made for the shape of the surface of the disc and for the remaining free functions in the problem. Instead of constructing one particular solution in this manner, one can reduce the problem to the solution of ordinary differential equations, while retaining considerable generality (Paper I).

One possibility is to impose self-similarity in the spherical radial coordinate, as is common in the study of winds and jets from discs (Blandford and Payne 1982; Königl 1989; Li 1995). This is expected to yield a solution that is valid in an intermediate asymptotic domain far from the radial boundaries, and provides a useful method for studying thick discs. An alternative method is to obtain the asymptotic solutions in the limit of thin discs, without imposing self-similarity. This approach may be considered as a careful generalization of the familiar thin-disc approximation for hydrodynamic discs (Pringle 1981). It is a powerful method which is useful not only for studying thin-disc equilibria and their stability, but also for studying winds and jets from discs, in each case clarifying the relevant physics.

The remainder of this paper is concerned with thin-disc equilibria and their stability. In Section 2 the basic properties of the equilibria are described. It is intended to present only a summary of the more detailed treatment of equilibria in Paper I, where a discussion of self-similar solutions can also be found. In Section 3 the results of a stability analysis are given. A full exposition will be given elsewhere (Ogilvie 1997b, hereafter 'Paper II').

2 Thin-disc equilibria

2.1 Asymptotic analysis

Let $z = H(r)$ be the location of the upper surface of the disc at radius r, where (r, ϕ, z) are the usual cylindrical polar coordinates. It is assumed that the density goes to zero at a finite height above the equatorial plane, so a polytropic relation between pressure and density should be used instead of an isothermal one. The small parameter ϵ of the problem is conveniently defined as either the maximum value, or a characteristic value, of $H(r)/r$. (One may reasonably expect that $H(r)/r$ is only weakly dependent on r except near the edges of the disc.) As in boundary-layer theory, the rapid variation of quantities with z is resolved by introducing a stretched vertical coordinate $\zeta = \epsilon^{-1}z$, which is $O(1)$ inside the disc. This is preferable to using vertically integrated equations or a strictly two-dimensional treatment, and is essential if a full stability analysis (to three-dimensional perturbations) is to be applied. The next stage is to determine the scalings of all physical quantities (the density, pressure, angular velocity and magnetic field) with ϵ in such a way that the equations of equilibrium are satisfied. Of the several possible ways to achieve this, there are two distinct sets of scalings that are capable of describing accretion discs. These may be described as 'weakly magnetized' and 'strongly magnetized', and their properties are summarized in Sections 2.2 and 2.3 below.

In each case there are two aspects to obtaining a solution at leading order. First, there is a set of non-linear ordinary differential equations in ζ which determine the vertical equilibrium of the disc at each radius separately. Second, there is an integral relation which determines the global magnetic structure of the disc. This is described in Section 2.4.

2.2   Weakly magnetized discs

The first family of solutions may be regarded as the natural generalization of the standard thin-disc approximation for hydrodynamic discs. The scalings are

$$u_\phi = O(1), \tag{1}$$
$$v_s = O(\epsilon), \tag{2}$$
$$v_A = O(\epsilon), \tag{3}$$
$$|\Omega - \Omega_K| = O(\epsilon), \tag{4}$$

where these quantities are the azimuthal fluid velocity, the sound speed, the Alfvén velocity and the deviation from Keplerian rotation, respectively. In a hydrodynamic thin disc, scalings (1) and (2) would apply, but $|\Omega - \Omega_K|$ would be $O(\epsilon^2)$ if self-gravitation were unimportant. The deviation from Keplerian rotation is now larger because of the radial Lorentz force associated with the bending of magnetic field lines in the disc.

The vertical equilibrium at each radius involves a balance between the pressure gradient, the Lorentz force and the gravitational force. Since all three are comparable, the equations governing the equilibrium do not have analytical solutions in general. If a polytropic relation

$$p = K\rho^\Gamma \tag{5}$$

is assumed locally in radius, then the equations, which form a third-order system, can be solved numerically. Apart from Γ, the two dimensionless parameters in the problem are B_{rs} (the radial component of the magnetic field at the upper surface $z = H$) and B_z (the vertical component of the magnetic field), measured in units of

$$\mu_0^{1/2}(GM)^{\Gamma/2(\Gamma-1)}K^{-1/2(\Gamma-1)}r^{-3\Gamma/2(\Gamma-1)}H^{\Gamma/(\Gamma-1)}. \tag{6}$$

The symmetries of the equations are such that these parameters may be taken to be non-negative without loss of generality.

The results of the calculation for the case $\Gamma = 5/3$ are shown in Figure 1. The existence of a solution depends strongly on the angle θ at which the magnetic field lines are inclined to the vertical at the surface of the disc. If $\theta < 30°$, then

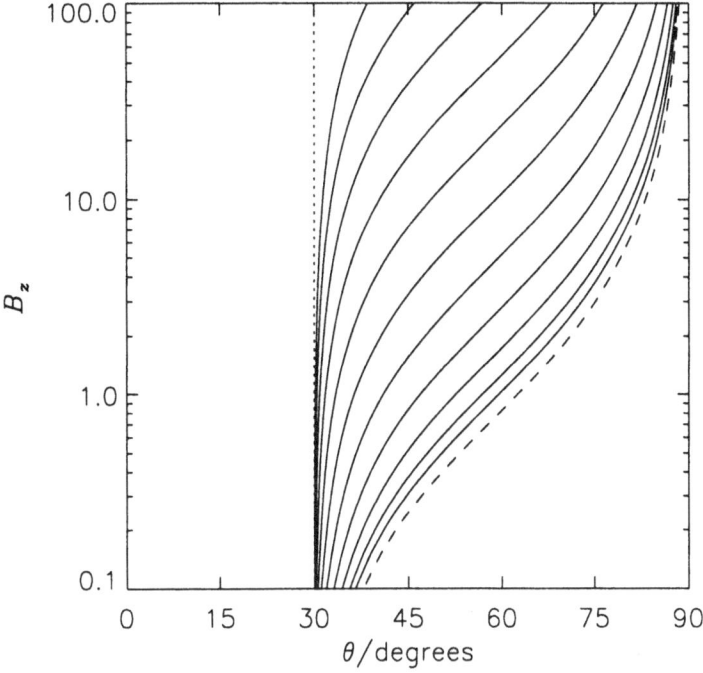

Fig. 1. Parameter space for weakly magnetized discs with $\Gamma = 5/3$. To the left of the straight dotted line, the angle of inclination $\theta < 30°$ and a solution exists at every point. To the right of the dotted line, $\theta > 30°$ and an acceptable solution exists only in the region filled with contour lines. These are contours of z_{sonic}, in units of H, with contour values $1.1, 1.2, 1.4, \ldots, 103.4$ (from right to left). The dashed line, equivalent to a contour value of 1, marks where the principal solution ceases to exist. (Other branches of physically unacceptable solutions are not included.)

at least one solution is found for every value of the magnetic field strength. If $\theta > 30°$, then a solution is found only if the magnetic field is sufficiently strong.

As discussed in Section 1.1, the angle of inclination also determines whether a predominantly centrifugally driven wind can be launched from the disc. If the equations governing the equilibrium are continued above the surface of the disc for a perfectly conducting atmosphere of negligible mass, it is found that the magnetic field lines are straight on the spatial scale H, since the field must be force-free. The dynamics of matter in this region is governed by the component $g_{\parallel}$ of the effective gravitational acceleration parallel to the magnetic field (measured downwards). This is a linear function of distance along the field line, and is positive at the surface of the disc if an equilibrium exists at all. If $\theta < 30°$, then $g_\parallel$ increases with distance along the field line. This may be compared with the linear increase of the (downward) vertical gravitational acceleration above a hydrodynamic thin disc. If $\theta > 30°$, however, then $g_\parallel$ decreases with distance and goes to zero at some height $z = z_{sonic}$. This name is appropriate because it is the expected location of the slow magnetosonic point of a wind. Contours of z_{sonic}, in units of H, are shown in the appropriate region of Figure 1.

The numerical results indicate that it is easier to satisfy the condition $\theta > 30°$ when the magnetic field is stronger. It should be emphasized that the angle of inclination varies with radius according to the solution of the exterior field equation (Section 2.4), and that vertical equilibrium must be satisfied at each radius separately.

2.3   Strongly magnetized discs

The second family of solutions is more closely related to models of solar prominences in which a sheet of matter is supported against gravity by a magnetic field that bends as it passes through the sheet (Kippenhahn and Schlüter 1957). The scalings are

$$u_\phi = O(1), \tag{7}$$

$$v_s = O(\epsilon^{1/2}), \tag{8}$$

$$v_A = O(\epsilon^{1/2}), \tag{9}$$

$$|\Omega - \Omega_K| = O(1). \tag{10}$$

The sound speed and the Alfvén velocity, although still comparable, are larger than in a weakly magnetized disc. The deviation from Keplerian rotation is also larger, because the radial Lorentz force is able to balance an $O(1)$ fraction of the radial gravitational force.

The vertical equilibrium at each radius involves a balance between the pressure gradient and the Lorentz force only. The vertical gravitational force does not appear at leading order because the disc is compressed by the Lorentz force rather than by gravity. The governing equations are therefore simpler than for a weakly magnetized disc and can be solved in terms of special functions if a polytropic relation is assumed.

Above the surface of a strongly magnetized disc, the effective gravitational acceleration parallel to the magnetic field is always directed downwards close to the disc, although it may possibly change sign at a height $z \gg H$, depending on the shape of the field lines. A strongly magnetized disc cannot have a predominantly centrifugally driven ('cold') wind whatever the angle of inclination.

The two families of magnetized thin discs represent physically different limits and must be carefully distinguished. It would appear that, in view of their closer relation to hydrodynamic thin discs and the possibility of their having predominantly centrifugally driven winds, the weakly magnetized discs are likely to be more important. Previous discussions of magnetized thin discs (in particular Heyvaerts and Priest 1989) have been based on two-dimensional treatments

which are equivalent to the vertically integrated equations for a strongly magnetized disc. A two-dimensional treatment is not sufficient to describe weakly
magnetized discs, however. One reason for this is that the deviation from Keplerian rotation at leading order $[O(\epsilon)]$ is not independent of ζ.

2.4 The exterior field equation

In the exterior of the disc the magnetic field is force-free, and must therefore
be current-free since it is also axisymmetric and purely poloidal. The potential
problem in the exterior of a thin disc, which in general has a non-zero inner radius
and a finite outer radius, is an example of a mixed boundary-value problem and
is conveniently solved using integral-transform techniques. If $B_{rs}(r)$ is specified
over the whole extent $r_1 \leq r \leq r_2$ of the disc, then the exterior field is completely
determined; in particular, the flux distribution $\psi(r)$ on the disc is then known
in the form

$$\psi(r) = \int_{r_1}^{r_2} K(r,r') B_{rs}(r') \, dr', \tag{11}$$

where the kernel $K(r,r')$ can be expressed in terms of special functions. The
vertical field in the disc follows immediately from

$$B_z = \frac{1}{r} \frac{d\psi}{dr}. \tag{12}$$

A more physical argument which leads to exactly the same result involves treating the disc as a current sheet with surface current density $(2/\mu_0) B_{rs}(r)$, and
deducing the exterior field using the Biot–Savart law (cf. Lubow, Papaloizou and
Pringle 1994). So far it has been assumed that the magnetic field is generated
only by the electric current flowing in the disc. If the disc is finite in extent, then
the field lines are closed and an O-type neutral point, at which $B_z = 0$, must
occur somewhere in the disc. This situation is unsatisfactory for several related
reasons, especially for weakly magnetized discs.

(a) In the neighbourhood of a neutral point, the angle of inclination θ is close
    to 90°. Figure 1 shows that an acceptable vertical equilibrium for a weakly
    magnetized disc cannot be found for any value of the magnetic field strength.
(b) If the exterior of the disc is treated as a conducting atmosphere rather
    than an insulating vacuum, then isorotation requires that pairs of points
    on the disc that are connected by magnetic field lines around the neutral
    point should have the same angular velocity. This is incompatible with the
    requirement that $|\Omega - \Omega_K| = O(\epsilon)$ for a weakly magnetized disc, although
    it could in principle be satisfied by a strongly magnetized disc.
(c) A wind cannot flow to infinity if the field lines are all closed.

These difficulties can be overcome by adding a uniform vertical field of sufficient
strength to eliminate the neutral point from the disc. This also opens the field
lines to infinity in a way conducive to the driving of a wind. The additional field

may be considered to be generated by a current at infinity; this is either the current in the distant interstellar medium that gives rise to the local magnetic field, or it represents currents beyond the Alfvén surface if the disc has a wind.

3   Stability to axisymmetric perturbations

The principal aim of constructing equilibria is to study the magnetorotational instability in more realistic situations than have been considered previously. When the magnetic field is purely poloidal, the instability is essentially axisymmetric and could therefore be investigated using the variational principle of Papaloizou and Szuszkiewicz (1992). However, this method is exceedingly difficult to apply in practice and does not in general yield a simple criterion. Since the equilibria are constructed on the basis of an asymptotic analysis for thin discs, it is both consistent and appropriate to develop an asymptotic theory for their normal modes. This problem also reduces to a consideration of ordinary differential equations at each radius separately. The details of this analysis can be found in Paper 2.

Attention may be restricted to weakly magnetized discs, not simply because they are likely to be more important, but because strongly magnetized discs are stable to the magnetorotational instability. This is because the typical frequencies of Alfvén waves for wavelengths comparable with H are $O(\epsilon^{-1/2})$ in a strongly magnetized disc, whereas the angular velocity is only $O(1)$. For a weakly magnetized disc, however, the angular velocity, the buoyancy frequency, and the typical frequencies of Alfvén and acoustic waves for wavelengths comparable with H are all $O(1)$, so this is clearly the appropriate scaling for the frequency eigenvalue ω of a normal mode. The form of a normal mode at leading order is found to be

$$\boldsymbol{\xi}(r, z) \sim \mathrm{Re}\left\{\boldsymbol{\xi}(r, \zeta)\, \exp\left[-\mathrm{i}\omega t + \mathrm{i}\epsilon^{-1}\int k(r)\,\mathrm{d}r\right]\right\}, \qquad (13)$$

where $\boldsymbol{\xi}$ is the Lagrangian displacement of fluid elements. This takes the form of a WKB function in r (although no additional approximation is made in deriving this form) with a complicated dependence on ζ. Strictly speaking, equation (13) represents a wave propagating in the r-direction (when ω is real), whereas the normal mode is a standing wave composed of two waves propagating in opposite directions. The wave is trapped between two radii which are either boundaries of the disc or turning points at which $k(r)$ vanishes. However, the local dispersion relation is insensitive to the sign of k and so it is sufficient to consider propagating waves of the form (13).

The leading-order equations for the Lagrangian displacement form a sixth-order set of ordinary differential equations in ζ at each radius separately. (The slow variation of the amplitude of the mode with r is not determined at this order.) For each value of k, which is real within the wave region, this is an eigenvalue problem for ω whose solution yields a local dispersion relation for the disc. It has the property that ω^2 is real whenever k is real, so that the onset of instability can occur only via a marginal mode with $\omega^2 = 0$.

The dispersion relation has many types of branches: f modes, p modes and g modes (as in a static body), and also r modes (due to inertial forces, introduced by the angular momentum gradient) and m modes (introduced by the magnetic field). The modes may be further classified according to their symmetry about the equatorial plane and their vertical mode number. The dispersion relation has been studied previously by Korycansky and Pringle (1995) for polytropic hydrodynamic discs and by Lubow and Pringle (1993) for isothermal discs. The m modes, however, are potentially unstable; they are the manifestation of the magnetorotational instability.

An example of the local dispersion relation is given in Figure 2. In this example the g modes are suppressed by taking the adiabatic exponent γ to be equal to the polytropic exponent Γ; this ensures that there is no entropy gradient which would otherwise have an effect, either stabilizing or destabilizing, on the disc. Also, the magnetic field is purely vertical rather than bending through the disc. Nevertheless, the dispersion diagram is particularly complicated because of the interactions between different branches of modes. In the absence of interactions, the f modes and p modes, which generally have greater group velocities $d\omega/dk$, would cross over the r modes and m modes; however, couplings between the modes lead to avoided crossings in which modes of different type exchange character as their branches approach and then diverge from each other. Avoided crossings occur only between modes of equal parity, since parity is a discrete property which cannot be exchanged smoothly between modes.

The case of a purely vertical magnetic field in a stratified disc has been considered previously by Gammie and Balbus (1994), who focused on the unstable side of the spectrum, although not specifically for a polytropic disc. For a general weakly magnetized disc, the magnetic field bends as it passes through the disc. A qualitatively similar dispersion diagram to Figure 2 is then obtained, although the avoided crossings are generally wider, indicating that the couplings between different types of mode are stronger.

In the example in Figure 2 the magnetic field is sufficiently weak that there are two unstable modes. These are the lowest m modes of odd and even symmetry, m_1^o and m_1^e. An examination of numerous different cases (restricted to $\gamma = \Gamma$) confirms the reasonable expectation that the last mode to be stabilized when the magnetic field strength is increased is m_1^o, and it is stabilized last at $k = 0$. This means that the overall condition for marginal stability to axisymmetric perturbations is that m_1^o should be marginal at $k = 0$. The stability boundary in the parameter space is therefore found by solving directly for an equilibrium possessing a marginal mode m_1^o at $k = 0$. The result of this calculation is shown in Figure 3. Above this marginal curve, equilibria are stable to the magnetorotational instability, but below it they are unstable. In the stable region, the equilibria are unique for given values of B_z and B_{rs}. In the unstable region, multiple solutions appear at a series of bifurcations. As well as being non-unique, the equilibria have peculiar properties; the magnetic field may bend

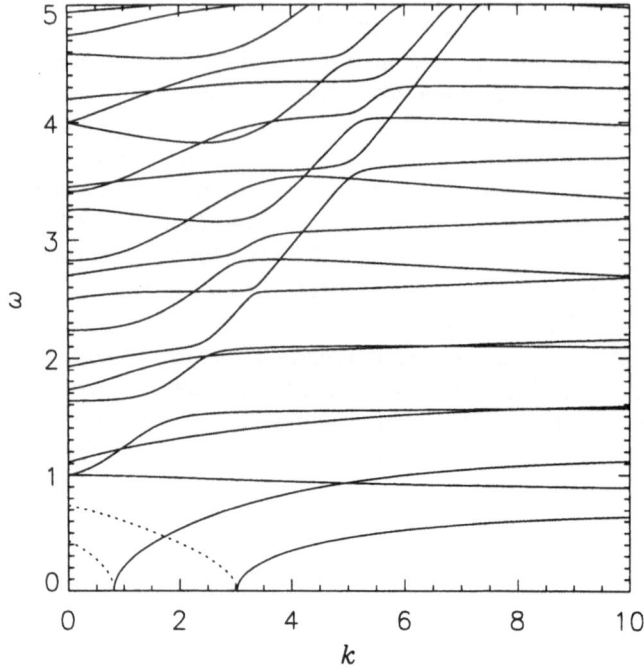

Fig. 2. Part of the local dispersion diagram for a weakly magnetized disc with $\gamma = \Gamma = 5/3$, $B_{rs} = 0$ and $B_z = 0.1$. The frequency eigenvalues of f, p, r and m modes, in units of Ω, are plotted against the radial wavenumber, in units of H^{-1}. Unstable branches, for which ω is imaginary, are shown as dotted lines. The most unstable mode is m_1^o, which achieves a growth rate of 0.7412 at $k = 0$, close to the Oort parameter $A = 0.75$.

several times as it passes through the disc, and there may be inversion layers. However, these solutions are all in the unstable part of the diagram; the stable equilibria do not have these peculiarities.

The results indicate that an equilibrium with a bending poloidal magnetic field, especially when the angle of inclination $\theta > 30°$, may be stable even when the corresponding equilibrium with a purely vertical magnetic field of the same value is unstable. They show that equilibria exist which are capable of driving a wind and are also stable to the magnetorotational instability. The instability therefore does not preclude the possibility of a wind-driven accretion disc from which most of the angular momentum is removed by a wind rather than being transported outwards by turbulent stresses. In Paper I the asymptotic method for thin discs is extended to include wind-driven accretion discs, by incorporating several additional features. Inside the disc, a slow accretion flow is included,

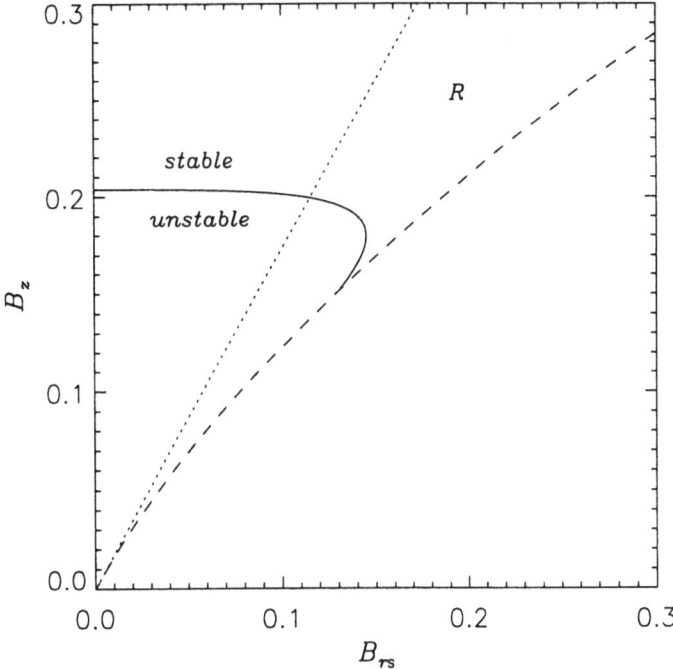

Fig. 3. Stability boundary in the parameter space for weakly magnetized discs with $\Gamma = 5/3$. The dotted line denotes an angle of inclination $\theta = 30°$. The dashed line is the critical curve for the existence of the principal solution branch. The solid line is the marginal curve, at $k = 0$, for the mode m_1^0, when $\gamma = \Gamma$. This is the overall stability boundary of the equilibria to axisymmetric perturbations. In region R, which extends indefinitely beyond the upper right of the figure, there exist equilibria which are capable of driving a wind and are also stable to the magnetorotational instability.

and Ohmic resistivity is required to allow a steady state in which the poloidal magnetic field slips outwards against its inward advection. A toroidal magnetic field is also needed to mediate the torque between the disc and the wind. Above the surface of the disc, the wind must be heated sufficiently to overcome the potential difference between the surface and the sonic point. For $z \gg H$, the flow is determined principally by the shape of the poloidal field lines until it reaches the Alfvén surface at a large distance from the disc. The most important feature of this analysis is that the basic disc solution of Section 2.2 and the exterior field solution of Section 2.4 are not affected at leading order by the additional terms. Only in the distant region at and beyond the Alfvén surface must a partial differential equation be solved for the wind.

The stability analysis described above, which of course applies only to the disc and not to any wind, is valid only for axisymmetric modes. It is expected that all models are subject to global non-axisymmetric instabilities which depend strongly on the radial boundaries of the disc. The equilibria described here as 'stable' here may therefore not exist as truly laminar flows but could have weak turbulence or at least fluctuations. However, that would be quite different from equilibria that are locally unstable to the magnetorotational instability, which are bound to degenerate into strong MHD turbulence.

Acknowledgements: I am grateful to Jim Pringle for helpful discussions. A research studentship from the Particle Physics and Astronomy Research Council is acknowledged.

References

Balbus S. A., Hawley J. F., 1991, Astrophys. J. 376, 214
Blandford R. D., Payne D. G., 1982, Mon. Not. R. Astr. Soc. 199, 883
Brandenburg A., Nordlund Å., Stein R. F., Torkelsson U., 1995, Astrophys. J. 446, 741
Brandenburg A., Nordlund Å., Stein R. F., Torkelsson U., 1996, Astrophys. J. 458, L45
Chandrasekhar S., 1960, Proc. Natl Acad. Sci. 46, 253
Curry C., Pudritz R. E., 1995, Astrophys. J. 453, 697
Curry C., Pudritz R. E., Sutherland P. G., 1994, Astrophys. J. 434, 206
Foglizzo T., Tagger M., 1995, Astron. Astrophys. 301, 293
Gammie C. F., Balbus S. A., 1994, Mon. Not. R. Astr. Soc. 270, 138
Hawley J. F., Gammie C. F., Balbus S. A., 1995, Astrophys. J. 440, 742
Hawley J. F., Gammie C. F., Balbus S. A., 1996, Astrophys. J. 464, 690
Heinemann M., Olbert S., 1978, J. Geophys. Res. 83, 2457
Heyvaerts J., Norman C., 1989, Astrophys. J. 347, 1055
Heyvaerts J. F., Priest E. R., 1989, Astron. Astrophys. 216, 230
Kippenhahn R., Schlüter A., 1957, Z. Astrophys. 43, 36
Königl A., 1989, Astrophys. J. 342, 208
Korycansky D. G., Pringle J. E., 1995, Mon. Not. R. Astr. Soc. 272, 618
Li Z.-Y., 1995, Astrophys. J. 444, 848
Lubow S. H., Pringle J. E., 1993, Astrophys. J. 409, 360
Lubow S. H., Papaloizou J. C. B., Pringle J. E., 1994, Mon. Not. R. Astr. Soc. 207, 235
Mestel L., 1968, Mon. Not. R. Astr. Soc. 138, 359
Ogilvie G. I., Pringle J. E., 1996, Mon. Not. R. Astr. Soc. 279, 152
Ogilvie G. I., 1997a, Mon. Not. R. Astr. Soc. (submitted) ('Paper I')
Ogilvie G. I., 1997b, Mon. Not. R. Astr. Soc. (in preparation) ('Paper II')
Papaloizou J. C. B., Szuszkiewicz E., 1992, Geophys. Astrophys. Fluid Dyn. 66, 223
Pringle J. E., 1981, Annu. Rev. Astron. Astrophys. 19, 137
Shu F. H., 1991, in The Physics of Star Formation and Early Stellar Evolution, C. J. Lada, N. D. Kylafis (eds), Kluwer, Dordrecht, 365
Stone J. M., Hawley J. F., Gammie C. F., Balbus S. A., 1996, Astrophys. J. 463, 656
Velikhov E. P., 1959, Sov. Phys. JETP 9, 995
Wardle M., Königl A., 1993, Astrophys. J. 410, 218

Magneto-viscous accretion discs

C.G. Campbell

Department of Mathematics, University of Newcastle upon Tyne, NE1 7RU, UK

Abstract: Magnetic models of accretion discs are presented and discussed. The magnetic field is generated and maintained in the disc by a dynamo mechanism. The relative magnitudes of the azimuthal viscous and magnetic forces depend on the diffusion mechanism considered. With buoyancy dominant, magnetic stresses control the radial advection of angular momentum. With buoyancy and turbulent diffusion, viscous and magnetic advection of angular momentum become comparable. All the magnetic disc solutions have eigenstate structures, corresponding to differing vertical variations of the field.

1 Introduction

Accretion discs play a central role in stellar astrophysics. They occur around the compact components in close binary stars, around young stellar objects in T Tauri stars, and are also believed to exist in active galactic nuclei. They are usually the main luminosity source in such systems, derived from the gravitational binding energy released by infalling matter.

The fundamental problem in disc theory is to explain the angular momentum advection required to allow inflow of material to the accreting object. Standard molecular viscosity generates azimuthal forces which are far too weak to account for this advection. An anomalous viscosity must therefore be invoked, and the standard disc model due to Shakura and Sunyaev (1973) uses a parameterized form of turbulent viscosity. However, until recently, no instability was known that could generate the required turbulence. Balbus and Hawley (1991) showed that Keplerian discs are unstable in the presence of a weak poloidal magnetic field, and suggested that this could generate turbulence. Subsequent local numerical simulations (e.g. Brandenburg *et al* 1995; Hawley, Gammie and Balbus 1995) confirmed such turbulent generation. The work of Brandenburg *et al* (1995) indicates turbulence with a finite mean helicity, $\langle \mathbf{v}_\mathrm{T} \cdot \nabla \wedge \mathbf{v}_\mathrm{T} \rangle$, and hence an α-effect. This effect converts large-scale toroidal field to poloidal field. The differential rotation in the disc shears radial field to generate toroidal field, this being the ω-effect. The large-scale field resulting from this $\alpha\omega$-dynamo leads to an azimuthal force, $F_{\mathrm{m}\phi}$, and hence to radial advection of angular momentum. The turbulence also leads to a viscous force, $F_{\mathrm{v}\phi}$.

The purpose of this paper is to investigate the relative magnitudes of $F_{\mathrm{m}\phi}$ and $F_{\mathrm{v}\phi}$ for self-consistent disc models. The nature of the magnetic diffusivity, η, plays a central role in the type of solution. Section 2 presents the magnetic disc equations. Section 3 contains a disc solution in which $F_{\mathrm{m}\phi}$ dominates $F_{\mathrm{v}\phi}$, while more general solutions are discussed in Section 4. The results are summarized in Section 5.

2 The Magnetic Disc Equations

An axisymmetric, steady disc is considered around a non-magnetic accreting star of mass M and radius R. Cylindrical polar coordinates (ϖ, ϕ, z) are used, with the origin at the centre of the star and the central plane corresponding to $z = 0$. The external medium is taken to be a vacuum.

The structure of the magnetic field must lead to an azimuthal force on rings of disc material, for net angular momentum advection. For a vacuum exterior, B_ϕ vanishes on the disc surfaces and it follows from (4) that a net $B_\varpi B_\phi$ stress is necessary for a finite magnetic torque. A quadrupolar-type field is the most easily generated in such a disc (e.g. Campbell 1997) and has the properties

$$|B_\varpi / B_\phi| \ll 1, \quad |B_\varpi / B_z| \sim \varpi / h \gg 1, \tag{1}$$

where $h(\varpi)$ is the disc height and the first relation follows for an $\alpha\omega$-dynamo, while the second is a consequence of $\nabla \cdot \mathbf{B}_p = 0$.

The two small quantities $|B_\varpi / B_\phi|$ and h/ϖ allow the fundamental equations to be reduced to ;

$$v_\phi = v_K = \left(\frac{GM}{\varpi}\right)^{\frac{1}{2}}, \tag{2}$$

$$\frac{z}{\varpi}\frac{v_K^2}{\varpi} + \frac{1}{\rho}\frac{\partial}{\partial z}\left(P + \frac{B_\phi^2}{2\mu_0}\right) = 0, \tag{3}$$

$$v_\varpi \frac{\partial}{\partial \varpi}(\varpi^2 \Omega) = \frac{1}{\mu_0 \varpi \rho}\frac{\partial}{\partial \varpi}(\varpi^2 B_\varpi B_\phi) + \frac{\varpi}{\mu_0 \rho}\frac{\partial}{\partial z}(B_z B_\phi) + \frac{1}{\varpi \rho}\frac{\partial}{\partial \varpi}\left(\rho \nu \varpi^3 \frac{\partial \Omega}{\partial \varpi}\right), \tag{4}$$

$$\eta \frac{\partial B_\varpi}{\partial z} = \alpha B_\phi, \tag{5}$$

$$\eta \frac{\partial^2 B_\phi}{\partial z^2} = -\varpi R_\varpi \frac{d\Omega_K}{d\varpi}, \tag{6}$$

$$\frac{1}{\varpi}\frac{\partial}{\partial \varpi}(\varpi \rho v_\varpi) + \frac{\partial}{\partial z}(\rho v_z) = 0, \tag{7}$$

with,

$$\nu = \epsilon^{\frac{3}{2}} c_s h, \tag{8}$$

$$\eta = \eta_B + \eta_T = \frac{\xi |B_{\phi c}|}{(\mu_0 \rho_c)^{\frac{1}{2}}} h + \bar{\epsilon} c_s h, \tag{9}$$

$$\alpha = \begin{cases} \tilde{\alpha}(\varpi), & 0 < z < h, \\ 0, & z = 0, \\ -\tilde{\alpha}(\varpi), & -h < z < 0, \end{cases} \tag{10}$$

where

$$\tilde{\alpha} = \epsilon \left(\frac{P_c}{\rho_c} \right)^{\frac{1}{2}}.$$ (11)

Equations (2)–(4) are the ϖ, z and ϕ-components of the momentum equation, (5) and (6) are the poloidal and toroidal components of the induction equation, and (7) is the continuity equation. The magnetic diffusivity, η, is due to buoyancy and turbulence, with $\xi < 1$ due to turbulent reconnection, and a simple form is taken for the turbulent α function, with $\epsilon < 1$.

It is noted that the viscosity coefficient ν, given by (8), contains a simple modification due to the effect of rotation. The Keplerian time of $\tau_K \sim 1/\Omega_K$ is shorter than the convective time $\tau_c \sim h/v_T$, where v_T is the rms turbulent speed. The mixing length h and v_T are each multiplied by the ratio τ_K/τ_c. It follows, using (11), that

$$\epsilon = \left(\frac{\tau_K}{\tau_c} \right)^2 = \left(\frac{v_T}{c_s} \right)^2,$$ (12)

and since the rotationally modified viscosity is

$$\nu = \left(\frac{\tau_K}{\tau_c} \right)^2 v_T h,$$ (13)

equation (8) follows.

3   A Magnetically-Controlled Disc

3.1   The reduced equations

The case of $\eta_B \gg \eta_T$ was considered in Campbell (1992). The magnetic diffusivity is

$$\eta = \frac{\xi |B_{\phi c}|}{(\mu_0 \rho_c)^{\frac{1}{2}}} h.$$ (14)

It can then be shown that the azimuthal force ratio is

$$F_{v\phi}/F_{m\phi} \sim \epsilon^{\frac{1}{2}} \xi^2 \ll 1,$$ (15)

The vertical equilibrium and azimuthal momentum equations reduce to

$$\frac{z}{\varpi} \frac{v_K^2}{\varpi} + \frac{1}{\rho} \frac{\partial P}{\partial z} = 0,$$ (16)

$$\varpi \rho v_\varpi \frac{\partial}{\partial \varpi} (\varpi^2 \Omega) = \frac{1}{\mu_0} \frac{\partial}{\partial \varpi} (\varpi^2 B_\varpi B_\phi) + \frac{1}{\mu_0} \frac{\partial}{\partial z} (\varpi^2 B_\phi B_z).$$ (17)

3.2 Vertical integrals

Combining (7) and (17) yields the angular momentum advection equation

$$\frac{\partial}{\partial\varpi}(\varpi\rho v_\varpi\varpi^2\Omega) = \frac{1}{\mu_0}\frac{\partial}{\partial\varpi}(\varpi^2 B_\varpi B_\phi) + \frac{1}{\mu_0}\frac{\partial}{\partial z}(\varpi^2 B_\phi B_z). \qquad (18)$$

A vacuum exterior leads to the surface conditions

$$B_\phi(\varpi,\pm h) = \rho(\varpi,\pm h) = 0. \qquad (19)$$

Vertical integration of (18) then gives

$$\frac{d}{d\varpi}\left(\frac{\dot{M}}{2\pi}\varpi^2\Omega_K + \frac{2\varpi^2}{\mu_0}\int_0^h B_\varpi B_\phi dz\right) = 0, \qquad (20)$$

where the mass inflow rate is

$$\dot{M} = -4\pi\int_0^h \varpi\rho v_\varpi dz. \qquad (21)$$

If Ω turns over at the edge of a boundary layer of width δ near the accretor, then $B_\varpi B_\phi = 0$ can be applied at $\varpi = R+\delta$ and radial integration of (20) yields

$$\int_0^h B_\varpi B_\phi dz = -\frac{\mu_0(GM)^{\frac{1}{2}}\dot{M}}{4\pi\varpi^{\frac{3}{2}}}\left[1 - \left(\frac{R}{\varpi}\right)^{\frac{1}{2}}\right]. \qquad (22)$$

Equation (5) gives

$$B_\varpi B_\phi = \frac{\eta}{2\alpha}\frac{\partial}{\partial z}(B_\varpi^2). \qquad (23)$$

A quadrupolar-type magnetic field has the surface conditions

$$B_\phi(\varpi, h) = B_\varpi(\varpi, h) = 0, \qquad (24)$$

the latter following from the thin disc approximation. Vertical integration of (23) then yields

$$\frac{\eta}{2\alpha}B_\varpi^2(\varpi,0) = \int_0^h B_\varpi B_\phi dz, \qquad (25)$$

and hence use of (22) gives

$$B_\varpi^2(\varpi,0) = \frac{\mu_0(GM)^{\frac{1}{2}}\dot{M}\tilde{\alpha}}{2\pi\varpi^{\frac{3}{2}}\eta}\left[1 - \left(\frac{R}{\varpi}\right)^{\frac{1}{2}}\right]. \qquad (26)$$

Vertical integration of (16) yields

$$P_c = \Omega_K^2\int_0^h z\rho dz. \qquad (27)$$

The thermal problem for an optically thick disc leads to

$$P_c^{\frac{15}{2}} = \frac{3\bar{K}}{4\sigma}\left(\frac{\mathcal{R}}{\mu}\right)^{\frac{15}{2}} hF_+\rho_c^{\frac{19}{2}},\tag{28}$$

where σ is the Stefan-Boltzmann constant, F_+ is the surface flux given by

$$F_+ = \frac{3GM\dot{M}}{8\pi\varpi^3}\left[1 - \left(\frac{R}{\varpi}\right)^{\frac{1}{2}}\right],\tag{29}$$

and $\bar{K}$ is a constant in the Kramer opacity

$$\kappa = \bar{K}\rho T^{-\frac{7}{2}}.\tag{30}$$

The gas equation of state is

$$P = \frac{\mathcal{R}}{\mu}\rho T,\tag{31}$$

where μ is the mean molecular weight.

3.3 The Dynamo

Combining (5) and (6) to eliminate B_ϖ gives

$$\frac{\partial^3 B_\phi}{\partial z^3} + \frac{\varpi\Omega'_K\tilde{\alpha}}{\eta^2}B_\phi = 0.\tag{32}$$

For a thin disc, the azimuthal magnetic field can be expressed in the separable form

$$B_\phi = \tilde{B}_\phi(\varpi)f_\phi(z/h).\tag{33}$$

Substitution in (32) leads to

$$f_\phi'''(\zeta) + Nf_\phi(\zeta) = 0,\tag{34}$$

where $\zeta = z/h$ and the dynamo number

$$N = \frac{\varpi\Omega'_K h^3\tilde{\alpha}}{\eta^2}.\tag{35}$$

The quadrupolar boundary conditions require

$$f_\phi''(1) = f_\phi(1) = f_\phi'(0) = 0.\tag{36}$$

The solution of (34) satisfying these conditions is

$$f_\phi = A\left[\exp[-K(\zeta-1)] - 2\exp\left[\frac{K}{2}(\zeta-1)\right]\cos\left(\frac{\sqrt{3}K}{2}(\zeta-1) - \frac{\pi}{3}\right)\right],\tag{37}$$

where A is a constant and the eigenvalues K satisfy

$$2\cos\left(\frac{\sqrt{3}K}{2}\right) + \exp\left(\frac{3K}{2}\right) = 0, \tag{38}$$

with $K^3 = N$.

The ratio of the field components follows from (6) and (33) as

$$\frac{B_\phi}{B_\varpi} = \frac{3\Omega_K h^2}{2\eta}\frac{f_\phi}{f_\phi''}. \tag{39}$$

Use of (35) then gives

$$B_\phi(\varpi,0) = |K|^3 \frac{f_\phi(0)}{f_\phi''(0)}\frac{\eta}{\tilde{\alpha}h}B_\varpi(\varpi,0), \tag{40}$$

where

$$\frac{f_\phi(0)}{f_\phi''(0)} = \frac{\exp K + 2\sqrt{3}\exp\left(\frac{-K}{2}\right)\sin\frac{\sqrt{3}}{2}K}{K^2\left[\exp K - 2\sqrt{3}\exp\left(\frac{-K}{2}\right)\sin\frac{\sqrt{3}}{2}K\right]}. \tag{41}$$

3.4   The Disc Solution

Equations (2), (26)–(35) and (40) can be used to derive the radial structure of the disc. The resulting solution is;

$$h(\varpi) = 2.3\times10^6\frac{\xi^{\frac{1}{5}}|K|^{\frac{3}{4}}}{\epsilon^{\frac{3}{20}}}\left|\frac{f_\phi(0)}{f_\phi''(0)}\right|^{\frac{1}{5}}\frac{\dot{M}_{-10}^{\frac{3}{20}}}{M_1^{\frac{3}{8}}}\varpi_8^{\frac{9}{8}}f^{\frac{3}{20}}\text{ m}, \tag{42}$$

$$B_\varpi(\varpi) = 6.6\times10^{-2}\frac{\epsilon^{\frac{13}{40}}|K|^{\frac{3}{8}}}{\xi^{\frac{1}{10}}}\left|\frac{f_\phi(0)}{f_\phi''(0)}\right|^{-\frac{1}{10}}M_1^{\frac{7}{16}}\dot{M}_{-10}^{\frac{17}{40}}\frac{f^{\frac{17}{40}}}{\varpi_8^{\frac{21}{16}}}\text{ T} \tag{43}$$

$$B_\phi(\varpi) = -9.5\times10^{-2}\frac{|K|^{\frac{15}{8}}}{\xi^{\frac{1}{10}}\epsilon^{\frac{7}{40}}}\left|\frac{f_\phi(0)}{f_\phi''(0)}\right|^{\frac{9}{10}}M_1^{\frac{7}{16}}\dot{M}_{-10}^{\frac{17}{40}}\frac{f^{\frac{17}{40}}}{\varpi_8^{\frac{21}{16}}}\text{ T}, \tag{44}$$

$$\rho(\varpi) = 9.3\times10^{-6}\frac{\xi^{\frac{7}{5}}|K|^{\frac{21}{4}}}{\epsilon^{\frac{21}{20}}}\left|\frac{f_\phi(0)}{f_\phi''(0)}\right|^{\frac{7}{5}}M_1^{\frac{5}{8}}\dot{M}_{-10}^{\frac{11}{20}}\frac{f^{\frac{11}{20}}}{\varpi^{\frac{15}{8}}}\text{ Kg m}^{-3}, \tag{45}$$

$$P(\varpi) = 3.4\times10^3\frac{\xi^{\frac{9}{5}}|K|^{\frac{27}{4}}}{\epsilon^{\frac{27}{20}}}\left|\frac{f_\phi(0)}{f_\phi''(0)}\right|^{\frac{9}{5}}M_1^{\frac{7}{8}}\dot{M}_{-10}^{\frac{17}{20}}\frac{f^{\frac{17}{20}}}{\varpi_8^{\frac{21}{8}}}\text{ N m}^{-2}, \tag{46}$$

$$T(\varpi) = 2.7\times10^4\frac{\xi^{\frac{2}{5}}|K|^{\frac{3}{2}}}{\epsilon^{\frac{3}{10}}}\left|\frac{f_\phi(0)}{f_\phi''(0)}\right|^{\frac{2}{5}}M_1^{\frac{1}{4}}\dot{M}_{-10}^{\frac{3}{10}}\frac{f^{\frac{3}{10}}}{\varpi_8^{\frac{3}{4}}}\text{ K}, \tag{47}$$

$$v_\varpi(\varpi) = -2.3 \times 10^2 \frac{\epsilon^{\frac{8}{5}}}{\xi^{\frac{8}{5}}|K|^6} \left| \frac{f_\phi(0)}{f''_\phi(0)} \right|^{-\frac{8}{5}} \frac{\dot{M}_{-10}^{\frac{3}{10}}}{M_1^{\frac{1}{4}}} \frac{1}{\varpi_8^{\frac{1}{4}} f^{\frac{7}{10}}} \text{ m s}^{-1}, \qquad (48)$$

where $M_1 = M/M_\odot$, $\dot{M}_{-10} = \dot{M}/10^{-10} M_\odot \, \text{yr}^{-1}$, $\varpi_8 = \varpi/10^8$ m and

$$f = 1 - \left(\frac{R}{\varpi} \right)^{\frac{1}{2}}. \qquad (49)$$

3.5 Discussion

The dimensionless quantity ϵ expresses the strength of the turbulence. Self-consistent solutions are possible for ϵ as small as 10^{-4} which, from (12), corresponds to $v_T \sim 10^{-2} c_s$ and hence to weak turbulence. Although the radial structure is similar to the standard viscous disc, the viscous force is negligible in the present case. The eigenvalue dependence of the magnetic disc allows different states. Typical values of B_ϕ in the central plane are ~ 100 G, these being weakly dependent on K. However, the central temperature increases with $|K|$. This property of the magnetic disc may be able to be related to the outburst behaviour of discs in binary stars.

4 Magneto-Viscous Discs

The foregoing disc solution is valid for $\eta_B \gg \eta_T$, leading to the magnetic stress dominating in the radial advection of angular momentum. The more general case with $\eta_B \sim \eta_T$ was considered by Campbell and Caunt (1997). Then η is given by (9). The radial magnetic force is still small, so the disc remains Keplerian up to an inner boundary layer at the stellar surface. The vertical and azimuthal momentum equations now become

$$\frac{z}{\varpi} \frac{v_K^2}{\varpi} + \frac{1}{\rho} \frac{\partial}{\partial z} \left(P + \frac{B_\phi^2}{2\mu_0} \right) = 0, \qquad (50)$$

$$v_\varpi \frac{\partial}{\partial \varpi} (\varpi^2 \Omega) = \frac{1}{\mu_0 \varpi \rho} \frac{\partial}{\partial \varpi} (\varpi^2 B_\varpi B_\phi) + \frac{\varpi}{\mu_0 \rho} \frac{\partial}{\partial z} (B_\phi B_z) + \frac{1}{\varpi \rho} \frac{\partial}{\partial \varpi} \left(\rho \nu \varpi^3 \frac{\partial \Omega}{\partial \varpi} \right), \qquad (51)$$

noting $\partial \Omega / \partial z = 0$. The continuity and induction equations are unchanged.

Combining the continuity and azimuthal momentum equations and integrating vertically through the disc, using the surface conditions (19), yields

$$\frac{d}{d\varpi} \left(\frac{\dot{M}}{2\pi} \varpi^2 \Omega + \nu \Sigma \varpi^3 \Omega' + \frac{2\varpi^2}{\mu_0} \int_0^h B_\varpi B\phi dz \right) = 0. \qquad (52)$$

Radial integration and application of the standard boundary layer condition then gives

$$\frac{\dot{M}}{2\pi}\left[\varpi^2\Omega_K - (GMR)^{\frac{1}{2}}\right] + \nu\Sigma\varpi^3\Omega_K' + \frac{2\varpi^2}{\mu_0}\int_0^h B_\varpi B_\phi dz = 0, \qquad (53)$$

valid for $\varpi > R+\delta$, where δ is the boundary layer width. The vertical equilibrium (50) yields

$$P_c + \frac{B_{\phi c}^2}{2\mu_0} = \Omega_K^2 \int_0^h z\rho dz. \qquad (54)$$

Since viscous dissipation is now significant, the surface heat flux becomes

$$F_+ = \frac{\eta}{\mu_0}\int_0^h \left(\frac{\partial B_\phi}{\partial z}\right)^2 dz + \frac{1}{2}\nu\Sigma(\varpi\Omega_K')^2. \qquad (55)$$

Use of (5), (53) and (55) gives (29) for F_+. Equation (28) relating P_c to ρ_c also still holds.

The vertical dynamo problem remains essentially unchanged, except that η is now given by (9). Equation (35), with $N = K^3$, yields

$$\eta^2 = (\eta_T + \eta_B)^2 = \frac{3\Omega_K h^3 \tilde{\alpha}}{2|K|^3}. \qquad (56)$$

Then, noting from (9) and (11)

$$\tilde{\alpha}h = \epsilon c_s h = \left(\frac{\epsilon}{\bar{\epsilon}}\right)\bar{\epsilon}c_s h = \frac{\epsilon}{\bar{\epsilon}}\eta_T, \qquad (57)$$

it follows that

$$(\eta_T + \eta_B)^2 = \frac{3}{2|K|^3}\frac{\epsilon}{\bar{\epsilon}}\Omega_K h^2 \eta_T. \qquad (58)$$

Using $\langle z\rho \rangle \simeq h\rho_c/2$ in (54), gives

$$\frac{P_c}{\rho_c} + \frac{B_{\phi c}^2}{2\mu_0\rho_c} = \frac{1}{2}h^2\Omega_K^2. \qquad (59)$$

Since $c_s^2 = P_c/\rho_c$, this yields

$$\frac{1}{\bar{\epsilon}^2}\eta_T^2 + \frac{1}{2\xi^2}\eta_B^2 = \frac{1}{2}h^4\Omega_K^2. \qquad (60)$$

Dimensionless diffusivities can be defined by

$$\bar{x} = \frac{\eta_T}{h^2\Omega_K}, \qquad \bar{y} = \frac{\eta_B}{h^2\Omega_K}. \qquad (61)$$

Equations (58) and (60) then become

$$(\bar{x} + \bar{y})^2 = \frac{3}{2|K|^3}\frac{\epsilon}{\bar{\epsilon}}\bar{x}, \qquad (62)$$

$$\frac{2}{\bar{\epsilon}^2}\bar{x}^2 + \frac{1}{\xi^2}\bar{y}^2 = 1. \tag{63}$$

It is noted that the coefficients in these equations are constants and hence $\bar{x}$ and $\bar{y}$ are spatially independent. Eliminating $(\bar{x} + \bar{y})$ and then $\bar{x}$ yields the quartic equation

$$4(\bar{\epsilon}^2 + 2\xi^2)^2\bar{y}^4 - 48(\bar{\epsilon}\epsilon\xi^2/|K|^3)\bar{y}^3 + \left(9\epsilon^2\xi^2/2K^6 - 2\bar{\epsilon}^2\xi^2(\bar{\epsilon}^2 + 2\xi^2)\right)\bar{y}^2$$
$$+12(\bar{\epsilon}\epsilon\xi^4/|K|^3)\bar{y} + \left(\bar{\epsilon}^4\xi^4 - 9\epsilon^2\xi^4/2K^6\right) = 0. \tag{64}$$

After solving this for $\bar{y}$, equation (63) gives $\bar{x}$. It follows from (61) that the total diffusivity is

$$\eta = (\bar{x} + \bar{y})\Omega_{\rm K}h^2. \tag{65}$$

The foregoing equations can be solved to obtain the radial disc structure. The solution is similar to (42)–(49), except with coefficients containing $\bar{x}(|K|, \epsilon, \bar{\epsilon}, \xi)$ and $\bar{y}(|K|, \epsilon, \bar{\epsilon}, \xi)$. Solutions are now possible with $F_{v\phi} \sim F_{m\phi}$, so the viscosity and magnetic stress make similar contributions to the radial advection of angular momentum.

5 Summary

Magnetic fields play a dual role in accretion discs. They generate the turbulence necessary for a self-sustaining dynamo, and large-scale fields lead to radial advection of angular momentum. It is particularly interesting that only weak turbulence ($v_{\rm T} \sim 10^{-2}$) is needed to generate the required poloidal field source leading to toroidal field creation and hence to B_pB_ϕ stress.

The similarity of the radial structure of the magnetic disc to that of the viscous disc is noted. This arises because of the Keplerian rotation and the similar thermal problems. However, the eigenstate structure is a distinct feature of the magnetic disc. The possible connection between the different disc states and outbursts in cataclysmic variable discs needs investigation.

References

Balbus, S.A. and Hawley, J.F., 1991. *Astrophys. J.*, **376**, 214.
Brandenburg, A., Nordlund, A., Stein, R.F. and Torkelsson, U., 1995. *Astrophys. J.*, **446**, 741.
Campbell, C.G., 1992. *Geophys. Astrophys. Fluid. Dynam.*, **63**, 197.
Campbell, C.G., 1997. *Magnetohydrodynamics in Binary Stars*, Kluwer Academic Publishers, in press.
Campbell, C.G. and Caunt, S.E., 1997. *Mon. Not. R. Astr. Soc.*, submitted.
Hawley, J.F., Gammie, C.F. and Balbus, S.A., 1995. *Astrophys. J.*, **440**, 743.
Shakura, N.I. and Sunyaev, R.A., 1973. *Astron. Astrophys.*, **24**, 337.

Precessing warped discs in close binary systems

J.C.B. Papaloizou[1], J.D. Larwood[1], R.P. Nelson[1], C. Terquem[1,2]

[1] Astronomy Unit, School of Mathematical Sciences, Queen Mary and Westfield College, Mile End Road, London E1 4NS

[2] Laboratoire d'Astrophysique, Observatoire de Grenoble,
Université Joseph Fourier/CNRS, BP 53X, 38041 Grenoble Cedex, France

Abstract: We describe some recent nonlinear three dimensional hydrodynamic simulations of accretion discs in binary systems where the orbit is circular and not necessarily coplanar with the disc midplane.

The calculations are relevant to a number of observed astrophysical phenomena, including the precession of jets associated with young stars, the high spectral index of some T Tauri stars, and the light curves of X–ray binaries such as Hercules X-1 which suggest the presence of precessing accretion discs.

1   Introduction

Protostellar discs appear to be common around young stars. Furthermore recent studies show that almost all young stars associated with low mass star forming regions are in multiple systems (Mathieu, 1994 and references therein). Typical orbital separations are around 30 astronomical units (Leinert et al. 1993) which is smaller than the characteristic disc size observed in these systems (Edwards et al. 1987). It is therefore expected that circumstellar discs will be subject to strong tidal effects due to the influence of binary companions.

The tidal effect of an orbiting body on a differentially rotating disc has been well studied in the context of planetary rings (Goldreich and Tremaine, 1981), planetary formation, and generally interacting binary stars (see Lin and Papaloizou, 1993 and references therein). In these studies, the disc and orbit are usually taken to be coplanar (see Artymowicz and Lubow, 1994). However, there are observational indications that discs and stellar orbits may not always be coplanar (see for example Corporon, Lagrange and Beust, 1996 and Bibo, The and Dawanas, 1992.)

In addition, reprocessing of radiation from the central star by a warped non coplanar disc has been suggested in order to account for the high spectral index of some T Tauri stars (Terquem and Bertout 1993, 1996).

A dynamical study of the tidal interactions of a non coplanar disc is of interest not only in the above contexts, but also in relation to the possible existence of precessing discs which may define the axes for observed jets which apparently precess (Bally and Devine, 1994).

Various studies of the evolution of warped discs have been undertaken assuming that the forces producing the warping were small so that linear perturbation theory could be used (Papaloizou and Pringle, 1983, Papaloizou and Lin, 1995 and Papaloizou and Terquem, 1995). The results suggested that the disc would precess approximately as a rigid body if the sound crossing time was smaller than the differential precession frequency.

We describe here some recent non linear simulations of discs which are not coplanar with the binary orbits using a Smoothed Particle Hydrodynamics (SPH) code originally developed by Nelson and Papaloizou (1993, 1994). We study the conditions under which warped precesing discs may survive in close binary systems and the truncation of the disc size through tidal effects when the disc and binary orbit are not coplanar. The simulations indicate that the phenomenon of tidal truncation is only marginally affected by lack of coplanarity. Also our model discs were able to survive in a tidally truncated condition while warped and undergoing rigid body precession provided that the Mach number in the disc was not too large. The inclination of the disc was found to evolve on a long timescale likely to be the viscous timescale, as was indicated by the linear calculations of Papaloizou and Terquem (1995).

2   Basic equations

The equations of continuity and of motion applicable to a gaseous viscous disc may be written

$$\frac{d\rho}{dt} + \rho \mathbf{\nabla} \cdot \mathbf{v} = 0, \tag{1}$$

$$\frac{d\mathbf{v}}{dt} = -\frac{1}{\rho} \mathbf{\nabla} P - \mathbf{\nabla} \Psi + \mathbf{S}_{visc} \tag{2}$$

where

$$\frac{d}{dt} \equiv \frac{\partial}{\partial t} + \mathbf{v} \cdot \mathbf{\nabla}$$

denotes the convective derivative, ρ is the density, $\mathbf{v}$ the velocity and P the pressure. The gravitational potential is Ψ, and $\mathbf{S}_{visc}$ is the viscous force per unit mass.

For the work described here, we adopt the polytropic equation of state

$$P = K\rho^{\gamma}$$

where

$$c_s^2 = K\gamma\rho^{\gamma-1}$$

gives the usual associated sound speed, c_s. Here we take $\gamma = 5/3$, and K is the polytropic constant. This corresponds to adopting a fluid that remains isentropic throughout even though viscous dissipation may occur. This means that an efficient cooling mechanism is assumed.

3   Orbital configuration

We consider a binary system in which the primary has a mass M_p and the secondary has a mass M_s. The binary orbit is circular with separation D. The orbital angular velocity is ω. We suppose that a disc orbits about the primary such that at time $t = 0$ it has a well defined mid-plane. We adopt a non rotating Cartesian coordinate system (x, y, z) centred on the primary star and we denote the unit vectors in each of the coordinate directions by $\hat{i}$, $\hat{j}$ and $\hat{k}$ respectively. The z axis is chosen to be normal to the initial disc mid-plane. We shall also use the associated cylindrical polar coordinates (r, φ, z).

We take the orbit of the secondary to be in a plane which has an initial inclination angle δ with respect to the (x, y) plane. For a disc of negligible mass, the plane of the orbit is invariable and does not precess. We denote the position vector of the secondary star by $\mathbf{D}$ with $D = |\mathbf{D}|$. Adopting an orientation of coordinates and an origin of time such that the line of nodes coincides with, and the secondary is on, the x axis at $t = 0$, the vector $\mathbf{D}$ is given as a function of time by

$$\mathbf{D} = D \cos \omega t \, \hat{i} + D \sin \omega t \cos \delta \, \hat{j} + D \sin \omega t \sin \delta \, \hat{k}. \qquad (3)$$

The total gravitational potential Ψ_{ext} due to the binary pair at a point with position vector $\mathbf{r}$ is given by

$$\Psi_{ext} = -\frac{GM_p}{|\mathbf{r}|} - \frac{GM_s}{|\mathbf{r} - \mathbf{D}|} + \frac{GM_s \mathbf{r} \cdot \mathbf{D}}{D^3}$$

where G is the gravitational constant. The first dominant term is due to the primary, while the remainder, $\equiv \Psi'_{ext}$, gives perturbing terms due to the secondary. Of these, the last indirect term accounts for the acceleration of the origin of the coordinate system. We note that a disc perturbed by a secondary on an inclined orbit becomes tilted, precesses and so does not maintain a fixed plane. Our calculations presented below are referred to the Cartesian system defined above through the initial disc mid-plane. However, we shall also use a system defined relative to the fixed orbital plane for which the 'x axis' is as in the previous system and the 'z axis' is normal to the orbital plane. If the disc were a rigid body its angular momentum vector would precess uniformly about this normal, as indicated below.

3.1   Disc response

The form of the disc response to the perturbing gravitational potential due to the secondary is determined by the properties of the free modes of oscillation. These are divided into two classes according to whether the associated density perturbation has even or odd symmetry with respect to reflection in the unperturbed disc midplane. The modes with even symmetry are excited when the binary orbit and disc are coplanar and have been well studied in the context of angular momentum exchange between disc and binary leading to tidal truncation of and wave excitation in the disc (see Lin and Papaloizou, 1993 and references therein). They are also excited at a somewhat reduced level in the non coplanar case where they produce similar effects.

Here we shall focus attention on the modes with odd symmetry and with azimuthal mode number $m = 1$. These are only excited in the non coplanar case. They are of interest because they are responsible for disc warping, twisting and precession.

3.2   Potential expansion

When the orbital separation is much larger then the outer disc radius such that for any disc particle, $r \ll D$, and $z \ll D$, we can expand Ψ'_{ext} in powers of r/D and z/D.

We are interested in bending modes which are excited by terms in the potential which are odd in z and which have azimuthal mode number $m = 1$ when a Fourier analysis in φ is carried out. The lowest order terms in the expansion of the potential which are of the required form are given by

$$\Psi'_{ext} = -\tfrac{3}{4}\tfrac{GM_s}{D^3}rz\left[(1 - \cos\delta)\sin\delta\sin(\varphi + 2\omega t)\right.$$
$$- (1 + \cos\delta)\sin\delta\sin(\varphi - 2\omega t)$$
$$\left. + \sin 2\delta\sin(\varphi)\right] \tag{4}$$

In linear perturbation theory, we can calculate the response of the disc to each of the three terms in Ψ'_{ext} separately and superpose the results. The general problem is then to calculate the response due to a potential of the form

$$\Psi'_{ext} = \text{const} \times rz\sin(\varphi - \Omega_P t)$$

or in complex notation

$$\Psi'_{ext} = \mathcal{R}\left(frze^{im(\varphi - \Omega_P t)}\right) \tag{5}$$

Here, the azimuthal mode number, $m = 1$, the pattern frequency Ω_P of the perturber is one of 0, 2ω or -2ω and f is the appropriate complex amplitude. We remark that the magnitude of the tidal perturbation acting on the disc is measured by the dimensionless quantity $GM_s/(\Omega^2 D^3)$, Ω being the angular velocity in the disc and for comparable primary and secondary masses this is of order ω^2/Ω^2.

4   Free bending modes

Bending waves with $m = 1$ are naturally excited by a perturbing potential of the form given by (5). For a binary with large orbital radius, we may consider the pattern frequency Ω_P to be small compared to the rotation frequency in the disc, Ω. Thus the forcing is at low frequency.

Bending waves in thin discs have been studied in the context of disc galaxies (Hunter and Toomre, 1969), planetary rings (Shu,1984) and gaseous accretion discs (Papaloizou and Lin, 1995, Papaloizou and Terquem, 1995).

In a self-gravitating disc with no pressure and of small enough mass that the unperturbed disc is in a state of near Keplerian rotation, the local dispersion relation for bending waves with $m = 1$ is given by (Hunter and Toomre, 1969)

$$(\Omega_P - \Omega)^2 = \Omega^2 + 2\pi G\Sigma|k|, \tag{6}$$

where Σ is the disc surface density. In the limit of small pattern speeds this takes the form

$$\Omega_P\Omega = -\pi G\Sigma|k|, \tag{7}$$

from which it follows that the waves propagate without dispersion with speed

$$c_g = \frac{\pi G\Sigma}{\Omega} = \frac{\langle c_s \rangle}{Q}, \tag{8}$$

where the Toomre $Q = \Omega\langle c_s \rangle/(\pi G\Sigma)$, where the angled brackets denote an apropriate vertical mean of the sound speed. For stability to axisymmetric modes, we require $Q \geq 1$, which implies that bending waves in a stable disc propagate with a speed not exceeding the maximum sound speed at a particular radial location. Papaloizou and Lin (1995) considered the case when pressure is included, giving the corresponding local dispersion relation for bending waves with $m = 1$ in the low frequency limit

$$\Omega_P\Omega = -\pi G\Sigma|k| - \frac{(1-\Theta)\Omega\langle c_s \rangle^2|k|^2}{4\Omega_P}, \tag{9}$$

where $\Theta < 1$ is a dimensionless parameter which vanishes when self-gravity is unimportant. This gives a quadratic equation for Ω_P with the two roots

$$\Omega_P = -\frac{\langle c_s \rangle|k|}{2}\left(\frac{1}{Q}\pm\sqrt{\frac{1}{Q^2}+1-\Theta}\right). \tag{10}$$

In this case there are fast and slow waves which in the case of a stable disc propagate with speeds comparable to the sound speed. When self-gravity is unimportant there is a single sonic like wave. The excitation by the forcing potential (5) , and angular momentum transport associated with these waves with non zero Ω_P has been considered by Papaloizou and Terquem (1995).

The secular term in the forcing potential (4) with zero pattern speed causes the disc to be subject to a precessional torque. The properties of the bending waves determine the form of the disc response. For the disc to precess approximately like a rigid body, the effects of the precessional torque which acts largely in the outer parts of the disc, must be communicated to the inner regions where it is weakest, within a precession period. This roughly corresponds to the condition that the disc sound crossing time be less than the precession period.

5   Precession frequency

In order to calculate the precession frequency, we consider the time independent, or secular term in the perturbing potential (4) as it is only this term which produces a non zero net torque after performing a time average. This is given by

$$\Psi'_{ext0} = -\frac{3}{4}\frac{GM_s}{D^3}rz\sin 2\delta \sin(\varphi). \tag{11}$$

For a conservative system, the Lagrangian displacement vector $\boldsymbol{\xi}$ will satisfy an equation of the form (see Lynden-Bell and Ostriker 1967)

$$\mathbf{C}(\boldsymbol{\xi}) = -\nabla\Psi'_{ext0}. \tag{12}$$

Here $\mathbf{C}$ is a linear operator, which needs to be inverted to give the response. When a barotropic equation of state applies, and the boundaries are free, $\mathbf{C}$ is self-adjoint with weight ρ. This means that for two general displacement vectors $\boldsymbol{\xi}(\mathbf{r})$ and $\boldsymbol{\eta}(\mathbf{r})$ we have

$$\int_V \rho\boldsymbol{\eta}^* \cdot \mathbf{C}(\boldsymbol{\xi})\,d\tau = \left[\int_V \rho\boldsymbol{\xi}^* \cdot \mathbf{C}(\boldsymbol{\eta})\,d\tau\right]^* \tag{13}$$

where * denotes complex conjugate and the integral is taken over the disc volume V.

Because of the spherical symmetry of the unperturbed primary potential, the unperturbed system is invariant under applying a rigid tilt to the disc. This corresponds to the existence of rigid tilt mode solutions to (12) when there is no forcing ($\Psi'_{ext0} = 0$). For a rotation about the x axis, the rigid tilt mode is of the form

$$\boldsymbol{\xi} = \boldsymbol{\xi}_T = (\mathbf{r}\times\hat{\varphi})\sin\varphi - \hat{\varphi}z\cos\varphi \tag{14}$$

where $\hat{\varphi}$ is the unit vector along the φ direction. For time averaged forcing potentials $\propto \sin\varphi$, the existence of the solution (14) results in an integrability condition for (12). When $\mathbf{C}$ is self-adjoint, this is

$$\int_V \boldsymbol{\xi}_T \cdot \nabla\overline{\Psi}_s\rho\,d\tau \equiv \int_V \hat{\mathbf{i}} \cdot \left(\mathbf{r}\times\nabla\overline{\Psi}_s\right)\rho\,d\tau = 0. \tag{15}$$

The above condition is equivalent to the requirement that the x component of the external torque vanishes. This will clearly not be satisfied in the problem we consider.

To deal with this one may suppose that the disc angular momentum vector precesses about the orbital angular momentum vector with angular velocity ω_p (Papaloizou and Terquem 1995). This in turn is equivalent to supposing our coordinate system rotates with angular velocity ω_p about the orbital rotation axis. Treating the Coriolis force by perturbation theory produces an additional term on the right hand side of (12) equal to $-2r\Omega\omega_p\times\hat{\varphi}$. Using the modified force in formulating (15) gives the integrability condition as

$$\omega_p \sin \delta \int_V r^2 \Omega \rho d\tau = \int_V \hat{\mathbf{i}} \cdot \left(\mathbf{r} \times \nabla \overline{\Psi}_s \right) \rho d\tau, \qquad (16)$$

with $\omega_p = |\boldsymbol{\omega}_p|$. Equation (16) gives a precession frequency for the disc that would apply if it were a rigid body. However, approximate rigid body precession is only expected to occur if the disc is able to communicate with itself, either through wave propagation or viscous diffusion, on a timescale less than the inverse precession frequency (see for example Papaloizou and Terquem 1995 and below). Otherwise, a thin disc configuration may be destroyed by strong warping and differential precession.

We comment that the situation described above in which the external perturbation produces a precessional torque in the x direction only is a consequence of the assumption of a conservative response for which the density and potential perturbations are in phase. However, if dissipative processes are included, there will be a phase shift between the perturbing potential and density response. This will result in a net torque in the y direction which can change the angle between the disc and orbital angular momentum vectors (see Papaloizou and Terquem, 1995) possibly leading to their alignment. Such a process is likely to occur on the long dissipative timescale. A torque in the y direction originating from a disc wind has been proposed by Schandl and Meyer (1994) in order to produce misalignment between the disc and binary orbit angular momentum vectors in HZ-Hercules.

6   Numerical simulations

Three dimensional simulations of warped precessing discs in close binary systems have been carried out by solving the set of basic equations (1) and (2) numerically using an SPH code (Lucy 1977, Gingold and Monaghan 1977), developed by Nelson and Papaloizou (1994), which uses a conservative formulation of the method that employs variable smoothing lengths. A suite of test calculations illustrating the accurate energy conservation obtained with this method is described by Nelson and Papaloizou (1994), and additional tests and calculations are presented in Nelson (1994).

Larwood et al (1996) considered circumprimary discs in close binary systems with mass ratio of order unity and Larwood and Papaloizou (1997) have considered circumbinary discs in systems with a variety of mass ratios.

In order to stabilize the calculations in the presence of shocks, the artificial viscous pressure prescription of Monaghan and Gingold (1983) has been used in the simulations. This induces a shear viscosity which leads to angular momentum transport and the standard viscous evolution of an accretion disc (Lynden-Bell and Pringle 1974) in which disc expansion is produced by outward transport of angular momentum as mass flows inwards. In the studies presented here, the discs undergo angular momentum loss through tidal interaction with orbiting secondaries. Then disc expansion arising from outward transport of angular momentum is halted by tidal truncation. This effect, well known in the coplanar

case (see for example Lin and Papaloizou 1993), also occurs here when the disc
and binary angular momenta are not aligned.

In the simulations reported here the shear viscosity, ν operating in our disc
models was well fitted by a constant value. To specify the magnitude of ν we
write $\nu = \alpha c_s^2(R)/\Omega(R)$, where α corresponds to the well known Shakura and
Sunyaev (1973) α parameterization and R denotes the outer radius of the disc.
However, it is applicable only at the outside edge of the disc. The discs were here
set up with a distribution of 17500 particles such that the surface density was independent of radius. Then the aspect ratio H/r, with H being the semi-thickness
was found to be approximately independent of radius. As a consequence of this
the Mach number $\mathcal{M} = (H/R)^{-1}$ is also approximately constant for a particular
simulation and can be used to parameterize it. Also the radial dependence of
the viscosity parameter is $\alpha \propto r^{-1/2}$.

We comment that a characteristic value of $\alpha = 0.03$ that we have in the simulations is about two orders of magnitude larger than that expected to be associated
with tidally induced inwardly propagating waves (Spruit 1987, Papaloizou and
Terquem 1995). Accordingly, it is expected that tidal truncation will instead occur through strong nonlinear dissipation near the disc's outer edge (Savonije,
Papaloizou and Lin 1994). The large viscosity of the disc models considered here
will damp inwardly propagating waves before they can propagate very far.

In order to deal with the central regions of the disc, the primary's gravitational
potential softened such that

$$\Psi_p = -\frac{GM_p}{\sqrt{r^2 + b^2}},$$

where b is the softening length.

We adopt units such that the primary mass $M_p = 1$, the gravitational constant
$G = 1$, and the outer disc radius $R = 1$. In these units the adopted softening
parameter $b = 0.2$ and the time unit is $\Omega(R)^{-1}$. The self-gravity of the disc
material has been neglected in the simulations presented here.

7   Numerical results

We have considered the evolution of disc models set up according to the procedure outlined above. The models were characterized only by the mean Mach
number, $\mathcal{M}$. After a relaxation period of about two rotation periods at the outer
edge of the disc, the time was reset to zero and the secondary was introduced in
an inclined circular orbit, moving in a direct sense, crossing the x axis at $t = 0$.
In some cases the full secondary mass was included immediately corresponding
to a sudden start. However, for strong initial tidal interactions such as those
that occur when $D/R = 3, M_s/M_p = 1$, this can result in disruption of the outer
edge of the disc with a small number of particles being ejected from the disc. It
was found that this could be avoided by using a 'slow start' in which M_s was
built up gradually (see Larwood et al,1996). However, subsequent results were
found to be indepent of the initiation procedure.

In the above discussion of bending modes we indicated that the disc should be able to approximately precess as a rigid body if the sound crossing time was short compared to the characteristic precesion period.

The general finding from the simulations was that a disc with an initial angular momentum vector inclined to that of the binary system tended to precess approximately as a rigid body, with a noticeable but small warp if $\mathcal{M}$ was not too large. In such cases only small changes in the inclination angle between the angular momentum vectors were found over the run time. This is consistent with the expectation from Papaloizou and Terquem (1995) that the timescale for evolution of the inclination in such cases should be comparable to the viscous evolution timescale of the disc, assuming outward disc expansion is prevented by tidal interaction.

We here describe simulations of three circumprimary disc models in a close binary system with $D/R = 3$ and $\delta = \pi/4$ initially. Models 1 ,2 and 3 had $\mathcal{M}$ equal to 20, 25 and 30 respectively and the total run times for these models initiated with a slow start was 310, 217.7, and 397.9 units respectively. Note that the viscosity is larger in the models with smaller Mach number so that this aids disc communication in these cases also.

The calculations presented here use a coordinate system which is based on the initial disc mid-plane. However, as the disc precesses, the mid-plane changes location with time. It is then more convenient to use a coordinate system (x, y_o, z_o) based on the fixed orbital plane, the z_o axis coinciding with the orbital rotation axis. We shall refer to these as 'orbital plane coordinates'. We locate the inclination angle ι (equal to δ at $t = 0$) between the disc and binary orbit angular momentum vectors through

$$\cos \iota = \frac{\mathbf{J}_D \cdot \mathbf{J}_O}{|\mathbf{J}_D||\mathbf{J}_O|}.$$

Here, $\mathbf{J}_O$ is the orbital angular momentum. The disc angular momentum is $\mathbf{J}_D = \sum_j \mathbf{J}_j$, where the sum is over all disc particle angular momenta $\mathbf{J}_j$.

A precession angle β_p, measured in the orbital plane can be defined through

$$\cos \beta_p = -\frac{(\mathbf{J}_D \times \mathbf{J}_O) \cdot \mathbf{u}}{|\mathbf{J}_D \times \mathbf{J}_O||\mathbf{u}|}$$

where $\mathbf{u}$ may be taken to be any fixed reference vector in the orbital plane. We take this to point along the y_o axis such that initially $\beta_p = \pi/2$ in all cases. For retrograde precession of $\mathbf{J}_D$ about $\mathbf{J}_O$ the angle β_p should initially decrease as is found in practice.

For a disc with constant Σ and radius R, the period of rigid body precession of a thin disc is $2\pi/\omega_p$ where, from equation (16) we obtain

$$\omega_p = -\left(\frac{3GM_s}{4D^3}\right)\cos\delta \frac{\int_0^R \Sigma r^3 dr}{\int_0^R \Sigma r^3 \Omega dr} = -\frac{15 M_s R^3}{32 M_p D^3}\Omega(R)\cos\delta. \qquad (17)$$

The condition for sound to propagate throughout the disc during a precession time is approximately that

$$\frac{H}{R} > \frac{|\omega_p|}{\Omega(R)}. \tag{18}$$

For models 1-3, equation (17) gives $\omega_p/\Omega(R) = 0.012$ corresponding to a precession period of 512 time units. Our results were consistent with the condition (18) to within a factor of two in that models 1 and 2 with $\mathcal{M} < 25$, showed modest warps and approximate rigid body precession while Model 3 with $\mathcal{M} = 30$ showed severe warping and a more complex precessional behaviour.

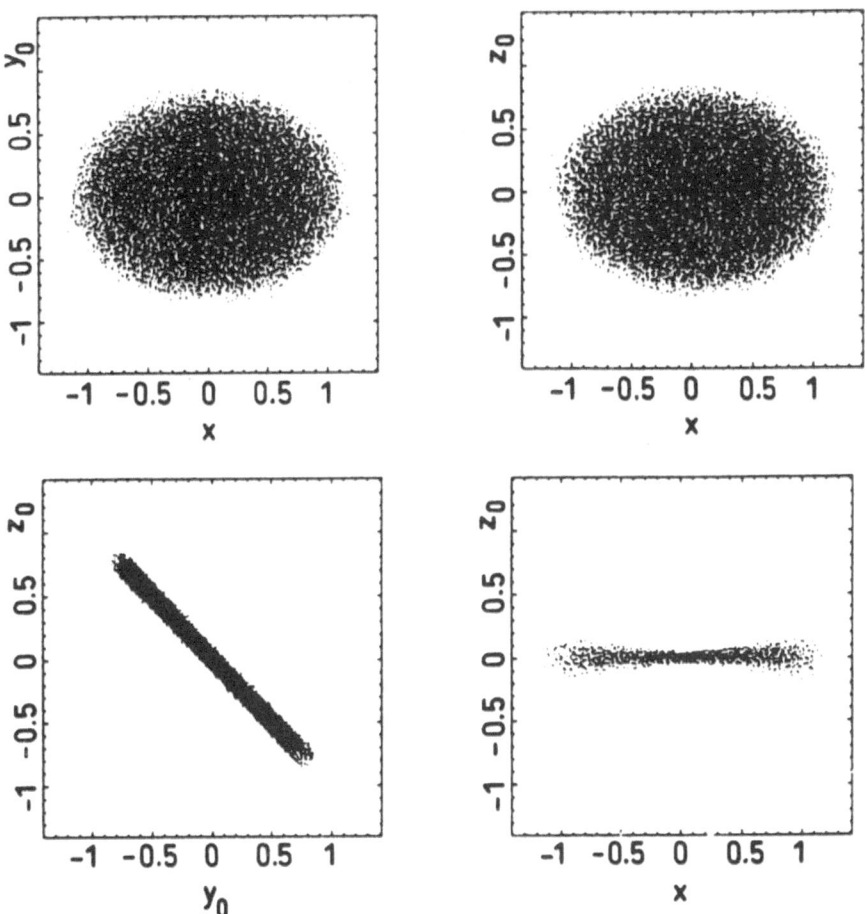

Fig. 1. Projection plots in orbital plane coordinates for model 1 at time $t \simeq 0$. The projections are in the (x, y_o) plane (top left), (x, z_o) plane (top right), (y_o, z_o) plane (bottom left) and (x, z_o) plane (bottom right).

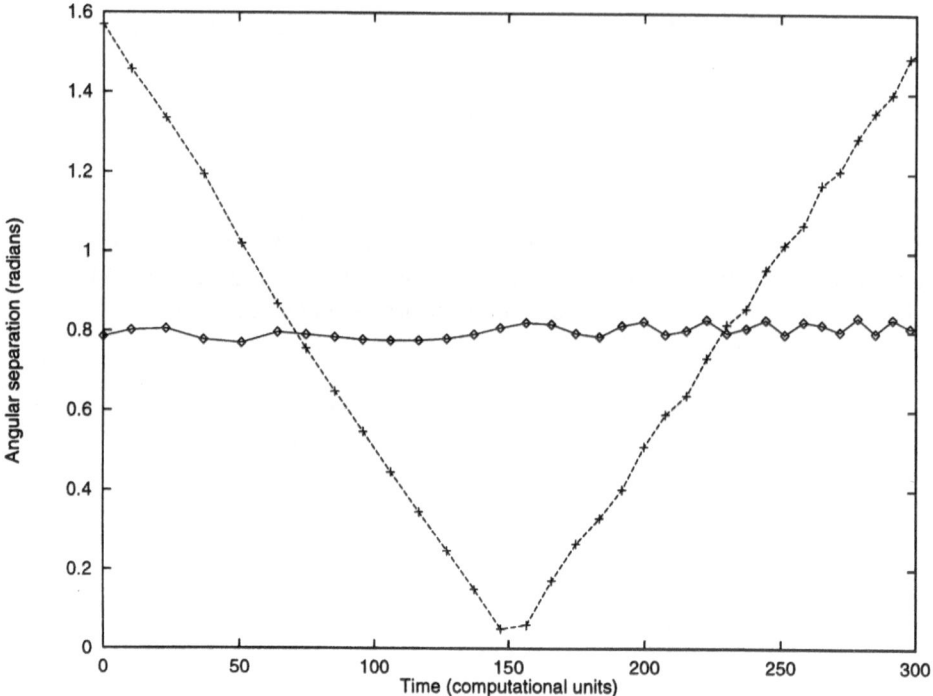

Fig. 2. The precession angle β_p (dashed line) and inclination angle ι (solid line) for model 1.

We now present particle projection plots for each of the Cartesian planes using obital plane coordinates. In all such figures a fourth 'sectional plot' is also included in which only particles such that $-0.05 < y_o < 0.05$ are plotted.

7.1   Model 1

A projection plot for model 1 is shown in Fig. (1) near $t = 0$ when the disc is unperturbed. Note that the disc appears as edge on and inclined at 45 degrees in the (y_o, z_o) plane. The time dependence of the angles ι and β_p is plotted in Fig. (2). It may be seen that there is little change in ι during the whole run. On the other hand β_p decreases approximately linearly, corresponding to uniform precession (note that β_p is plotted as positive rather than negative for the latter section of this plot). The inferred precession period is around 600 units, in reasonable agreement with the value of 512 units obtained from equation (17). Fig. (3) shows a projection plot at $t = 297.8$, near the end of the run when the disc has precessed through about 180 degrees. This amount of precession is demonstrated by the fact that the disc appears almost edge on in the (y_o, z_o) plane just as it did at time $t = 0$. However, its plane is inclined at about 90 degrees to the original disc plane. At this stage our results indicate that the disc

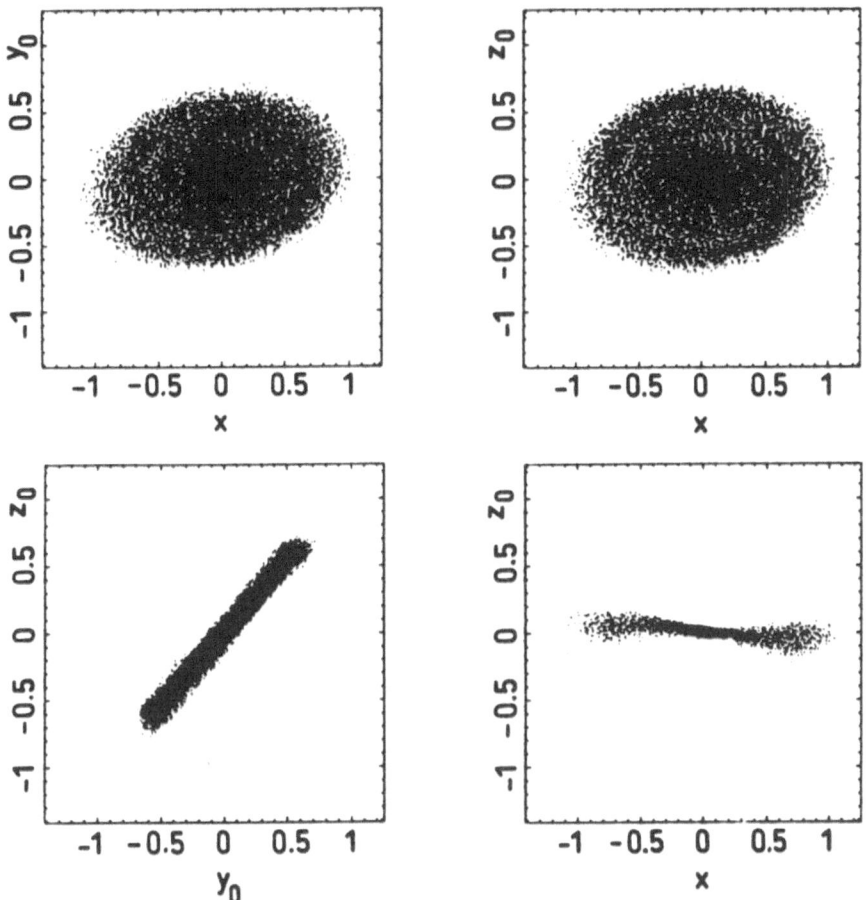

Fig. 3. Projection plots in orbital plane coordinates for model 1 at time $t = 297.8$.

has attained a quasi-steady configuration as viewed in a frame that precesses uniformly about the orbital rotation axis. The disc develops a warped structure that initially grows in magnitude but then levels off. The sectional plot in Fig. (3) indicates that the disc has developed a modest warp in this case.

7.2   Models 2, and 3

The behaviour of model 3 with $\mathcal{M} = 30$ is considerably more complex than that of model 1. This disc develops a strong warp such that the inner and outer parts of the disc try to separate. A projection plot is shown at $t = 397.9$ in Fig. (4). The inner part of the disc seems to occupy a different plane from the outer part. The outer part was found to precess like a rigid body at the expected rate, and it tended to drag the inner section behind it. This is indicated in Fig. (5) where we

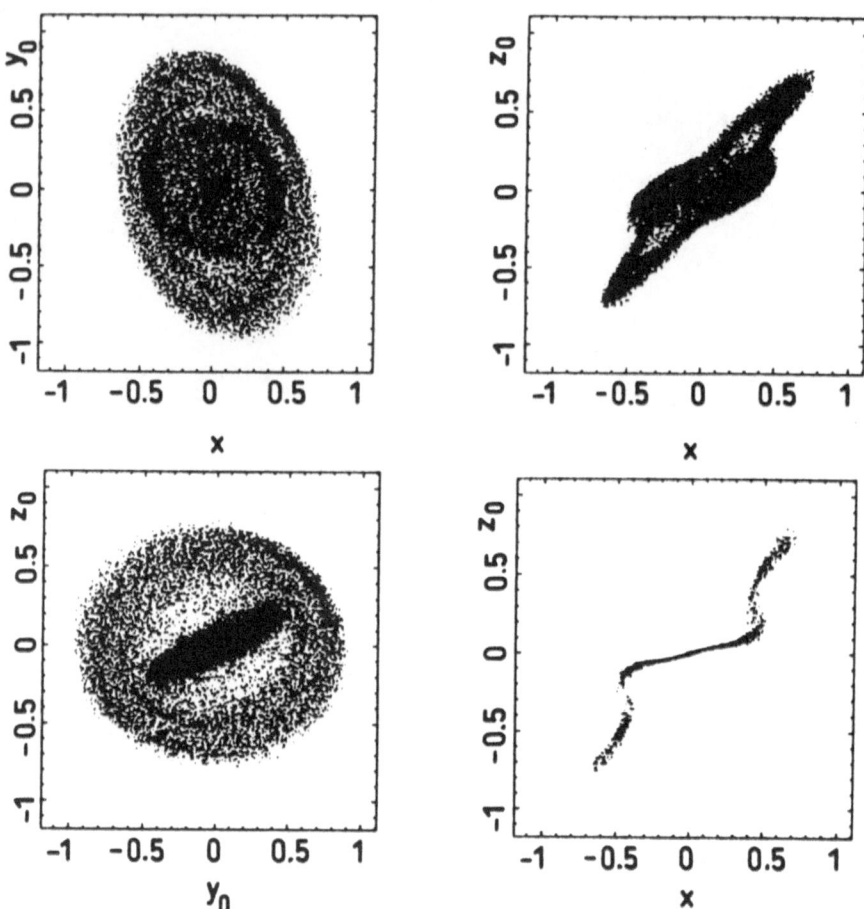

Fig. 4. Projection plots in orbital plane coordinates for model 3 at time $t = 397.9$.

plot the angle β_p calculated using the outer disc section with $r > 0.5$ only and also the same angle calculated using only the inner section with $r < 0.3$. The angle associated with the inner segment progresses at a variable rate indicating coupling to the outer section. The angle ι associated with the two sections is also plotted. This becomes significantly smaller for the inner segment consistent with the existence of a large amplitude warp. We note that the relatively larger inclination associated with the outer segment enables a closer matching of the precession frequencies associated with the two segments and aids coupling, due to the presence of large pressure gradients induced by the strong warping. Each segment of this model remained thin throughout the run.

Fig. (6) shows the evolution of β_p and ι for model 2, which has $\mathcal{M} = 25$ and is intermediate between model 1 and model 3. There is an indication of differential precession initially. But the inner part of the disc couples to the outer part in

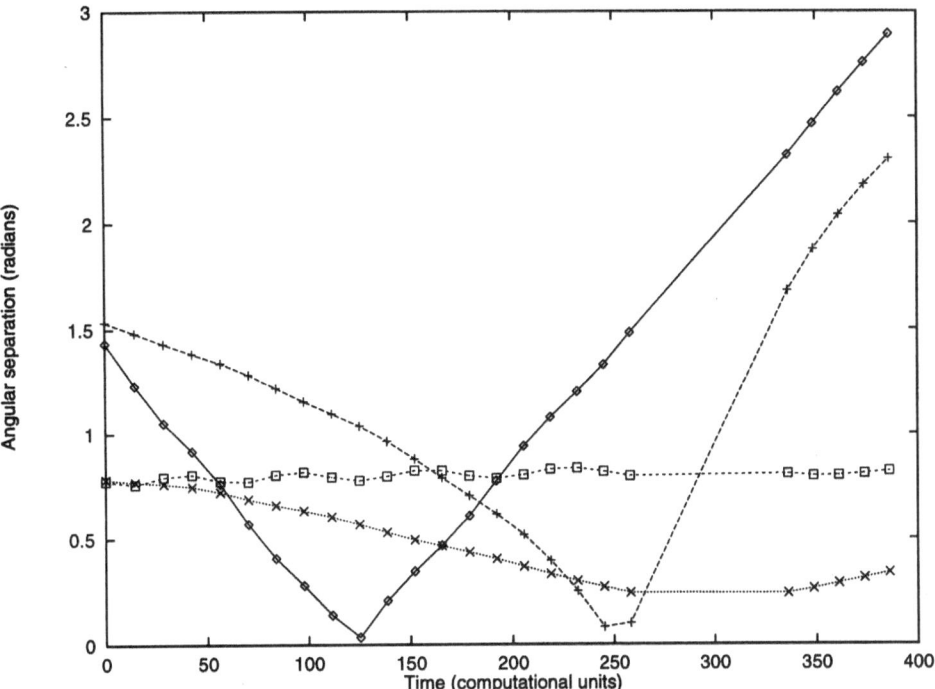

Fig. 5. The precession angle β_p (solid and long-dashed lines) and inclination angle ι (short-dashed and dotted lines) for model 3. The solid and short-dashed lines correspond to the outer disc section with $r > 0.5$, and the long-dashed and dotted lines correspond to the inner disc section with $r < 0.3$.

such a way that the precession becomes uniform after about 150 time units. It appears tha the inner part is able to adjust its precession frequency to the outer part by changing slightly its relative inclination. In this way the dependence of the precession frequency on inclination is exploited to remove the differential precession.

8 Discussion

We have described nonlinear simulations of an accretion disc in a close binary system when the disc midplane is not necessarily coplanar with the plane of the binary orbit. For our constant viscosity SPH models we found the tidal truncation phenomenon to be only marginally affected by non coplanarity. We found that modestly warped and thin discs undergoing near rigid body precession may survive in close binary systems. However, extremely thin discs may be severely disrupted by differential precession depending on the magnitude of the characteristic Mach number, $\mathcal{M}$. The crossover between obtaining a warped, but coherent disc structure, and disc disruption occurs for a value of the Mach number

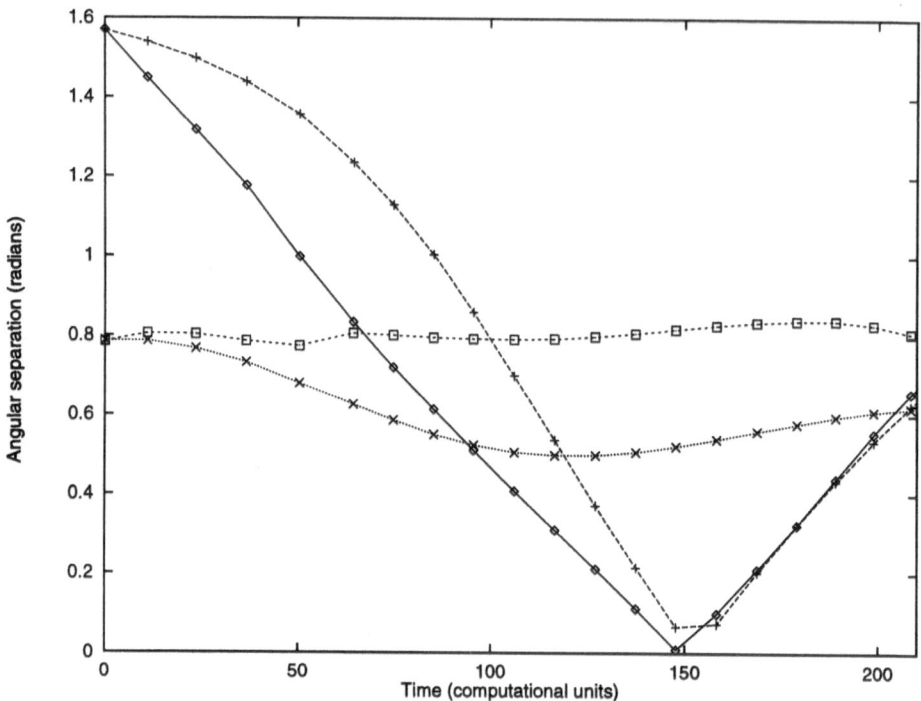

Fig. 6. The precession angle β_p (solid and long-dashed lines) and inclination angle ι (short-dashed and dotted lines) for model 2. The solid and short-dashed lines correspond to the outer disc section with $r > 0.5$, and the long-dashed and dotted lines correspond to the inner disc section with $r < 0.3$.

$\mathcal{M} \sim 30$. We also found that the inclination evolved on a long timescale, likely to be the viscous timescale, as indicated by the linear calculations of Papaloizou and Terquem (1995).

A class of models formulated to explain the generation of jets in young stellar objects assumes that a wind flows outwards from the disc surface. This is then accelerated and collimated by the action of a magnetic field (see Königl and Ruden, 1993 and references therein). It is reasonable to assume that a precessing disc may lead to the excitation of a precessing jet. The precession period obtained from our calculations with a mass ratio of unity is about 500 units. When scaled to a disc of radius 50 AU, surrounding a star of $1M_\odot$, the unit of time is $\Omega(R)^{-1} \simeq 56\ yr$, leading to a precession period of $3.10^4 yr$.

Bally and Devine (1994) suggest that the jet which seems to be excited by the young stellar object HH34* in the L1641 molecular cloud in Orion precesses with a period of approximately $10^4\ yr$. This period is consistent with the source being a binary with parameters similar to those we have used in our simulations with a separation on the order of a few hundred astronomical units.

Some of the results presented here demonstrate how a warped disc can present a large surface area for intercepting the primary star's radiation. The effect that the consequent reprocessing of the stellar radiation field can have on the emitted spectral energy distribution has been investigated by Terquem and Bertout (1993, 1996). They find that it may account for the high spectral index of some T Tauri stars or even for the spectral energy distribution of some class 0/I sources. The Model 3 simulation indicates that the required strongly warped disc could be physically realisable.

Finally, there is evidence from the light curves of X–ray binaries such as Hercules X-1 and SS433, that their associated accretion discs may be precessing in the tidal field of the binary companion (Schandl and Meyer, 1994). Larwood et al (1996) have demonstrated that the disc precession periods seen in simulations are in reasonable agreement with those that are inferred observationally (Petterson, 1975, Gerend and Boynton, 1976, Margon, 1984).

Acknowledgement: This work was supported by PPARC grant GR/H/09454, JDL is supported by a PPARC studentship.

References

Artymowicz P., Lubow S.H., 1994, ApJ, 421, 651

Bally J., Devine D., 1994, ApJ, 428, L65

Bibo E.A., The P.S., Dawanas D.N., 1992, A&A, 260, 293

Corporon P., Lagrange A.M., Beust H., 1996, A&A, 310, 228

Edwards S., Cabrit S., Strom S.E., Heyer I., Strom K.M., Anderson E., 1987, ApJ, 321, 473

Gerend D., Boynton P., 1976, ApJ, 209, 562

Gingold R.A., Monaghan J.J., 1977, MNRAS, 181,375

Goldreich P., Tremaine S., 1981, ApJ, 243, 1062

Hunter, C. & Toomre, A. 1969, *Ap. J.*, **155**, 747

Königl A., Ruden S.P., 1993, in Levy E.H., Lunine J., eds, Protostars and Planets III (Univ. Arizona Press, Tucson), p. 641

Larwood J.D., Nelson R.P., Papaloizou J.C.B., Terquem C., 1996, MNRAS, 282, 597

Larwood J.D., Papaloizou J.C.B., 1997, MNRAS, In press

Leinert C., Zinnecker H., Weitzel N., Christou J., Ridgway S.T., Jameson R.F., Haas M., Lenzen R., 1993, A&A, 278, 129

Lin D.N.C., Papaloizou J.C.B., 1993, in Levy E.H., Lunine J., eds, Protostars and Planets III (Univ. Arizona Press, Tucson) p.749

Lucy L.B., 1977, AJ, 83, 1013

Lynden-Bell D., Ostriker J.P., 1967, MNRAS, 136, 293

Lynden-Bell D., Pringle J.E., 1974, MNRAS, 168, 603

Margon B., 1984, ARA&A, 22, 507

Mathieu R.D., 1994, ARA&A, 32, 465

Monaghan J.J., Gingold R.A., 1983, J. Comp. Phys., 52, 374

Monaghan J.J., Lattanzio J.C., 1985, A&A, 149, 135

Nelson R.P., Papaloizou J.C.B., 1993, MNRAS, 265, 905

Nelson R.P., Papaloizou J.C.B., 1994, MNRAS, 270, 1

Nelson R.P., 1994, Ph.D Thesis, University of London

Papaloizou J.C.B., Lin D.N.C., 1995, ApJ, 438, 841

Papaloizou J.C.B, Pringle J.E., 1983, MNRAS, 202, 1181

Papaloizou J.C.B., Terquem C., 1995, MNRAS, 274, 987

Petterson J.A., 1975, ApJ, 201, L61

Savonije G.J., Papaloizou J.C.B., Lin D.N.C., 1994, MNRAS, 268, 13

Shakura N.I., Sunyaev R.A., 1973, A&A, 24, 337

Schandl S., Meyer F., 1994, A&A, 289, 149

Shu, F. S. 1984. In Planetary Rings, Greenberg, R. & Brahic, A., eds., University of Arizona Press.

Spruit H.C., 1987, A&A, 184, 173

Terquem C., Bertout C., 1993, A&A, 274, 291

Terquem C., Bertout C., 1996, MNRAS, 279, 415

Super-Eddington luminosity in the bursting pulsar GRO J1744-28. GRANAT/WATCH results

S.Y. Sazonov[1,2], R.A. Sunyaev[1,2], N. Lund[3]

[1] Space Research Institute, Moscow, Russia
[2] Max-Planck Institut für Astrophysik, Garching, Germany
[3] Danish Space Research Institute, Kopenhagen, Denmark

Abstract: We present the results of GRANAT/WATCH observations of the BATSE bursting pulsar GRO J1744-28 in January – March 1996, during an intense outburst of this transient. The observations started in mid-January, exactly at the peak of the source light curve. High, ~ 3.7 Crab, photon flux measured in these first observations in the energy band 8 – 20 keV indicates that the bolometric persistent X-ray luminosity of GRO J1744-28 was at that moment reaching 10^{39} erg/s, a value that exceeds the Eddington luminosity for a neutron star by a factor of ~ 5. This estimate is made assuming that the source is at the distance of the Galactic Center (8.5 kpc). 70 bursts have been detected from GRO J1744-28 by WATCH, during which the luminosity of the source increased 7 to 22 times relative to the persistent level, reaching over $10^{40}(D/8.5\text{ kpc})^2$ erg/s, thus exceeding the neutron star Eddington luminosity by two orders of magnitude. All of the bursts have similar profiles characterized by FWHM ~ 4 sec. The frequency of bursts also remained constant through the observations, 41 ± 5 day^{-1} (after correction for the observations' windowing). The WATCH flux measurements are completely unaffected by any dead-time or pile-up effects, therefore the luminosity estimates we have obtained are quite reliable. We believe that the thin-wall column geometry of the accretion flow near the neutron star magnetic poles outlined by Basko & Sunyaev (1976) makes it possible for the GRO J1744-28 luminosity to overcome the Eddington limitation.

1 Introduction

GRO J1744-28 is a unique X-ray transient source, which shows both pulsations and bursts. Its first bursts coming at intervals of ~ 3 min were detected by the CGRO/BATSE instrument on 2 December 1995 (Fishman et al., 1996a). A few weeks after that discovery a new hard X-ray pulsar with a period of 467 ms, located in the direction of the Galactic Center, was detected by BATSE (Paciesas et al., 1996; Finger et al., 1996a). It was shortly shown that the burster and the transient pulsar GRO J1744-28 were the same source (Kouveliotou et al., 1996a). The object is a neutron star in a binary system with an orbital period of 11.8 days and an X-ray mass function of $1.31 \times 10^{-4} M.$ (Finger et al., 1996b). The GRO J1744-28 very close celestial position to the Galactic Center,

$l = +0.°1$, $b = +0.°3$ (Kouveliotou et al., 1996b), strongly suggests that the source is approximately as far from us as the Galactic Center, i.e. ~ 8.5 kpc away. This possibility is further justified both by X-ray measurements of the absorbing column along the line of sight to GRO J1744-28, which give values for N_H around 5×10^{22} cm^{-2} (Dotani et al., 1996), typical for the objects in the Galactic Center region, and by the non-detection of any safe counterpart in optical and infra-red.

On 13 January GRO J1744-28 for the first time appeared in the field of view of one of the WATCH instruments aboard the GRANAT satellite. The photon flux, ~ 4 Crab, measured from the source on 14-15 January at 8 – 20 keV indicated that its luminosity was $\sim 4 \times 10^{38}$ erg/s only in this restricted energy band (Sazonov & Sunyaev, 1996). We continued to observe the source with WATCH repeatedly throughout most of its outburst in January - March.

2   Observations and Results

The observations of GRO J1744-28 presented here were carried out with the WATCH all-sky monitor in the period 13 January to 12 March, 1996, when the GRANAT observatory was monitoring the sky in scanning mode. In this regime the spacecraft spins with a period of ~ 20 min, and the spinning axis roughly follows the ecliptical motion of the Sun in the course of a year. The modulation assembly of the WATCH instrument is halted for the scanning observations, in contrast to the normal pointed observations when it is kept rotating with a frequency of about 1 sec^{-1} by means of a motor. Therefore the modulation of the fluxes of the field of view X-ray sources is created by the rotation of the satellite itself, thus allowing one to determine the intensities and locations of the sources. We have analyzed the data taken in the instrument's lowest energy band 8 – 20 keV. Throughout the whole cycle of observations, the Crab Nebula, the standard candle in X-ray astronomy, was in the instrument's field of view, enabling continuous calibrating of the GRO J1744-28 flux.

We specially note that during the first months of 1996 it was very difficult to implement precise measurements of the luminosity of GRO J1744-28, especially during its bursts, with high-throughput pointed telescopes, because photon fluxes from the source were extremely high and various "dead-time" effects inevitably interfered in the observations. In the case of WATCH such problems do not exist at all, because the contribution of GRO J1744-28 to the count flux on the detector is for example of the order of the usual WATCH background flux.

2.1   Fraction of the GRO J1744-28 Luminosity in the WATCH Energy Range

From XTE measurements (Giles et al., 1996) it is known that the GRO J1744-28 energy spectrum both in quiescence and during the bursts can be fitted by a power law with a slope of ~ 1.2 and a cutoff above ~ 20 keV with an e-folding energy of ~ 15 keV. This means that GRO J1744-28 radiates in the WATCH low energy band (8 – 20 keV) ~ 35 % of its total X-ray luminosity. We will make use of this value in calculations of the source bolometric luminosity based on WATCH flux measurements.

2.2 Persistent Flux

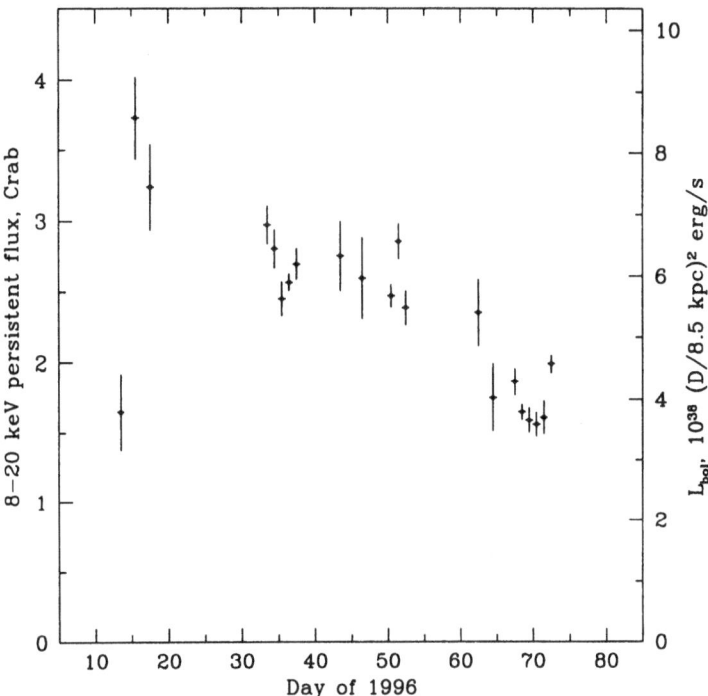

Fig. 1. GRO J1744-28 light curve measured by GRANAT/WATCH in January – March 1996. Each point corresponds to one day of observations.

Fig. 1 shows the evolution of the GRO J1744-28 persistent X-ray flux during the period January 13 to March 12. One can see an almost linear decline of the intensity at $\sim 25\%$ per month over all the period described, the flux maximum of $\sim 3.7 \pm 0.3$ Crab occurring on January 15. We note that during the very first WATCH observation on 13 January GRO J1744-28 was very close to the edge of the field of view, therefore we do not consider the measured on that day flux 1.6 ± 0.3 (the first point in Fig. 1) reliable. Already in the subsequent observation, on 14-15 January, GRO J1744-28 had moved to the inner region of the field of view, which is well calibrated, and in all the following observations the source flux could be determined safely. Earlier BATSE observations indicate that the flux rise lasting for approximately one month did complete by the middle of January (Fishman et al., 1996b).

The right vertical scale of Fig. 1 measures the source bolometric (2 – 100 keV) luminosity estimated from the WATCH flux values applying for the 8 – 20 keV fraction of the total source energy (see above). We infer that the persistent luminosity of GRO J1744-28 peaked at $\sim 9 \times 10^{38}$ erg/s assuming an isotropic radiation, if the source is at the distance of the Galactic Center, which exceeds the Eddington luminosity for a spherically accreting neutron star of mass $\sim 1.4 M_\odot$ by a factor of 5.

We ought to mention that there are several relatively bright X-ray sources in the vicinity of the Galactic Center, which should influence our calculation of the GRO J1744-28 flux, however, the cumulative photon flux from these sources is usually below 200 mCrab (see e.g. Pavlinsky et al., 1994). Also, at the end of February close to the Galactic Center and only 1° away from GRO J1744-28 an X-ray Nova, GRS 1739-238, outburst, which was discovered and localized by the GRANAT/SIGMA (Paul et al., 1996) and MIR-KVANT/TTM (Borozdin et al., 1996) telescopes. It is not possible to separate the contributions of GRO J1744-28 and GRS 1739-238 to the flux measured with WATCH, but GRS 1739-283 seems to have not ever exceeded the level of 300 mCrab at 8 – 20 keV according to the observations by the GRANAT and MIR/KVANT observatories. Therefore, we conclude that the light curve shown in Fig. 1 is only slightly contaminated by the GRO J1744-28 nearby sources.

2.3   Bursts

Data reduction procedure   As we have already mentioned, the GRO J1744-28 observations took place when the GRANAT satellite was spinning. Hence, the count rate on the WATCH detectors recorded with a time resolution of either 2.1 sec or 4.2 sec was persistently highly variable on the time scale of tens of seconds due to the modulation of the photon fluxes from the bright X-ray sources in the field of view by the instrument's grid collimator. The problem was to extract the burst signal from the complex raw count rate history. This problem was successfully solved, and Fig. 2 shows the scheme of data reduction that was used for producing burst light curves that give the photon flux incident on the collimator.

A burst was considered detected if an increase in the count rate of more than 8 standard deviations over the background level had been registered at one of the following time scales: 2 sec, 4 sec etc. We then had to single out GRO J1744-28 bursts from the whole sample of burst events detected, including those of solar or magnetospheric origin. This selection was based on three conditions: 1) the spectrum of the burst must be much softer than the typical spectra of cosmic gamma-ray bursts, 2) the burst must last more than a few but less than ~ 100 seconds, and 3) the burst must occur at a rotation phase when GRO J1744-28 is in the field of view. The first two conditions are based on information available

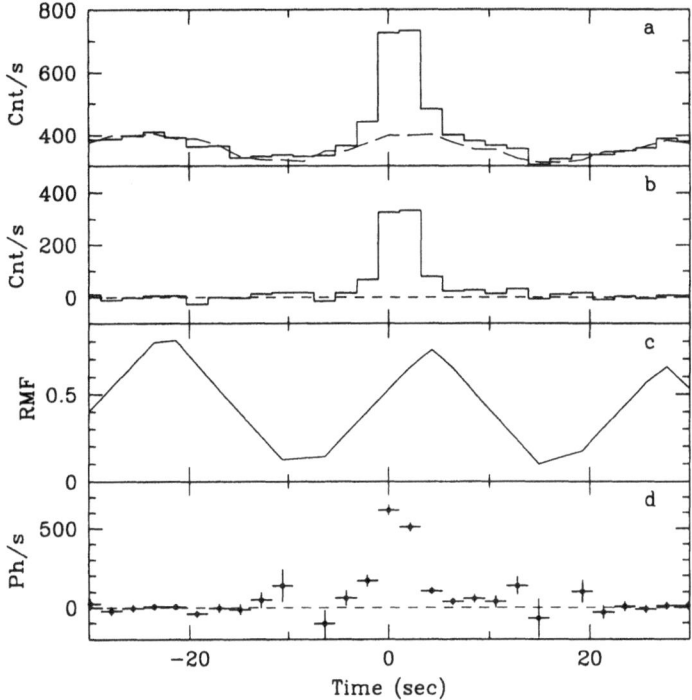

Fig. 2. Stages of the reconstruction of the light curve of a burst: a) Time history of the raw count rate on the detector (solid line). It includes the contribution of the background and persistent sources (dashed line). b) Count rate history upon the subtraction of the persistent component. c) Evolution of the rotation modulation function calculated for GRO J1744-28. d) The final light curve of the burst obtained by dividing the burst count rate history (b) by the RMF (c).

a-priory from other X-ray instruments. The fact that all but a few events selected by these two properties pass successfully the third geometrical test implies that more than 90 % of the bursts in our final list are indeed originated by GRO J1744-28. We also note that the bursts that have not passed the geometrical test are somewhat shorter than the GRO J1744-28 bursts.

Burst frequency Our list of GRO J1744-28 bursts includes 70 events, which were detected by WATCH with a mean frequency of 9 events per day. Taking into account the fact that the source was in the field of view during roughly 30% of each satellite rotation and the more complicated effect that the effective observational time is smaller for weaker bursts than for stronger ones, we have found that GRO J1744-28 on average emitted 41 + / − 5 events per day. Our observations reveal no significant difference between the burst frequency in the first half of the period January – March and that in the second half.

Burst profiles The profile of individual bursts remained strikingly stable over the whole period of observations. A histogram of the distribution of burst durations, defined here as full width at quarter-maximum, is shown in Fig. 3. The mean burst duration is ∼ 7 sec, and the scatter of individual measures around this mean can be almost totally attributed to the measurement errors. We have averaged the profiles of the bursts, aligned on the burst peak to obtain a mean burst light curve, which is shown in Fig. 4. Its FWHM is ∼ 4 sec, and it is asymmetric, the rise time being shorter than the decay time.

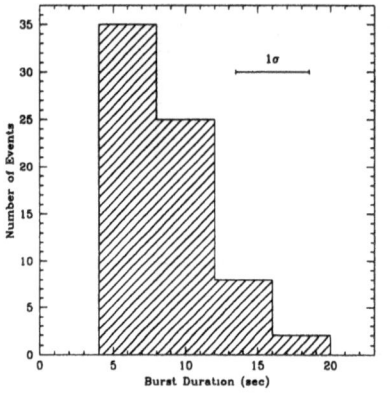

Fig. 3. Histogram of the GRO J1744-28 burst duration distribution constructed from the 70 bursts observed by WATCH. The typical 1 σ uncertainty of an individual measurement is also shown.

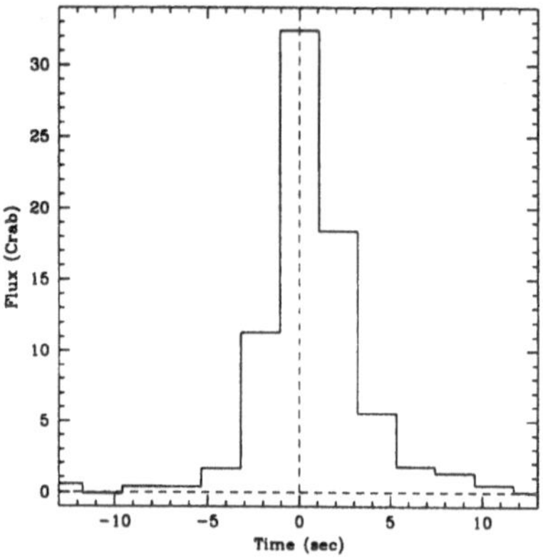

Fig. 4. Averaged GRO J1744-28 burst light curve. Individual burst light curves were aligned on the peak time.

Burst peak fluxes, fluences and luminosities The time history of the burst fluence (left vertical scale) and correspondingly the burst total energy (right vertical scale) is shown in Fig. 5. Along with jumps between individual measures, a decline of the fluence with a characteristic slope similar to that of the persistent light curve (Fig. 1) is clearly seen. Thus, the strength of the bursts was roughly proportional to the source persistent flux. Given that the spectrum of GRO J1744-28 during bursts is not markedly different from its persistent spectrum we find that the total energy emitted by the source in X-rays during a burst for some of the events amounts to ∼ 10^{41} erg. The ratio of the peak burst flux to the persistent near burst flux ranges between 7 and 22 for the bursts observed. The burst peak bolometric luminosity averaged over the 10 strongest bursts is

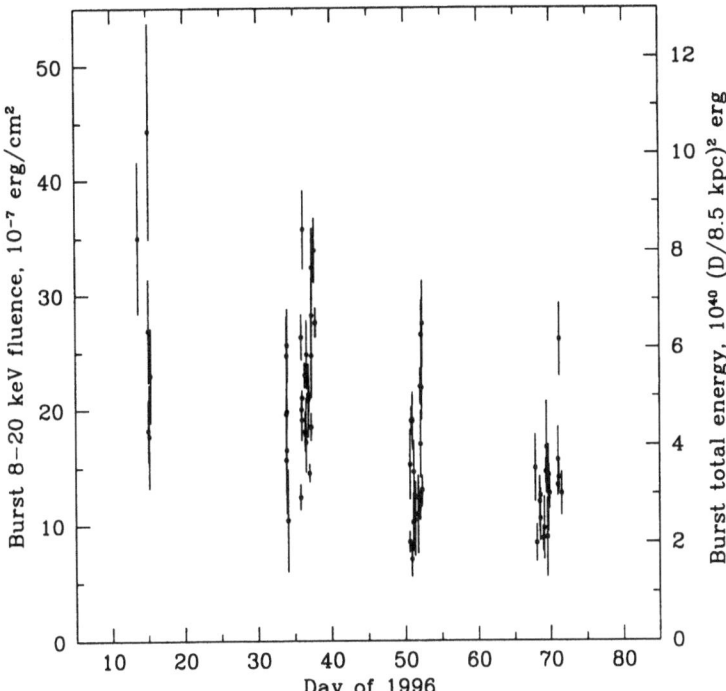

Fig. 5. GRO J1744-28 bursts' fluences and total energies as a function of date during the period January – March 1996.

$(1.1 \pm 0.1) \times 10^{40}(D/\ 8.5 \text{ kpc})^2$ erg/s, where D is the source distance, a value that is some 60 times as high as the neutron star Eddington luminosity. Note that the best time resolution available in our observations was 2.1 sec, which is compared with the burst profile top width (see e.g. Giles et al., 1996), therefore we consider that the source luminosity at the very maximum of its most powerful bursts was actually some 1.5 times the value quoted above.

3 Discussion

3.1 The Luminosity vs. Period Change

One of the most important results of the WATCH observations is the detection of the highly super-Eddington persistent luminosity in GRO J1744-28. An independent estimate of the source luminosity is provided by measurements of the pulsar period change. The standard theory of accretion by rotating magnetic neutron stars yields the following general relation between the pulsar luminosity L and the period change rate $\dot{P}$:

$$\dot{P} = -6 \times 10^{-4} \epsilon^{1/2} n R_6^{6/7} M^{-3/7} I_{45}^{-1} P^2 \mu_{30}^{2/7} L_{38}^{6/7} \text{ s yr}^{-1}, \tag{1}$$

where R is the neutron star radius measured in units of 10^6 cm, M is the mass of the neutron star in units of solar masses, I is the neutron star moment of inertia in units of 10^{45} g cm^2, P is the pulsar period in units of sec, μ is the dipole magnetic moment in units of Gauss cm^3, L is the luminosity in units of 10^{38} erg/s; ϵ and n are model-dependent coefficients describing the radius of the magnetosphere and the net torque on the neutron star due to the accretion flow. For example in the model of Ghosh & Lamb (1979) these parameters are equal to ~ 0.5 and 1.4 (for slow rotator) respectively. According to observations by BATSE (Finger et al., 1996c) and XTE (public data archive) during the outburst of GRO J1744-28 the period change followed a $L^{6/7}$ trend over at least an order of magnitude change in luminosity, indicating that the pulsar was a slow rotator (otherwise n would have been changing with luminosity). However the observed X-ray luminosity, $\sim 9 \times 10^{38}$ erg/s, according to formula (1) implies a period change about an order of magnitude higher than the observed $\dot{P} = -8 \times 10^{-5}$ s yr^{-1} at the maximum of the light curve, if we take $R \sim 1$, $M \sim 1.4$, $I \sim 1$ and $\mu \sim 1$, values that are believed to be typical for neutron stars. This apparent discrepancy between the observed GRO J1744-28 luminosity and period change rate, earlier noticed by Daumerie et al. (1996) (we became aware of this paper when our independent analysis was almost finished), in our opinion may be due to at least five different causes:

1) The source is considerably closer than the Galactic Center. As was argued above this situation is very unlikely.

2) The parameters of the neutron star (R, M, I) are very unusual. This may provide a tangible modification factor due to the rather strong dependence of $\dot{P}$ on these parameters.

3) The disc-magnetosphere transition zone has a smaller radius than assumed ($\epsilon < 0.5$). This factor is very unlikely to yield more than a factor of ~ 1.5 correction.

4) The radiation is beamed towards the observer. The beaming factor is indeed expected to be about 1.5 even for the trivial case that the radiation is emitted isotropically by a small spot on the surface of the neutron star, because we are apparently observing the binary system almost face-on (Daumerie et al., 1996). Possible anisotropy effects due to the presence of the strong magnetic field may somewhat increase the beaming factor.

5) The magnetic field is highly non-dipolar (see Daumerie et al., 1996). This would yield a strong ($\sim 10^{12}$ Gauss) magnetic field on the surface of the neutron star required to explain strong regular X-ray pulsations observed in the flux of GRO J1744-28 and a steep decline of the field outwards from the neutron star and consequently a smaller radius of the magnetosphere. It seems that this situation has been hardly studied so far, and one of the questions such a model should give an answer to is: how shall the accretion flow and the pulsation profile look like in such a case?

Although none of the above factors is likely to reconcile the observed luminosity and the period change alone, we believe that a combination of at least some of them may do so.

3.2   The Mechanism for the Super-Eddington Luminosity

It is well known that the radiation of luminosities in excess of the Eddington value $L_{Ed} \sim 1.3 \times 10^{38} M$ erg s^{-1} is impossible in the case of spherically symmetric accretion onto a compact star, because at this critical luminosity the pressure of the outgoing radiative flux equals the gravitational pressure of the accreting matter, inhibiting any further increase of the accretion rate. However, as was first shown by Basko & Sunyaev (1976), one may well expect to observe such high luminosities from the rotating pulsars possessing strong magnetic fields ($\sim 10^{12}$ Gauss on the surface of the neutron star). At high accretion rates the accreting matter is likely to liberate its gravitational energy inside high accretion columns standing above the neutron star magnetic poles. The radiation is then radiated by the side walls of these columns, and may escape the system without interacting with the accreting flow. The maximum luminosity achievable in this model is given by:

$$L^{**} \sim 8 \times 10^{38} \frac{l/d}{40} \frac{M}{M_\odot} \text{erg s}^{-1}. \tag{2}$$

Therefore if the cross-section of the accretion channel has a form of a thin annular arc, as suggested by Basko and Sunyaev, i.e. its length l is much larger than its width d, than the observed GRO J1744-28 luminosity can be accounted for. Whether the considered mechanism is responsible for the production of the X-ray bursts as well is still to be understood.

Acknowledgement: This work has been supported in part by RFFI grant No. 95-02-05938.

References

Basko M.M. Sunyaev R.A., 1976, MNRAS 175, 395
Borozdin K., Alexandrovich N. and Sunyaev R., 1996, IAUC 6350
Daumerie P., Kalogera V., Lamb F.K. and Psaltis D., 1996, Nat. 382, 141
Dotani T., Ueda Y., Ishida M., et al., 1996, IAUC 6337
Finger M.N., Wilson R.B., Harmon B.A., et al., 1996a, IAUC 6285
Finger M.N., Wilson R.B. and van Paradijs J., 1996b, IAUC 6286
Finger M.N., Koh D.T., Nelson R.W., et al., 1996c, Nat. 381, 291
Fishman G.J., Kouveliotou C., van Paradijs J., et al., 1996a, IAUC 6272
Fishman G.J., Harmon B.A., Kouveliotou C., et al., 1996b, IAUC 6290
Giles A.B., Swank J.H., Jahoda K., et al., 1996, ApJ 469, L25
Ghosh P. and lamb F.K., 1979, ApJ 234, 296
Kouveliotou C., Kommers J., Lewin W.H.G., et al., 1996, IAUC 6286
Kouveliotou C., Greiner J., van Paradijs J., et al., 1996, IAUC 6369

Paciesas W.S., Harmon B.A., Fishman G.J., et al., 1996, IAUC 6284
Paul J., Bouchet L., Churazov E., et al., 1996, IAUC 6348
Pavlinsky M.N., Grebenev S.A. and Sunyaev R.A., 1994, ApJ 425, 110
Sazonov S. and Sunyaev R., 1996, IAUC 6291

On viscous disc flows around rotating black holes

J. Peitz[1]*, S. Appl[2]**

[1] Landessternwarte Königstuhl, D-69117 Heidelberg, Germany
[2] Observatoire Astronomique, Université Louis Pasteur, 11 rue de l'Université, F-67000 Strasbourg, France

Abstract: The stationary hydrodynamic equations for transonic viscous accretion discs in Kerr geometry are derived and solved for a polytropic equation of state. The viscous angular momentum transport and the boundary conditions on the horizon of a central black hole are consistently treated. A refined expression for the scale-height of the disc is obtained from the vertical Euler equation for general accretion flows with vanishing vertical velocity. Different solution topologies are identified, characterized by a sonic transition close to or far from the marginally stable orbit. Global polytropic solutions for the disc structure are calculated by a new numerical method, covering each topology and a wide range of physical conditions. These solutions generally possess a sub-Keplerian angular momentum distribution and have maximum temperatures in the range $10^{11} - 10^{12}$ K. Accretion discs around rotating black holes are hotter and deposit less angular momentum on the central object than accretion discs around Schwarzschild black holes.

1 Introduction

Optically thin advection dominated accretion flows (ADAFs) yield encouraging fits to the observed hard X-ray/γ-ray spectral components of black hole X-ray binary and underluminous active galactic nuclei sources (e.g. models for soft X-ray transients (SXTs) A0620-00, V404 Cyg and Nova Muscae 1991 by Narayan, McClintock & Yi (1996) and the model for Sgr A* by Narayan, Yi & Mahadevan (1995)). These early models contained various simplifications. Narayan & Yi (1994, 1995ab) assumed a self-similar disc structure to calculate local thermal equilibrium solutions, whereas Abramowicz et al. (1995) calculated global solutions within the pseudo-Newtonian potential approximation for a Schwarzschild black hole (Paczyński & Wiita 1980). These simplifications are justified as long as the properties of ADAFs at intermediate radii are concerned, but they do not allow to properly treat the boundaries of the disc. In particular the presence of an event horizon in any black hole accreting system, which is the key feature for the existance of an ADAF, requires a relativistic treatment at the inner edge of the disc. Furthermore, the bounday conditions at the horizon imply that any accretion flow onto a black hole be transonic, a feature which cannot be

---

* e-mail: jpeitz@lsw.uni-heidelberg.de
** e-mail: appl@astro.u-strasbg.fr

accounted for in selfsimilar solutions. The equations governing viscous transonic accretion discs in Kerr geometry have been given by Lasota (1994). Global solutions have been calculated for a polytropic equation of state by Peitz (1994) and for optically thin flows cooled by non-relativistic bremsstrahlung emission by Abramowicz et al. (1996).

Here we report on a refined stationary model (Peitz & Appl 1997) for viscous transonic disc flows around rotating black holes.

2   Formulation

The disc is treated as a stationary axialsymmetric (1+1)d problem, i.e. the radial and vertical structure equations are assumed to decouple. The four-velocity of the fluid is of the form $u = (u^t, u^\phi, u^r, 0)$. The half-thickness h is assumed to satisfy $h/r \lesssim 1$.

The radial disc structure is calculated for vertically integrated thermodynamic variables $P \simeq 2hp$ and $\Sigma \simeq 2h\rho$, using the Kerr metric in cylindrical coordinates $\{t, \phi, r, z\}$ expanded to zeroth order in z/r (Novikov & Thorne 19973). A polytropic relation between P and Σ is assumed, $P = K\Sigma^\Gamma$. Neglecting contributions due to viscosity and heat flux, the radial Navier-Stokes equation can be cast into the form

$$\frac{du^r}{dr} = \frac{\mathcal{N}}{\mathcal{D}} . \tag{1}$$

Accretion discs around black holes are necessarily transonic. The flow encounters a critical point at the sonic radius r_s, where $\mathcal{D} = 0$. Regularity at r_s requires

$$\mathcal{N}(\mathcal{D} = 0) = 0 , \tag{2}$$

which provides an *internal* boundary condition. The radial disc structure is completed by the equation of angular momentum transport, which has the structure

$$\frac{\dot{M}}{2\pi}(\mu l - L_0) = 2hrt_\phi^r . \tag{3}$$

Here μ is the chemical potential, $l \equiv u_\phi$ the specific (per unit mass) angular momentum and $\dot{M}$ the (rest mass) accretion rate. Neglecting bulk viscosity, the viscous torque t is proportional to the shear σ of the flow, $t = -2\eta\sigma$. At the horizon, the coefficient of kinetic viscosity, $\nu = \eta/\rho_0$, vanishes for causality reasons,

$$\nu(r_h) = 0 , \tag{4}$$

independent of the specific physical processes responsible for viscosity. Thus the fluid penetrates the horizon on ideal trajectories, which have been shown (Anderson & Lemos 1988) to yield finite shear component σ_ϕ^r everywhere. Equation (4) then represents a *no-torque* condition $t_\phi^r(r_h) = 0$, i.e. viscous stresses vanish at the horizon and the angular momentum L_0 accreted by the black hole is equal to the angular momentum of the fluid at the horizon, $L_0 = L(r_h)$. In our model condition (4) is guaranteed by an appropriate parametrization with α as a viscosity parameter. We apply the full expression for σ^r_ϕ, expressed as a function of variables u^r, l and du^r/dr, dl/dr. Equations (1), (3) then yield a coupled system of two differential equations in u^r and l, which we integrate between r_h and $r_\mathrm{out} = 100r_\mathrm{g}$ ($r_\mathrm{g} = GM/c^2$ being the gravitational radius), respecting the regularity condition (2).

The half-thickness of the flow is calculated assuming vertical hydrostatic equilibrium, $u^z = 0$. For that purpose the vertical Euler equation is expanded to first order in z/r along the lines of Riffert & Herold (1995).

3   Global transonic solutions

The regularity condition (2) enforces sub-Keplerian profiles at the sonic point. Two distinct types of transonic solutions are identified (compare Abramowicz & Zurek 1981; Chakrabarti 1990). In type I solutions the sonic point is located in the vicinity of the marginal stable orbit r_ms ($r_\mathrm{ms} = 6r_\mathrm{g}$ for a Schwarzschild black hole and $r_\mathrm{ms} = 1.94r_\mathrm{g}$ for a rapidly rotating black hole with spin parameter $a = 0.95$). In type II solutions the sonic point is located further away from the horizon of the black hole. Whether a particular solution is type I or type II depends on the properties at the sonic point r_s, e.g. on the specific energy $E(r_\mathrm{s})$ and angular momentum $\lambda(r_\mathrm{s})$ with $E = -\mu u_t$ and $\lambda = -u_\phi/u_t$ respectively. In non-viscous flows E, λ are conserved and the solution type is known from the boundary conditions.

The situation is similar for viscous solutions. Type I solutions are generally found for moderate viscosity parameters α and higher 'modified accretion rates' $\dot{\mathcal{M}} \equiv (\Gamma K)^N (\dot{M}/2\pi)$. Figure 1 shows type I solutions around a non-rotating black hole, which differ only in α. They show globally sub-Keplerian angular momentum profiles. Variation of α allows to match any reasonable angular momentum $\lambda(r_\mathrm{out})$. There exist solutions which approach a Keplerian angular momentum distribution λ_k globally or even exceed λ_k locally. Figure 2 shows two examples of such trans-Keplerian solutions, again for a non-rotating black hole. The distribution of adiabatic sound speed, c_ad, and thus the temperature distribution possess a maximum outside r_s in any solution of type I .

Type II solutions are found for higher α and lower $\dot{\mathcal{M}}$. In such flows the angular momentum L_0 deposited onto the black hole is generally lower. Figure 3 shows a sequence of type II solutions around a non-rotating black hole for fixed value of 'modified accretion rate' $\dot{\mathcal{M}}$. Type II solutions generally do not allow to match any reasonable angular momentum at r_out by variation of α only. Type II solutions have generally lower temperature than type I , with a maximum at the horizon.

4   Summary

We obtain prograde disc solutions for any Kerr parameter a between $a = 0$ and $a \simeq 0.99$. Except from quantitative differences, solutions show the same

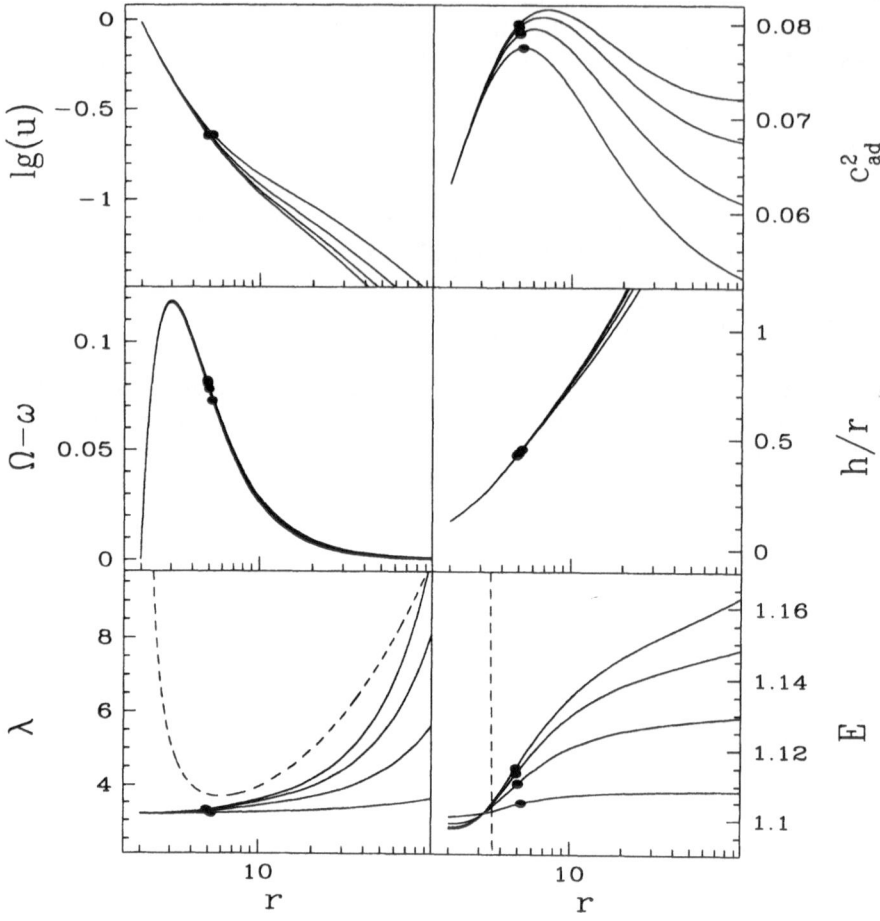

Fig. 1. Sequence of type I viscous disc solutions around a non-rotating black hole. Solutions are for $a = 0$, $\dot{M} = 0.032$ and $L_0 = 3.5$, with α decreasing (from top to bottom) as $\alpha = 0.045$, 0.04, 0.03, 0.01. Dashed curves represent Keplerian distributions. The sonic point is marked by •. Adopted from Peitz & Appl (1997).

qualitative behavior for any a. Accretion discs around rotating black holes are hotter and deposit less angular momentum on the central object than accretion discs around Schwarzschild black holes. Since the event horizon of a rotating black hole is smaller than the horizon of the corresponding (equal mass M) non-rotating black hole, an accretion disc around a rotating hole reaches deeper into the gravitational potential well. Thus the virial temperature of the gas at the inner edge is higher in discs around rotating black holes. Indeed, it is a general

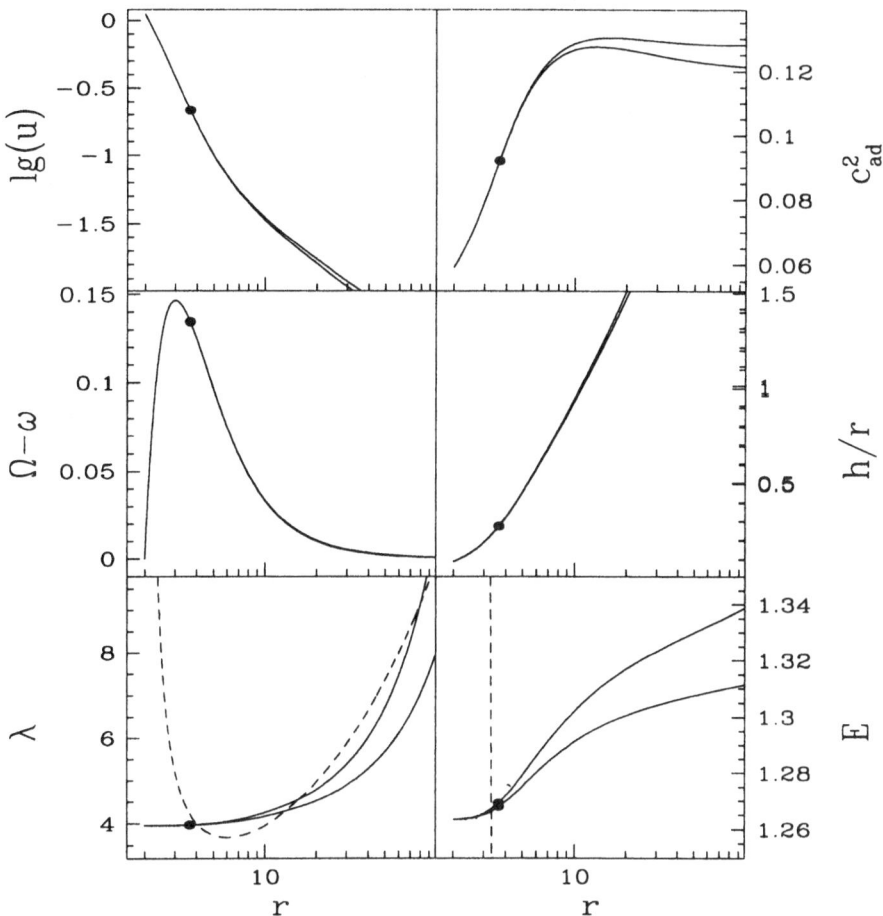

Fig. 2. Sequence of type I viscous disc solutions around a non-rotating black hole. Solutions are for $a = 0$, $\dot{M} = 0.032$ and $L_0 = 5.0$, with α decreasing (from top to bottom in the λ-diagram) as $\alpha = 0.008$, 0.006. Dashed curves represent Keplerian distributions. The sonic point is marked by •. Adopted from Peitz & Appl (1997).

feature of optically thin ADAFs that the ions are nearly virial (Narayan & Yi 1994, 1995ab). As a concequence the morphology of the disc is quasi-spherical, $h/r \simeq 1$. Since these general features of hot optically thin ADAFs are qualitatively and quantitatively reproduced by our global solutions, we conclude that a polytropic equation of state represents a close approximation to the properties of optically thin ADAFs.

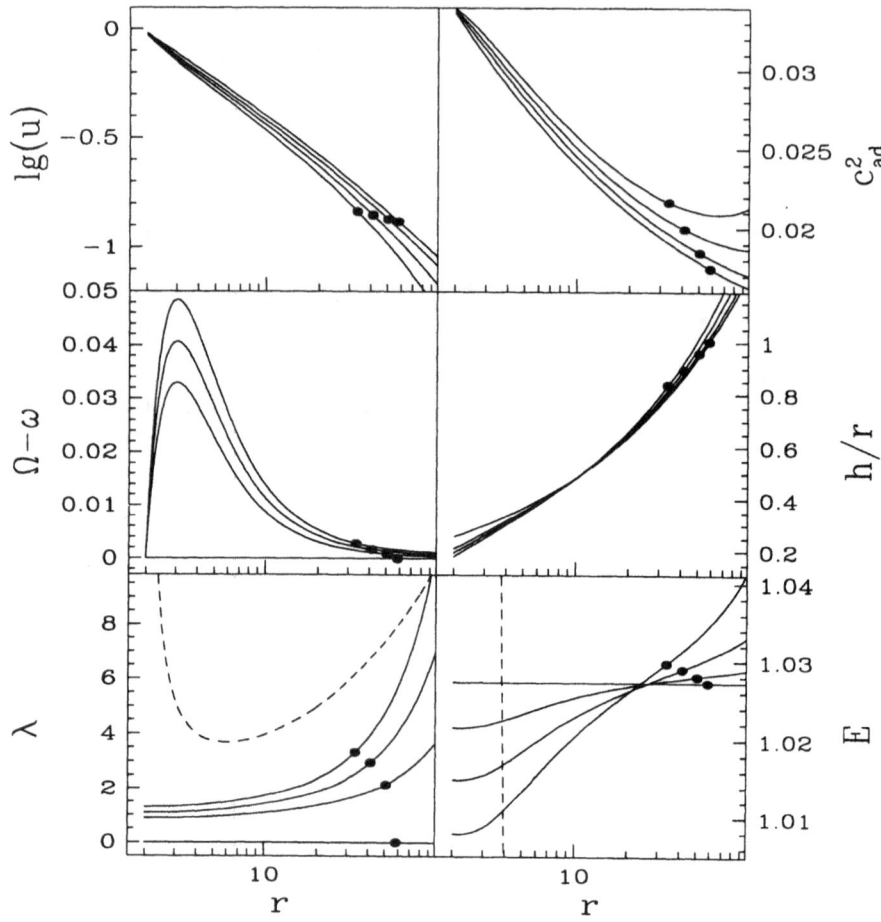

Fig. 3. Sequence of type II viscous disc solutions around a non-rotating black hole. Solutions are for $a = 0$ and $\dot{\mathcal{M}} = 0.011$. The upper two solutions have $\alpha = 0.6$ and (from top to bottom in the λ-diagram) $L_0 = 1.3, 1.1$. The lower two solutions have $\alpha = 0.5$ and (from top to bottom in the λ-diagram) $L_0 = 0.9$, 0.0. Dashed curves represent Keplerian distributions. The sonic point is marked by •. Adopted from Peitz & Appl (1997).

Acknowledgement: The work reported in this contribution was initiated by M. Camenzind. We thank him and R. Khanna for many fruitful discussions. The support by Deutsche Forschungsgemeinschaft SFB 328 (J.P.) and by the French Ministery of Foreign Affairs (S.A.) is acknowledged.

References

Abramowicz M.A., Chen X., Granath M., Lasota J.P., 1996, ApJ in press

Abramowicz M.A., Chen X., Kato S., Lasota J.P., Regev O., 1995, ApJ 438, L37

Abramowicz M.A., Zurek W.H., 1981, ApJ 246, 314

Anderson M.R., Lemos J.P.S., 1988, MNRAS 233, 489

Chakrabarti S.K., 1990, Theory of Transonic Astrophysical Flows, World Scientific, Singapore

Lasota J.P., 1994, in Duschl W.J., Frank J., Meyer F., Meyer-Hofmeister E., Tscharnuter W.M., eds, Theory of Accretion Disks 2.Kluwer, Dordrecht

Narayan R., Yi I., 1994, ApJ 428, L13

Narayan R., Yi I., 1995a, ApJ 444, 231

Narayan R., Yi I., 1995b, ApJ 452, 710

Narayan R., Yi I., Mahadevan R., 1995, Nature 374, 623

Narayan R., McClintock J.E., Yi I, 1996, ApJ 457, in press

Novikov I.D, Thorne K.S., 1973, in DeWitt C., DeWitt B.S., eds., Black Holes, Gordon and Breach, New York

Paczyński B., Wiita P.J., 1980, A&A 88, 23

Peitz J., 1994, Diploma Thesis, University of Heidelberg, Germany

Peitz J., Appl S., 1997, MNRAS in press

Riffert H., Herold H., 1995, ApJ 450, 508

Comparison of different models for the UV-X emission of AGN

S. Collin-Souffrin[1][2], A.-M. Dumont[1]

[1] DAEC, Observatoire de Paris, Section de Meudon, 92195 Meudon, France
[2] Institut d'Astrophysique, 98bis Bd Arago, 75014 Paris, France

Abstract: After briefly recalling some important observational results concerning the UV and X–ray spectrum of AGN, we discuss the radiation mechanisms which are involved and describe different models proposed to account for them. We schematize these models in a simple way, and compute the emitted spectra, insisting on their main characteristics and on their differences.

1   Introduction and historics

Since the discovery of quasars and of the whole generic class of AGN our ideas about their radiation mechanisms have strongly evolved. At the beginning, and mainly because the first quasars were radio loud objects, the spectrum from radio to X-ray was attributed to the synchrotron (or the synchro-Compton) process. Rapidly it was proposed that active nuclei are fueled by accretion onto a massive black hole (cf. Rees, 1984), so the seventies were devoted to build models accounting for synchrotron radiation (or cyclotron in some cases) in this framework.

A breakthrough occurred with the discovery, first in 3C 273, then in other quasars, of the "blue bump", which was attributed to the thermal emission of an accretion disk (Shields, 1978, Malkan & Sargent, 1982). Thanks to IUE, the blue bump was then also detected in local AGN, and was observed to vary rapidly. Meanwhile observations made with Einstein and EXOSAT lead to the discovery of the soft X-ray excess (Wilkes & Elvis, 1987). EXOSAT allowed also to discover the very short time variability of the X-ray flux in Seyfert galaxies and in a few quasars. It became then evident that a large fraction of the luminosity of AGN is emitted in the UV and soft X–ray range, has a thermal origin, and is produced in a small region close to the black hole, while the infrared is most probably dominated by thermal dust emission produced much further out, at least in radio quiet quasars (Sanders & al, 1989). This paper will concentrate on the part of the spectrum emitted in the very central regions, namely UV and X-rays.

Another important insight onto the emission mechanism came with the observations made by GINGA which lead to the detection of a "reflection" component in the X-ray spectrum od AGN (Pounds & al, 1990). Finally a last surprise (and contrary to results obtained almost two decades ago in the archetypal object NGC 4151) was hold in store for us by the GRANAT and GRO missions: at high energies the X-ray spectral distribution of radio quiet AGN has a break much below 1 MeV. This signed the death sentence of non thermal models for the X–ray emission and validated thermal Compton models which were already envisioned before.

The dominance of thermal luminosity is certainly also valid in a majority of radio loud AGN. Besides the presence of radio emission (which does not contribute to an important fraction of the luminosity), there are however small differences in the spectrum of radio loud and radio quiet AGN. Radio loud objects radiate more energy in the X–ray range than radio quiet AGN (typically by a factor 3), so it is possible that a non thermal mechanism such as the synchro-Compton contributes to the X–ray emission. Finally we do not consider at all in this paper the class of "radio-gamma loud" AGN, characterized by an intense gamma ray emission, which are most probably radio loud AGN seen in the direction of a relativistic jet. They constitute a minority of the radio loud class, which itself represents only 10% of all AGN.

In the following section we briefly recall some observational facts on which the most recent models are based. Deductions concerning the emission mechanisms are summarized in Section 3. Section 4 describes a few models integrating these results, and the characteristics of the predicted spectrum for each model are shown and discussed in Section 5.

2   Observational facts that any model should explain

We recall here some important results, which are based on a considerable number of papers, that we cannot refer to. A more extensive review can be found for instance in Collin-Souffrin & al, 1996.

2.1   The optical-UV range

The main feature in the optical-UV range is the "Big Blue Bump" (so called to distinguish it from the "Small Bump", which is attributed to a mixture of Balmer continuum and FeII blends). There is much confusion about this feature. Sometimes it corresponds only to the rise of the continuum in the optical-UV range (i.e. in a $\log \nu F_\nu$ versus $\log \nu$ diagram the slope of the continuum becomes positive), and sometimes it takes into account the soft X-ray excess (cf. later), and is therefore a much bigger feature linking the UV to the X-ray continuum. This is for instance the conclusion of Walter & al, 1994, after the study of 8 AGN observed simultaneously in the UV and in the soft X–ray ranges. However a thorough discussion by Collin-Souffrin & al, 1996, based on several samples of AGN shows that the UV continuum steepens below 2000Å, precluding a smooth junction with the soft X-ray continuum. In the following, and for an easy comparison with predicted spectra, we shall adopt an energy spectral index equal to unity at 2000Å.

A second important issue concerning the UV continuum is the absence of any Lyman discontinuity. A weak absorption edge is observed at the emission redshift in high-z quasars (the only objects observable in this range) but this discontinuity may be produced by the Broad Line Region or by an intervening intergalactic cloud not linked with the quasar. This raises a severe problem. Indeed all models for the UV emission of AGN predict a discontinuity in absorption or in emission, depending on the structure of the surface layers of the emission region. Its complete absence requires either fine tuning or a mechanism which suppresses it.

The variability of the spectral distribution and of the flux is also a strong constraint for models. The problem is reviewed by M-H. Ulrich in these proceedings. The following results should be recalled.

- The variability time scale (defined as the e-folding time for flux variations) scales with the luminosity, although the exact dependence is not well known. In Seyfert galaxies with a bolometric luminosity of the order of 10^{45} ergs s^{-1} the variation time scale is typically of the order of a few days, while it is several months in high luminosity quasars. The deduced maximum dimension of the emission region is about 10^3 R_G, R_G being the gravitational radius (note that this dimension implies a causal link between different parts of the emission region provided by a phenomenon propagating with the speed of light; if the velocity is less, the deduced dimension is also smaller).

- Although a correlation has been observed between UV and soft X-ray fluxes, it does not seem to be present in "high states" of the X-ray flux, indicating probably a larger size for the UV emission region.

- The continuum "hardens when it brightens". This trend is observed not only in the optical-UV but also up to the EUV range in one object (Marshall & al, 1996). It proves the existence of a tight relation between the UV and the EUV continuum.

- Finally an important constraint is set in well monitored Seyfert nuclei by the absence of a measurable time lag between the optical and UV light curves.

2.2   The soft X–ray range: the "soft X–ray excess"

Below 1 KeV the continuum displays an excess when compared to the extrapolation of the continuum in the 1-10 KeV range, where the energy distribution is close to a power law of slope 0.7. The variability time scale is smaller than in the UV range, and corresponds to a dimension smaller than 10^2 R_G (the same restriction holds than for the UV variation time scale concerning the velocity of propagation of the perturbation). Several spectral features are present: an absorption edge close to 1 KeV and two absorptions identified with resonance lines of OVII and OVIII (in one object these lines as well as several others are present in emission).

2.3   The hard X–ray range

In the hard X-ray range, the spectrum consists of an underlying power law with an energy spectral index ~ -0.9 and superimposed features, mostly a broad emission line at 6.4 KeV identified with a Kα line of weakly ionized iron, an absorption edge around 8 KeV, attributed to the same ions, and frequently traces of a "hump" above 10 keV. The variation time scale in this range is comparable to the soft X–ray time scale and the variations seem to be correlated. Recent observations with ASCA show that the line varies very rapidly, implying that it is emitted in a very small radius close to the black hole. The typical equivalent width of the line is a few hundreds eV. The profile of the line is characterized by a very extended red wing (cf. Fabian's paper in these proceedings).

Above 40 KeV the spectrum is characterized by a cut-off with an e-folding energy of 100 to 200 KeV. It does not display any emission feature around 500 KeV which could be attributed to the electron-positron annihilation line.

A last important result is that the UV/soft X–ray luminosity is larger than the hard X–ray luminosity, or at least of the same order. An easy way to figure out the UV to X-ray luminosity is the spectral index α_{UV-X} which is the slope of an hypothetical energy power law extending from 2500Å up to 2 KeV. This index varies -1.2 in local Seyfert 1 nuclei and in radio loud quasars, to -1.5 in radio quiet quasars.

3   Emission mechanisms

3.1   UV and soft X–ray emission

The distinction between the two assumptions concerning the Big Blue Bump is not without consequence. If the EUV flux is simply the interpolation between the broad UV and soft X-ray bands, it implies that the EUV contains a major fraction of the bolometric luminosity, and that the same mechanism is giving rise to the whole spectrum from UV to soft X-rays, which is not trivial. The last point is particularly well illustrated in the poster contribution of Staubert & al: in their fits of the soft X-ray spectrum by accretion disk models, the computed UV continuum is much flatter than the observed one.

Whatever the assumption, there is not a big choice for the emission mechanism: the shape of the continuum implies a thermal mechanism with the emission regions spanning a temperature from 10^5 to 10^6 K (by thermal emission we mean that the bulk of the gas participates to the emission, and not only a small fraction of particles). This can be achieved in two different cases:

1. an optically thick medium, radiating locally like a black body (or a modified black body);

2. a medium optically thin down to a frequency well below the peak of the Planck curve, so the emission is dominated by optically thin free-free emission (here the optical thickness does not include the diffusion coefficient).

In the first case, one finds:

$$R_{10} \sim 3 \left(\frac{\Omega}{4\pi}\right)^{-1/2} T_5^{-2} M_8^{-1/2} \left(\frac{L_{UV}}{L_{Edd}}\right)^{1/2} \tag{1}$$

where L_{UV} is the luminosity of the Big Blue Bump (eventually including the soft X–ray excess), L_{Edd} the Eddington luminosity, Ω the solid angle covered by the emission region (assuming a spherical emitting shell), T_5 the effective temperature in units of $10^5 K$, R_{10} the radius in units of $10 R_G$, and M_8 the black hole mass in $10^8 M_\odot$.

One deduces from Sections 2.2 and 2.3 and from this equation that the variations are associated with a phenomenon propagating with a velocity close to that of light, therefore most probably by electromagnetic transport. For instance in the standard accretion disk model the causal link between the inner regions of the disk (producing the UV continuum) and the outer regions (producing the optical continuum) should be provided at the speed of light to account for the absence of time lag between the optical and UV fluxes. In other words the disk is radiatively and not viscously heated.

In the second case, the emission is dominated by thin free-free emission and it can be easily shown that:

$$\frac{L_{ff}}{L_{bol}} \sim 0.3 \ 10^{-27} \frac{\Omega}{4\pi} \left(\frac{L_{bol}}{L_{Edd}}\right)^{-1} n^2 f R_{10}^3 \ T_5^{1/2} \tag{2}$$

where n is the density, f is the volumic filling factor, L_{ff} and L_{bol} are respectively the free-free and the bolometric luminosities. On the other hand, we know that the medium is optically thin down a frequency of at most 10^{15} Hz. Using the Rayleigh-Jeans expression of the free-free opacity, the previous equation, and the fact that L_{ff}/L_{bol} should be of the order of unity, $n^2 f$ can be eliminated and one finds the following relation:

$$R_{10} \geq 20 M_8^{-1/2} T_5^{-1} \left(\frac{L_{bol}}{L_{Edd}}\right)^{1/2} \left(\frac{\Omega}{4\pi}\right)^{-1/2} \tag{3}$$

which shows that the dimension of the UV emission region is larger than in the previous case. So again in these optically thin models the optical and UV emission regions are most probably radiatively heated (cf. Collin-Souffrin & al, 1996).

3.2   Hard X–ray emission

Owing to the absence of the electron-positron annihilation line and to the existence of the cut-off at about 100 KeV, it is now widely admitted that the emission mechanism is thermal, with $\Theta_e = kT_e/mc^2$ of the order of 0.1. Three thermal mechanisms can compete in this hot medium: free-free, cyclotron, and inverse Compton emission.

It can be shown that the time scale for inverse Compton cooling is smaller than the time scale for free-free cooling when the "compacity" l defined as :

$$l = \frac{L}{R} \frac{\sigma_T}{mc^3} \sim 10^3 \frac{L_{UV}}{L_{Edd}} R_{10}^{-1} \tag{4}$$

is larger than $0.03\,\Theta_e^{1/2}$ (σ_T is the Thomson cross section).

Using the dimension of the emission region inferred from the variability, one deduces that the condition is fulfilled, so inverse Compton cooling dominates over free-free cooling. It dominates also over cyclotron cooling unless the magnetic pressure is of the order of the density of radiation, which implies a very high value of the magnetic field.

We have seen that the hard X–ray spectrum is actually made of several components: a power law with a spectral index equal to 0.9, and two emission features, the 6.4 KeV line and the hump. These features are attributed respectively to continuum fluorescence and to Compton reflection of the power law spectrum on a "cold" medium (i.e. with a temperature smaller than a few 10^6 K) covering at least 2π of the source. The power law spectrum is often called the "primary", although it can itself be the result of a reprocessing mechanism (as it is the case in models 1 and 4 discussed below). In the accretion disk model for the UV emission, the cold medium is identified with the disk. In the "optically thin" case for the UV emission the existence of the hump requires a Thomson optical thickness of the order of unity.

In the past the 6.4 KeV line was sometimes ascribed to reflection on the molecular torus invoked in the "Unified Scheme" of AGN. The discovery of its very rapid variations in Seyfert 1 nuclei proves that it is most probably linked with the UV emission region and not with the torus which is located much further away. This means that the edge observed near 7 KeV is the signature of the absorption process leading to the fluorescence mechanism, and is also produced at least partly in very central regions. As we shall see below, the same explanation does not hold for the absorption edges observed near 1 KeV, which are therefore attributed to a warm dilute medium surrounding the primary source. The status of this "warm absorber" is still not clear, although it is often believed to be spatially associated with the Broad Line Region, at about $10^4 R_G$ from the center. The warm absorber is therefore out of the scope of the present discussion.

A "consistent" phenomenological model comes thus out, made of two different media. One medium is "cold" and emits UV radiation mainly by reprocessing the hard X–rays. The other is hot and is cooled by inverse Compton process on soft photon seeds. These soft photons are likely provided by the UV emission region, but they can also be produced by other mechanisms, such as thermal cyclotron process in the hot gas itself.

4   The models

We describe here a few models which are phenomenological more than physical ones, since they do not include a physical description of the accretion process. All these models share a common property (this is a result of our previous discussion): the UV emission region is illuminated by the X–ray continuum.

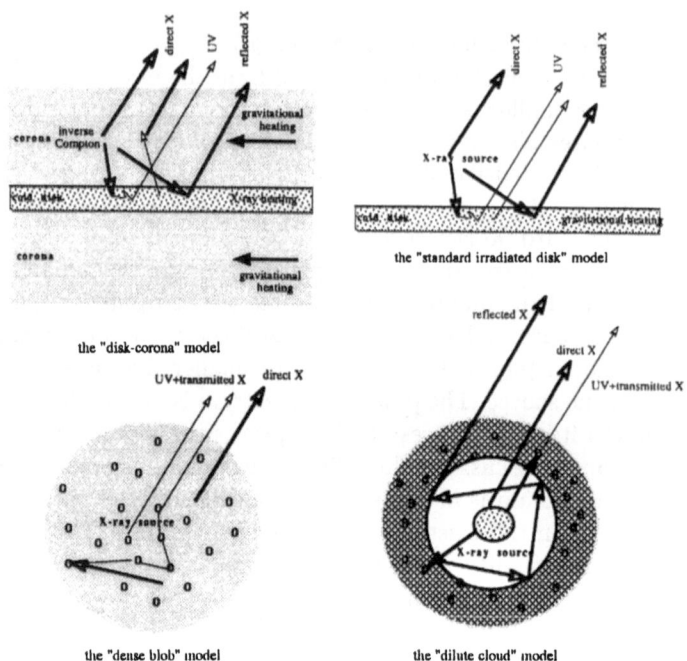

Fig. 1. Schemes of the 4 models described in the text.

This radiation is absorbed in a Thomson depth of a few, which is therefore photoionized. If thermal and ionization equilibrium is reached in this region (and this is the case, owing to its high density which implies very small time scales for the atomic processes), its ionization state and temperature depend on the "ionization parameter". We choose for the definition of this parameter (Krolik, McKee & Tarter, 1981):

$$\Xi_X = \frac{L_X}{4\pi R^2 nckT} \tag{5}$$

where T and n are again respectively the temperature and the density of the UV emission region, whose distance from the X–ray source is R. One finds that to get T of the order of 10^5 K, Ξ should be of the order of unity. It leads to the following relation:

$$n_{17}R_{10}^2 \geq \frac{L_X}{L_{\mathrm{Edd}}} M_8^{-1} T_5^{-1} \mathrm{cm}^{-3}. \tag{6}$$

This means that the smaller the distance, the higher the density, and vice versa. The relation is actually an inequality because in a thick medium the deep layers contributing to the bulk of the emission have a smaller temperature than the illuminated surface where the equality applies.

Note that another ionization parameter will be also used in the following:

$$\xi_X = \frac{L_X}{nR^2} = 5T_5 \Xi_X. \tag{7}$$

Eqs. 5 and 7 show that the temperature of a photoionized medium depends also on Ξn, i.e. on the illuminating flux. In particular if the medium is optically thick, the fraction of the irradiating flux which is not reflected is reprocessed as a quasi black body spectrum whose temperature is given by Eq. 3 (as the albedo is small).

Finally one should recall the emission properties of an irradiated medium:

1. If the medium is Thomson thick, it produces two different spectra. The illuminated side produces a **reflection** (inward) spectrum, and the other side produces an **emission** (outward) spectrum. These spectra are different: the "reflection spectrum" depends mainly on the ionization parameter, and the "emission spectrum" depends mainly on the flux.

2. If the medium is optically Thomson thin, it does not produce a reflection spectrum, but only an "emission spectrum" which depends on the ionization parameter.

4.1   Models involving an accretion disk for the UV emission

An important characteristic of these models is that their geometry allows an external X–ray source to be seen directly. The observed spectrum is thus the sum of 3 components: the disk emission (dominating in the UV and EUV), and the primary and reflection spectra, dominating in the X–ray range.

1. The disk-corona model. A model coupling the UV and X-ray spectra and satisfying several of the above specifications was proposed a few years ago by Haardt & Maraschi, 1991, and Haardt & Maraschi, 1993. It consists in an optically thick cold disk sandwiched by an optically thin hot corona where a large fraction of the dissipation is assumed to take place. The corona is cooled by inverse Compton on the soft UV photons coming from the disk and it emits X-rays. The disk is heated by X-ray photons from the corona, and reprocesses this radiation as thermal UV. The model is schematized on Fig. 1.

The model is "self regulated" and can therefore account for a universal spectrum. The ratio between the inverse Compton luminosity and the soft luminosity is indeed fixed and implies a Comptonization factor y of the order of unity. The model predicts then the correct spectral index of the hard X-ray spectrum α_X, given by (cf Rybicki & Lightman, 1979):

$$\alpha_X = \frac{-ln\tau_T}{-ln\tau_T + lny}. \tag{8}$$

where τ_T is the Thomson thickness of the corona. We see that α_X is close to unity almost independently of τ_T. If one assumes that gravitational energy is dissipated locally in the corona, the model predicts also a value of T_e which corresponds to the correct high energy cut-off, and finally it predicts that the UV flux be roughly equal to the X–ray flux (the UV/X-ray flux ratio can be increased by assuming an inhomogeneous corona (cf Haardt, Maraschi & Ghisellini, 1994).

The physics of the model has been studied in several papers (Nakamura & Osaki, 1993, Kusunose & Mineshige, 1994, Zycki, Collin-Souffrin & Czerny, 1994, Svenson & Zdiarski, 1995). Since the disk has no internal heat generation, it is almost isothermal and has a weak radiation pressure. It is therefore supported only by gas pressure and is very dense. According to the way angular momentum is transported in the disk and in the corona, several descriptions of the disk-corona structure can be given. For instance following Sincell & Krolik, 1996, one finds for a vertically averaged disk (neglecting boundary conditions and assuming a non rotating black hole):

$$H \sim 10^{11} T_5^{1/2} \, M_8 R_{10}^{3/2} \text{ cm} \qquad (9)$$

$$n \sim 10^{21} T_5^{-3/2} M_8^{-1} \frac{L_{\text{UV-X}}}{L_{\text{Edd}}} \alpha^{-1} R_{10}^{-3} \text{ cm}^{-3} \qquad (10)$$

$$T_5 \sim \left(\frac{L_{\text{UV-X}}}{L_{\text{Edd}}} \right)^{1/4} M_8^{-1/4} R_{10}^{-3/4} \text{ K} \qquad (11)$$

where α is the viscosity parameter and H is the thickness of the disk. These values are very different from those of a standard disk (cf. below).

Actually the fraction of the disk where X–ray photons are absorbed has a lower average density, since it corresponds only to an irradiated "skin", located about 3 scale heights above the midplane of the disk (Sincell & Krolik, 1996). A typical density for this skin is 10^{18} cm^{-3}. These authors have computed the UV continuum emitted by the X–ray heated skin, integrating the emission on the radius, neglecting heavy element cooling, assuming hydrostatic equilibrium, and using a two stream Eddington approximation for the transfer of radiation.

2. The standard irradiated disk model. Ross & Fabian, 1993, Matt, Fabian & Ross, 1993a, Zycki & al, 1994, studied the emission of "standard" irradiated accretion disks. The inner regions of these disks are radiation pressure supported and dominated by Thomson scattering opacity (Shakura & Sunayev, 1973). Their structure is given by the following equations:

$$H \sim 10^{14} \frac{L_{\text{UV}}}{L_{\text{Edd}}} \text{ cm} \qquad (12)$$

$$n \sim 10^{10} \left(\frac{L_{\text{UV}}}{L_{\text{Edd}}} \right)^{-2} M_8^{-1} \alpha^{-1} R_{10}^{3/2} \text{ cm}^{-3}. \qquad (13)$$

Contrary to the previous one, this model assumes the presence of an a priori given primary X–ray continuum. It is schematized on Fig. 1.

The computations of the emitted spectrum were made with different methods and different schematizations of the model. They all assume that the disk has a constant density in the vertical direction. It is irradiated on the top side by a power law X-ray continuum and on the bottom by an underlying black body (or by a Wien spectrum for Ross & Fabian, 1993) to approximate the effect of viscous heating. In Zycki & al, 1994, the computation is performed for a slab of constant density ($n = 10^{14}$ cm^{-3}) and different irradiation fluxes, and the transfer is treated with a Monte Carlo method taking into account Compton diffusions. In Matt, Fabian & Ross, 1993a, the emission is integrated on the whole disk, irradiated by a point source located at a given height above the disk, and the transfer is treated with a one stream approximation and the Kompaneets diffusion equation.

Special care was given to the computation of the iron line intensity and profile by Matt, Fabian & Ross, 1993b, and Matt, Fabian & Ross, 1996, within the framework of this model. Since it is the subject of Fabian's review in the same proceedings, we shall not devote a long discussion to this - however crucial - problem. One finds that it is indeed difficult to reconcile the large observed equivalent width of the line, which requires a strongly ionized medium, with the centroid of the line profile, which implies a low ionization state. This was called the "emission reflection paradox" by Collin-Souffrin & al, 1996. Integration on a large disk can help to solve the problem, since lines of FeXXIV to FeXXVI are emitted by the innermost regions, and are therefore Compton scattered and strongly Doppler broadened (in particular in the case of a rotating black hole). They are thus more difficult to detect than lines of iron in lower ionization states, emitted further away. However, if the very broad red wings observed in a few objects (which may be a universal feature) is due to a gravitational effect, it means that the bulk of the line emission takes place very close to the black hole, where the disk is found to be strongly ionized in this model.

4.2   Models not involving an accretion disk for the UV emission

In these models UV radiation is provided by free-free and free-bound processes in a medium which is optically thin in the UV range and covers the X-ray source. Therefore a characteristic of the models is to provide transmitted X-ray photons besides direct and reflected X-ray photons, as the geometry is no more disklike. In the following computations this transmitted spectrum is included in the outward emission.

This type of models requires a large covering factor, of the order of unity, to produce enough UV luminosity (cf Eq. 3). The problem is then that the X-ray source would be covered and heavily absorbed in the soft and even hard X-ray range in a majority of objects, contrary to what is observed. This can be avoided in two ways: either the UV reprocessing material is intimately mixed with the X-ray photons, or the UV reprocessing material covers almost completely the X-ray source, but X-ray reflection photons are provided in sufficient amount and are seen through holes in the medium. Two models satisfy these two requirements.

The "dense blob" model. This model was first proposed by Guilbert & Rees, 1988, and Ferland & Rees, 1988, and thereafter discussed by Celotti, Fabian, Rees, 1992, and Kuncic, Celotti & Rees, 1996. It is made of very dense blobs located **inside** a power-law X–ray source (which is a priori given and is not produced by a hot plasma). The blobs are very dense (typically $n = 10^{18}$ cm^{-3}) and each blob has a small thickness (typical column density of the order of 10^{21} cm^{-2}), but there are several clouds on one line of sight. So each cloud "sees" a primary radiation already reprocessed by the other clouds on its line of sight. The model is schematized on Fig. 1.

There are two consequences of this geometry.

1. The fraction of primary radiation seen by an external observer is small (it is equal to the ratio of the mean distance between the blobs to the size of the emission region).

2. For a given ionization parameter, the equilibrium temperature of the blobs is higher - and the blobs are more homogeneous - than for a unique cloud with a thickness equal to the sum of all clouds on a line of sight. In particular, if the power law spectrum extends down to small frequencies, induced Compton heating is important, and it affects a larger fraction of the blobs than in a "one blob" model.

The computations of the spectrum (cf. Kuncic, Celotti & Rees, 1996) have been made with the photoionization code CLOUDY (which treats the transfer in the one stream approximation) for one blob, and iterated in order to take into account the fact that the blobs are irradiated by a mixture of primary and reprocessed radiation.

The "dilute cloud" model. This model was proposed by Collin-Souffrin & al, 1996. It consists in a spherical shell made of dilute clouds illuminated by a central X–ray source. The size of the X–ray source is finite and larger than the size of one cloud. The total column density of the clouds is large, typically 10^{26} cm^{-2} (whether there is one or several clouds on the same line of sight does not make any difference in this model except if there are strong velocity gradients, cf below). The system of clouds is therefore Thomson thick, although optically thin for free-free process in the optical-UV range. The coverage factor of the central source by the clouds is close to unity. It is proposed that the X–ray source is a very hot gas (almost virialized) cooled by inverse Compton process on the UV photons emitted by the clouds: the model is then quite similar to the disk-corona from the point of view of radiation processes. However it presents some peculiarities.

1. The fraction of primary radiation seen by an external observer is small, unless the line of sight is free of clouds, but this has smaller probability to occur; moreover if the clouds are small, no line of sight will be completely free of clouds, and the situation will be that of a "partial absorber".

2. Since the cloud shell is almost completely closed, a light ray undergoes many reflections on the inner face of the shell before emerging outwards. This phenomenon has several consequences:

- first it increases the radiation density inside the shell, with respect to its value given by the primary source only. The medium is hotter than in the absence of reflections. This is similar to what happens in the "dense blob model";

- second it changes the quality of the illuminating radiation, which is partially reprocessed (this is also similar to the "dense blob" model, except that it is a reflection and not an emission process); this can help to solve the above mentioned "emission/reflection paradox";

- third it increases the fraction of reflection in the resulting spectrum so it is larger than $1-\Omega/4\pi$;

- finally it increases the number of Compton diffusions undergone by a photon; this leads to a larger amount of Compton diffusion for X-ray lines, which are therefore broadened and redshifted; it could explain the observed shape of FeK lines (cf. Section 2).

Computations have been performed using a code specially designed for photoionized, dense, thick and hot media. Its main characteristic is to treat lines with complete transfer in the same way as the continuum (avoiding therefore the use of the local escape probability formalism which is not adapted to these media). It solves the transfer in the Eddington two stream approximation (for a short description of the code, cf Collin-Souffrin & al, 1996; a more extensive description is in preparation Dumont & Collin-Souffrin, 1997).

5   The resulting spectra

5.1   The transfer problem

In order to best compare the different models we have run a set of computations using our code. Before discussing the results, let us explain why it is important to treat correctly the line transfer.

Lines are generally important in these problems for two reasons:
- they can be intense and then be used as diagnostics for the models,
- at temperatures of the order of 10^5 K, they carry a large fraction of the energy, so they strongly influence the thermal balance.

All previous photoionization models of AGN treat line transfer with the escape probability formalism, which uses the probability that a line photon emitted at a given point can escape from the cloud, assuming that the rest of the cloud is homogeneous and has the same properties as the emitting layer. Even if it is achieved the escape probability can be computed exactly only in the case of complete frequency redistribution, absence of line interlocking, and negligible opacity of the underlying continuum. Therefore more or less sophisticated estimates of this probability are used to account for partial redistribution, line or continuum interlocking, but none is valid for a continuum optical thickness of the order of or larger than unity, and when the properties of the medium vary considerably from one point to another. Here the continuum underlying many lines is optically thick. Several recipes to correct the escape probabilities for continuum absorption have been given in the literature, but they cannot account correctly for the fact that a line photon emitted by a given ion at a given point is absorbed in ionizing another atom far from its emission place.

Actually models involving thermal and ionization equilibrium, non LTE, Compton diffusions, for a thick and dense medium, are extremely difficult to handle. The problem is indeed similar to building non LTE atmosphere models, with the additional difficulties of inelastic diffusions and of an external illumination by non thermal photons creating a large range of ionization states. Moreover, as the medium is very dense and/or thick, some transitions and frequency ranges are very close to LTE, while some others are very far. In our opinion no code, including ours, is presently satisfactory (for the moment our code does not take into account the spherical geometry nor the subordinate lines of heavy elements, and it treats correctly the Compton diffusions in the energy balance but not in the transfer).

5.2   Schematization of the models

Our intention is not to perform detailed computations taking into account all subtleties of the models, which by nature would be strongly model dependent, but rather to try to find out the main characteristics of the resulting spectra, and in particular to define some tests which could be used to distinguish between the models.

We schematize the models by plan-parallel slabs with constant density illuminated by an isotropic primary power law spectrum with an energy index equal to unity, a low energy cut-off at 0.1 eV and a high energy cut-off at 100 KeV: from the point of view of the UV-soft X emission, it is quite similar to illumination by a comptonized spectrum, like in the disk-corona model. In the case of disk models, we add an underlying blackbody as it was done in the computations mentioned previously. For the disk-corona, the temperature of the black body is only 10^4K, as there is very little viscous dissipation in the disk, while in the standard irradiated disk case, it is 10^5K, corresponding to a flux comparable to that of the external source. However this representation is not satisfactory, and it would be better to compute a real disk model taking into account both viscous dissipation and external irradiation. Note that in the disk model what is called the "reflection spectrum" includes here also the reprocessed underlying blackbody.

Model 2bis has been added for a direct comparison with model 4, which has similar physical conditions.

The spectra are computed with a velocity dispersion due only to thermal motions, i.e. we assume a small turbulent velocity within a slab, and that one slab does not reabsorb the radiation of another one or there are no large scale velocities. This is clearly a bad approximation for the dense blob model, and it could also not be adapted to the dilute cloud model, if the clouds are small entities and there are many clouds on a given line of sight. On the contrary it is well adapted to the disk model, where radiation transfer is mostly in the vertical direction where velocity gradients are small. Large velocity gradients could increase the intensities of strong lines, allowing them to escape from deeper regions.

Tab. 1 gives the parameters - i.e. the density, the thickness and the ionization parameter - of the runs.

n° run	density (cm^{-3})	thickness cm	ξ (power-law)	underlying black body	comments	hv(eV)	ion	W(eV)
1	10^{18}	10^8	10^{-1}	10^4K	mimics the disk-corona	654	OVIII	60
2	10^{14}	10^{12}	10^3	10^5K	mimics the standard irradiated disk	6700	FeXXV	50
2bis	10^{12}	10^{14}	10^3	10^5K	mimics the standard irradiated disk	6700	FeXXV	67
3	10^{18}	10^3	10^{-1}		mimics the dense blob model	574	OVII	498
4	10^{12}	10^{14}	10^3		mimics the dilute cloud model	6700	FeXXV	176 [*]
						6700	FeXXV	67 [**]

[*] resulting spectrum equal to (in+out)/2
[**] resulting spectrum equal to (0.1in+0.9out)+0.1primary

Table 1. Parameters of the models and corresponding equivalent widths.

5.3   Results

Figs. 2, 3 and 4, display the resulting spectra for models 1 to 4, convolved with a gaussian corresponding to a resolution of 30. This is to account for the dispersion of velocity due to radial, rotational, or turbulent motions. Actually a smaller resolution of the order of 5 to 10 would be more adapted to a region close to the black hole, but a large value helps to better understand the origin of the spectral features (whether they are due to free-bound transitions or lines, whether they are in absorption or in emission). The figures show also a typical AGN broad band spectrum.

The resulting spectrum for model 1 is simply the sum of the incident and reflection spectra. In the EUV range the spectrum is dominated by reflection (which in this case is reprocessing more than reflection). The spectrum is actually close to a black body, owing to the very high thickness and density of the medium. In the soft X-ray range it displays strong lines, such as the OVIII line (cf. Table 1). In the hard X-ray range it is smooth and dominated by primary radiation. Sincell & Krolik, 1996, found that for the range of L_{UV-X}/L_{Edd} ratio spanned by their computations (from 0.003 to 3) the Lyman edge appears in emission and its amplitude increases with the mass and with the disk inclination. Contrary to them we do not find a strong Lyman discontinuity, perhaps owing to the different structure of the irradiated slab (they assume hydrostatic equilibrium, which implies that the deeper layers are much denser than the surface layers).

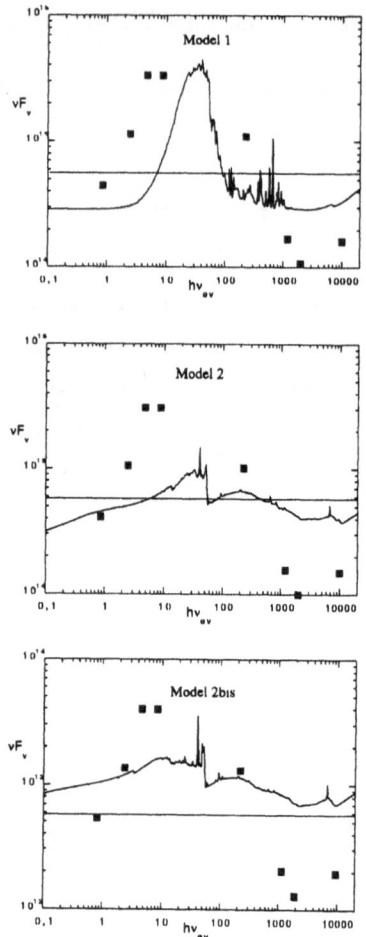

Fig. 2. The resulting spectra for models 1, 2 and 2bis. The straight line represents the incident continuum. The big squares correspond to a typical broad band AGN spectrum.

In any case this is not a severe problem, since an advantage of the disk-corona model is that the edge should be erased by Compton scattering in the corona (which is not taken into account).

Comparison with the AGN spectrum shows that this model is viable as far as the overall shape of the continuum is concerned, if the reflection spectrum could be shifted towards smaller wavelengths, which implies a smaller external flux. A detailed computation of the emission taking into account the vertical and the radial structure (which unfortunately would be quite arbitrary), as well as solving the self consistent problem of the coronal emission, would be worthwhile

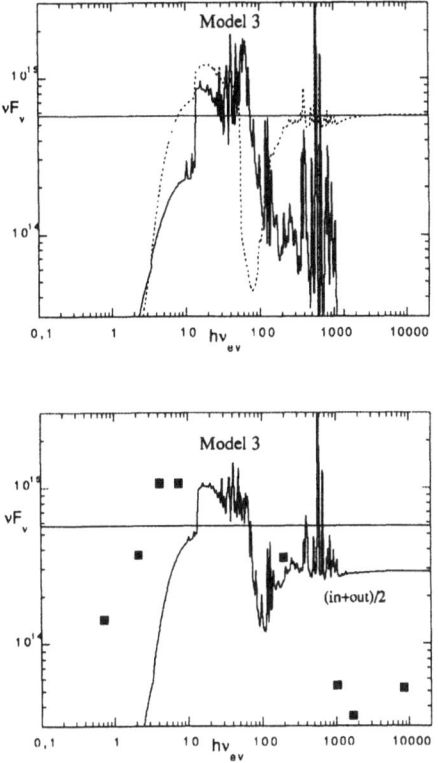

Fig. 3. Model 3. In the upper figure the dotted line corresponds to the emission spectrum and the solid line to the reflection spectrum. The lower figure shows the resulting spectrum.

in the future.

The resulting spectrum for models 2 and 2bis is also the sum of the incident and reflection spectra. Although the external flux is two orders of magnitude larger in model 2 than in model 2bis, the resulting spectrum is quite similar as it is dominated by the incident spectrum. To fit the observations the contribution of the disk emission should be much larger, which amounts to say that the ionization parameter should be much smaller.

For model 3 (the dense blob model), the resultant spectrum is assumed equal to half the sum of the emission and reflection spectrum (the emission spectrum includes transmission). We do not take into account the primary spectrum, according to our previous discussion. Note that a consistent computation would require to iterate on the spectrum incident on a cloud like Kuncic, Celotti & Rees, 1996.

Due to its high density this model is optically thick for free-free process up to a frequency larger than 10^{15} Hz. Moreover since the temperature is high the

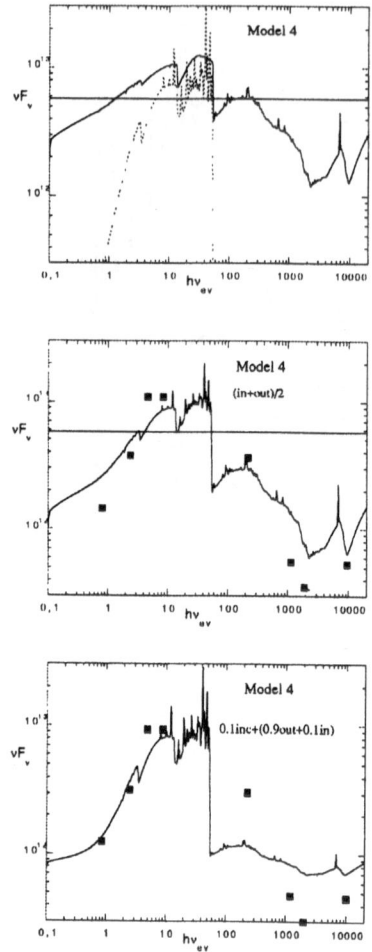

Fig. 4. Model 4. In the upper figure the dotted line corresponds to the emission spectrum and the solid line to the reflection spectrum. The two lower figures show resulting spectra.

maximum of the flux occurs around 10^{16} Hz. So the spectrum in the optical-UV range is close to a Rayleigh-Jeans law, and therefore very different from the observed one (energy spectral index -1, cf. Section 2). If we would have iterated on the reprocessed spectrum the temperature would have even be higher. Moreover, the X-ray "hump" is not reproduced, as the medium is Thomson thin. Note the very intense lines which should be easily observed. This model could perhaps account for objects with intense soft X-ray lines in emission.

There are several possible combinations of the inward, outward, and primary spectra for model 4. The emission (outward) spectrum is completely cut above

54eV owing to the large thickness of the cold outer layers of the clouds. So the X-ray spectrum is only made of reflection, and possibly of a small fraction of primary spectrum. It is steeper than in the case of the irradiated disk with the same physical parameters (model 2bis). Fig. 4 shows two examples of resulting spectra: the first corresponds to a small coverage factor, so one sees as much inward as outward radiation, and the primary source is not seen, and the second corresponds to a large coverage factor ($\Omega/4\pi = 0.9$), but the primary source is partly seen. The second case is closer to reality as the coverage factor should be large. One sees that the overall shape of the resulting spectrum fits roughly the observations, except that the Lyman edge (and in the first case the iron edge) is too strong. This is one of the problems we expect to solve in the near future by taking into account the effect mentioned previously of multiple reflections in the spherical geometry (Collin-Souffrin & al, 1996).

It is interesting to note that in the two "dense" models 1 and 3 the hard X-ray spectrum is dominated by the primary, and therefore does not display any feature. On the other hand the other "dilute" models 2, 2bis and 4, presents an Fe line in emission, but it is always the FeXXV line (actually a combination of several lines) at 6.7KeV, and not the "cold" iron line at 6.4KeV which is observed. We can also notice that the equivalent width of the line is smaller than observed, except for model 4. These are difficult problems which should be addressed in the future.

Finally in none of the models the OVII and OVIII edges are seen in absorption, so it is indeed necessary to invoke the presence of the "warm absorber" to account for these features.

6   Conclusion

We have tried to mimic in an oversimplified way different models proposed to account for the UV-X spectrum of AGN. Our aim was to show the potentiality of this kind of computation: the overall shape of the continuum, as well as the intensity of the spectral features (lines, and absorption or emission edges), could in principle help to discriminate between different models. Although the models correspond all to small dimensions (~ 10 to $100R_G$ for the UV emission region, $\sim 10R_G$ for the X-ray emission region) they sould be associated to different variation patterns of the lines and of the continuum (in particular according to the number of reflections undergone by the photons). At the present time it is not possible to discriminate between them, but one could expect that it will be feasible in the future with the new generation of X-ray experiments. In the case of organized motions, line profiles should also help to choose between the models. However we want to stress that all this work is still in infancy, as no existing code is able to treat correctly the whole complexity of these models. Moreover the physical constraints are still too loose to hope that a detailed modeling could be really conclusive at this stage.

Acknowledgement: This work has been supported by funds of the CNRS.

References

Celotti A., Fabian A.C. & Rees M.J., 1992, MNRAS 255, 419

Collin-Souffrin S., Dumont S., 1986, A&A 166, 13

Collin-Souffrin S., Czerny B., Dumont A-M., Zycki P., 1996, A&A 314, 393

Czerny B., Collin-Souffrin S., Dumont A-M., 1997, in preparation

Dumont A-M., Collin-Souffrin S., 1997 , in preparation

Ferland G.J., Rees M.J., 1988, ApJ 332, 141

Guilbert P.W., Rees M.J., 1988, MNRAS 233, 475

Haardt F., Maraschi L., 1991, ApJ 380, L51

Haardt F., Maraschi L., 1993, ApJ 413, 507

Haardt F., Maraschi L., Ghisellini G., 1994, ApJ 432, L95

Krolik J.H., McKee C.F., Tarter C.B., 1981, ApJ 249, 422

Kusunose M., Mineshige S., 1994, ApJ 423, 600

Kuncic Z., Celotti A., Rees M.J., 1996, in press in MNRAS

Malkan M.A., Sargent W.C.W., 1982, ApJ 254, 122

Matt G., Fabian A.C., Ross R.R., 1993, MNRAS 264, 839

Matt G., Fabian A.C., Ross R.R., 1993, MNRAS 262, 179

Matt G., Fabian A.C., Ross R.R., 1993, MNRAS 278, 1111

Marshall H.L., 1996, in press in ApJ

Nakamura K., Osaki, Y., 1993, PASJ 45, 775

Pounds K.A., Nandra K., Stewart G.C., George I.M., Fabian A.C., 1990, Nature 344, 132

Rees M.J., 1984, Ann. Rev. Ast. Ap. 22, 471

Ross R.R., Fabian A.C., 1993, MNRAS 258, 189

Rybicki G.B., Lightman A.P., 1979, "Radiative processes in Astrophysics", John Wiley & sons, Inc. New York

Sanders D.B., Phinney E.S., Neugebauer G., Soifer B.T., Matthews K., 1989, ApJ 347, 29

Shakura N.I., Sunyaev R.A. 1973, A&A 24, 337

Shields G.A., 1978, Nature 272, 423

Sincell M.W., Krolik J.H., 1996, in press in Ap.J.

Svenson R., Zdiarski A.A., 1995, ApJ 436, 599

Walter, R., Orr, A., Curvoisier, T.J.-L., Fink, H.H., Makino, F., Otani, C., Wamsteker, W., 1994, A&A 285, 119

Wilkes B., Elvis M., 1987, ApJ 323, 243

Zycki P., Collin-Souffrin S., Czerny B., 1995, MNRAS 277, 70

Zycki P.T., Krolik J.H., Zdziarski A.A., Kallman T.R., 1994, ApJ 437, 597

Broad X-ray emission lines from accretion disks

A.C. Fabian

Institute of Astronomy, Madingley Road, Cambridge, CB3 OHA

Abstract: The profile and variability of the broad iron lines seen in the X-ray spectra of Active Galactic Nuclei are briefly reviewed. Such lines imply the existence of dense, thin, accretion disks with abrupt density changes at their surfaces extending down to within a few gravitational radii of the central black hole.

1   Introduction

Most of the X-ray emission from radio-quiet active galactic nuclei (AGN) emerges from within a few gravitational radii of the central massive black hole. Although this is required on energetic arguments from the accretion process, we can now be fairly specific about it on observational grounds from measurements of a broad, skewed, iron emission line in the X-ray band. The observations also indicate that an accretion disk is involved and reveal some of the properties of that disk.

The shape of the X-ray continuum from AGN is not very informative in terms of its origin. It appears to be a power-law of approximately unit energy index up to a few hundred keV, above which it cuts off. Variability on times scales down to 1000 s or in some cases 100 s does indicate a compact source but does not in itself argue for say an accretion disk.

A breakthrough occurred with GINGA spectra of AGN which revealed small deviations from the power-law in the form of an iron emission line around 6.4 keV and a continuum bump around 20-30 keV (Pounds et al 1990; Matsuoka et al 1990). Both the emission line and the continuum bump are due to X-ray 'reflection' by relatively cold matter (the reflection component is the backscattered and fluorescent emission; for details see George & Fabian 1991; Matt, Perola & Piro 1991); their strength is consistent with the cold matter subtending about 2π sr at the source of the continuum. Overall, a picture emerges in which the X-ray continuum is due to flares (probably magnetic) above and below an optically-thick accretion disk.

The iron emission line is a very important ingredient. It occurs because the combination of elemental abundance and fluorescent yield are mazimized for iron (see Fig. 1). It essentially provides us with a clock (or at least a known frequency) in orbit at a few gravitational radii. Frequency shifts due to the Doppler effect, and to transverse and gravitational redshifts then cause the line to be distorted in shape, developing a characteristic skewed profile (Fabian et al 1989).

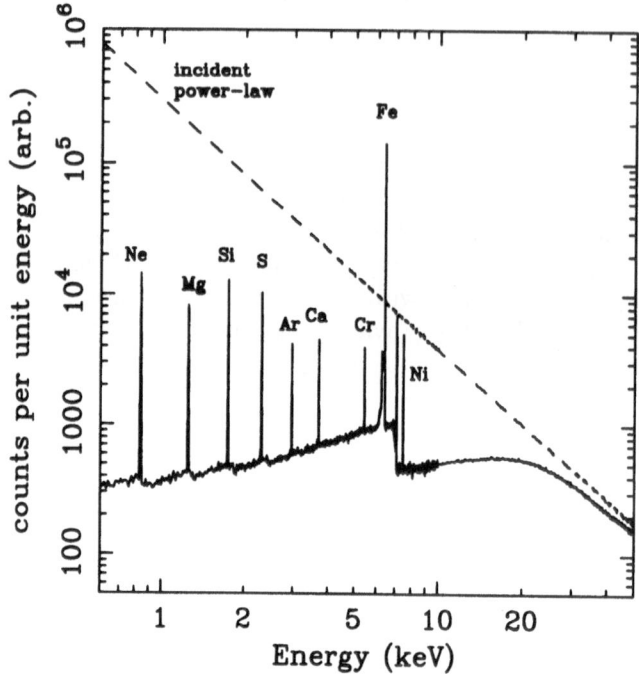

Fig. 1. A Monte Carlo simulation of reflection from a slab of gas, kindly provided by C. Reynolds.

2   Observations of the iron line profile

The proportional counters on GINGA did not clearly resolve the iron lines from AGN. Only when the CCDs on ASCA were available could this be achieved. The first look at the spectrum of the Seyfert 1 galaxy MCG–6-30-15 showed that the line is broad (Fabian et al 1994), as demonstrated by several other similar objects (Mushotzky et al 1995). The quality of the spectra from such one-day observations was not however sufficent to lead to an unambiguous statement on the line profile. It required a 4.5 day observation, also of MCG–6-30-15, to rectify that (Tanaka et al 1995).

    That observation showed clearly for the first time the skewed line profile (Fig. 2) expected from an emission source on a disk at about 20 gravitational radii (i.e. $\sim 20GM/c^2$). The line has a sharp edge to its blue wing (blue and red here mean high and low energies) at about 6.5 keV and a gradual red wing extending down to below 5 keV. The sharp blue wing means that most of the matter is moving at no more than about 30 deg from the line of sight. The gradual red wing means that most of the matter has high velocities and is deep

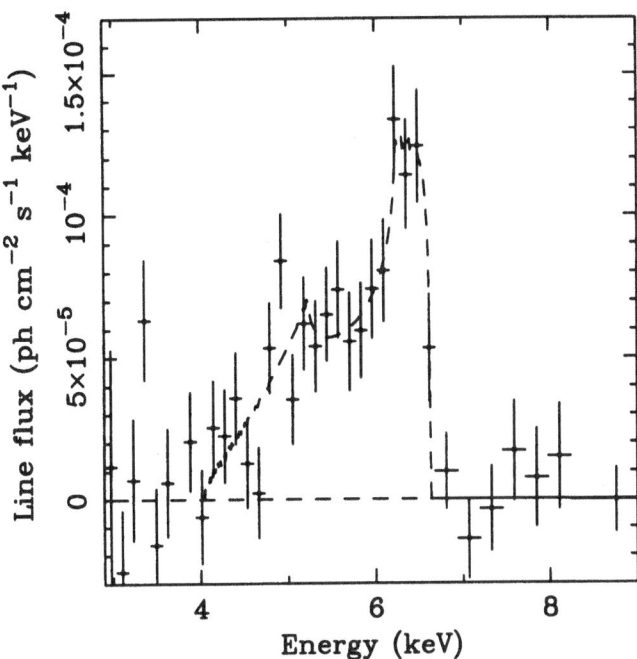

Fig. 2. The broad iron line from MCG–6-30-15 (Tanaka et al 1995). The observed energy of the line, if at rest in that galaxy, is 6.35 keV.

in a gravitational well, consistent with the emitting region being the inner parts of an accretion disk about a massive black hole (Fabian et al 1995).

Many other AGN have now been examined with ASCA and show skewed, broad iron lines (see Reynolds 1997 and Nandra et al 1997 and references therein). It appears to be a common phenomenon in Seyfert 1 galaxies. Few quasars appear to show a strong iron line and some powerful ones definitely seem to have little or no reflection (Nandra et al 1995; Cappi et al 1996). Some radio-loud objects do have broad iron lines, in particular the powerful, jetted, FRII galaxy 3C109 shows one (Allen et al 1997) indicating that a disk is present at 100 gravitational radii or closer.

A detailed examination of the long observation of MCG–6-30-15 by Iwasawa et al (1996) has revealed that some changes in the line profile took place. Most of the time the line appeared similar to that shown in Fig. 1 but during the brightest phase it just appeared to have only a blue horn, and during the faintest phase only a red wing. This last result requires that the disk extends within 6 gravitational radii which in turn requires that the black hole is spinning (i.e. is a Kerr black hole). This is because the orbiting disk can extend in to only

6 gravitational radii for a non-spinning (Schwarzschild) black hole but frame-dragging causes it to extend inward for a spinning hole. The result is tentative for the moment because the continuum is difficult to define accurately when the source is faint and it is to be hoped that it will be confirmed by further observations. It requires that the continuum source moves about on the surface of the disk.

3   Implications for accretion disks

The broad X-ray iron line from AGN now allows us to study the innermost regions about black holes and accretion disks.

• The line shape shows that matter is both approaching and receding in an inclined plane or flattened mass of gas which is best thought of as the accretion disk. The emission generally peaks at about 20 gravitational radii. This suggests a stable dense disk of matter.

• The variable X-ray continuum is emitted above a relatively cold region subtending about $\sim 2\pi$ sr which is also consistent with a disk. Much of the power is in this component (although about one half is reprocessed by the cold gas) indicating that there is good energy transport from the disk to the flare region (or corona). It may be that this stabilizes the disk (i.e. causes the disk to be dominated by gas rather than radiation pressure). Magnetic fields presumably are the medium by which the energy is transferred.

• The disk extends down to at most 6 gravitational radii and probably much closer, implying that a Kerr metric is relevant. (There are also many relativistic effects important for the observed phenomena such as variability).

• The 6.4 keV line dominates the iron emission. This means that the iron in the surface of the disk (i.e. the outer Thomson depth) is at most partially ionized so that the ions have full K and L shells. (If more ionized then the line energy increases and the line can be resonanly scattered and destroyed by the Auger effect – this may be the case for the Galactic Black Hole; Ross et al 1996.) In turn, this implies that the disk has a sharp surface, going from high density to low density in much less than a Thomson depth.

Acknowledgement: ACF thanks the Royal Society for support.

References

Allen S.W., Fabian A.C., Idesawa E., Kii T., Inoue H., 1997, MNRAS, in press
Cappi M. et al 1996, ApJ, in press
Fabian A.C., Rees M.J., Stella L., White N.E., 1989, MNRAS, 238, 729
Fabian A.C. et al, 1994, PASJ, 46, L59
Fabian A.C. et al 1995, MNRAS, 277, L11
George I.M., Fabian A.C., 1991, MNRAS, 249, 352
Iwasawa K. et al 1996, MNRAS, 282, 1038
Matsuoka M. et al 1991, ApJ, 361, 440

Matt G., Perola G.C., Piro L., 1991, AaA, 245, 75

Mushotzky R.F. et al 1995, MNRAS, 272, L9

Nandra K. et al 1995, MNRAS, 276, 1

Nandra K. et al 1997, ApJ, in press

Pounds K.A. et al 1991, Nature, 344, 122

Reynolds C.S., 1997, MNRAS, in press

Ross R.R., Fabian A.C., Brandt W.N., 1996, MNRAS, 278, 1082

Tanaka Y. et al 1995, Nature, 375, 659

An advection-dominated flow in the nucleus of M87

T. Di Matteo[1], C.S. Reynolds[1,2], A.C. Fabian[1], U. Hwang[3], C.R. Canizares[4]

[1] Institute of Astronomy, Madingley Road, Cambridge, CB3 OHA
[2] JILA, University of Colorado, Boulder, CO 80309-0440, USA
[3] NASA Goddard Space Flight Center, Greenbelt, MD 20771, USA
[4] Center for Space Research, Massachusetts Institute of Technology, Massachusetts Avenue, Cambridge, MA 02139, USA

Abstract:

   Most large elliptical galaxies should now possess massive black holes left over from an earlier quasar phase. Bondi accretion from the interstellar medium might then be expected to produce quasar-like luminosities from the nuclei of even quiescent elliptical galaxies. Such luminosities though are *not observed*. Motivated by this problem, Fabian & Rees have recently suggested that the final stages of accretion in these objects occurs in an advection-dominated mode with a correspondingly small radiative efficiency. Despite possessing a long-known active nucleus and dynamical evidence for a black hole, the low radiative and kinetic luminosities of the core of M87 provide the best illustration of this problem. Here, we examine an advection-dominated model for the nucleus of M87 and show that accretion at the Bondi rate is compatible with the best known estimates for the core flux from radio through to X-ray wavelengths. The success of this model prompts us to propose that FR-I radio galaxies and quiescent elliptical galaxies accrete in an advection dominated mode whereas FR-II type radio-loud nuclei possess radiatively efficient thin accretion disks.

1   Introduction

There is strong evidence that most nearby, large elliptical galaxies should host a massive black hole left over from an earlier quasar phase (Fabian & Canizares 1988; Fabian & Rees 1995). Quasar counts and integrated luminosities suggest masses above $10^8 - 10^9 \, M_\odot$. Giant elliptical galaxies also contain an extensive hot halo which provides a minimum fuelling level for accretion. This leads to a problem. Fabian & Canizares (1988) have determined a limit to the accretion rate for a sample of bright elliptical galaxies produced using the classical Bondi (1952) formula. They concluded that if the radiative efficiency of accretion is ~ 10 per cent, then all such nuclei would appear much more luminous than observed. The nearby giant elliptical galaxy M87 (NGC 4486) might be considered a counter example because it has long been known to host an active nucleus that powers a jet and the giant radio lobes of Virgo A. Furthermore, *HST* observations have now provided a direct dynamical determination of the nuclear black hole mass of $M \approx 3 \times 10^9 \, M_\odot$ (Ford et al. 1995; Harms et al. 1995). In fact, M87 illustrates the problem of quiescent black holes in giant ellipticals and, we suggest, illuminates the solution. Qualitative evidence for the relative quiescence of M87 comes from a comparison to the quasar 3C273, which presumably contains a black hole of comparable mass. While both have core, jet and lobe emission, the luminosity of M87 in all wavebands falls 5 orders of magnitude below that of 3C273 (see below). The contrast between M87 and 3C273 cannot be completely ascribed to a smaller mass accretion rate in the former, as can be seen by an estimate of the Bondi accretion rate in M87. Imaging X-ray observations provide information on the hot interstellar medium (ISM). A deprojection analysis of data from the *ROSAT* High Resolution Imager (HRI) shows that the ISM has a central density $n \approx 0.5 \, \mathrm{cm}^{-3}$ and sound speed $c_s = 500 \, \mathrm{km \, s}^{-1}$ (C. B. Peres, private communication). The resulting Bondi accretion rate onto the central black hole is $\dot{M} \sim 0.15 \, \mathrm{M}_\odot \, \mathrm{yr}^{-1}$. Following standard practice, we define a dimensionless mass accretion rate by

$$\dot{m} = \frac{\dot{M}}{\dot{M}_{\mathrm{Edd}}}, \tag{1}$$

where $\dot{M}$ is the mass accretion rate and $\dot{M}_{\mathrm{Edd}}$ is the Eddington accretion rate assuming a radiative efficiency of $\eta = 0.1$. For M87, the Eddington limit is $L_{\mathrm{Edd}} \approx 4 \times 10^{47} \, \mathrm{erg \, s}^{-1}$ corresponding to $\dot{M}_{\mathrm{Edd}} = 65 \, \mathrm{M}_\odot \, \mathrm{yr}^{-1}$. The Bondi accretion rate corresponds to $\dot{m} \sim 2 \times 10^{-3}$ and so, assuming a radiative efficiency $\eta = 0.1$, would produce a radiative luminosity of $L \sim 8 \times 10^{44} \, \mathrm{erg \, s}^{-1}$. Observationally, the nucleus is orders of magnitude less active. The observed radiative power does not exceed $L_{\mathrm{obs}} \sim 10^{42} \, \mathrm{erg \, s}^{-1}$ (Biretta, Stern & Harris 1991; also see Section 2 of this letter) and the time-averaged kinetic luminosity of the jet cannot exceed much more than $L_{\mathrm{K}} \sim 10^{43} \, \mathrm{erg \, s}^{-1}$ (Reynolds et al. 1996).

The recent interest in advection-dominated accretion disks (Narayan & Yi 1995; Abramowicz et al. 1995; Narayan, Yi & Mahadevan 1995) prompted Fabian & Rees (1995) to suggest that such disks exist around the nuclear black holes in quiescent giant elliptical galaxies. In this mode of accretion, the accretion flow is very tenuous and so a poor radiator. (The possibility of similarly tenuous 'ion-supported tori' had been discussed in the context of radio galaxies by Rees et al. 1982 and for the Galactic centre by Rees 1982). Much of the energy of the accretion flow cannot be radiated and is carried through the event horizon. Fabian & Rees (see also Mahadevan 1996) realised that the resulting low radiative efficiency provides a possible solution to the elliptical galaxy problem described above. They identify the weak parsec-scale radio cores seen in most elliptical galaxies (Sadler et al. 1989; Wrobel & Heeschen 1991; Slee et al. 1994) with synchrotron emission from the plasma of the advection-dominated disks (ADD).

Here we present a detailed examination of the possibility that the massive black hole in M87 accretes via an ADD. In particular, we compute the spectrum of the ADD and show that it is consistent with the observations for physically reasonable mass accretion rates. In Section 2 we compile data from the literature on the full-band spectrum of the core of M87 and present some additional data on the X-ray flux from the core. Care is taken to limit the effect of contaminating emission from the jet and/or the galaxy. Despite this, the spectrum we obtain must still be considered as a set of upper limits on the spectrum of the accretion flow with the non-thermal emission from the jet representing the main contaminant. We make a direct comparison to the quasar 3C 273. Section 3 describes some details of our ADD model spectrum calculation. Section 4 compares this model spectrum with the data and finds that accretion rates comparable with the Bondi rate do not overproduce radiation and are thus acceptable. Section 5 discusses some further astrophysical implications of this result.

2   The spectrum of the core emission

2.1   The M87 data

Table 1. Summary of the core data

Frequency ν (Hz)	Resolution (milliarcsecs)	νF_ν $(10^{-14}\,\mathrm{erg\,s^{-1}\,cm^{-2}})$	reference	notes
1.7×10^9	5	1.65	Reid et al. (1989)	VLBI
5.0×10^9	0.7	1.0	Pauliny-Toth et al. (1981)	VLBI
2.2×10^{10}	0.15	4.8	Spencer & Junor (1986)	VLBI
1.0×10^{11}	0.1	8.7	Bääth et al. (1992)	VLBI
7×10^{14}	50	200	Harms et al. (1994)	HST
2.4×10^{17}	4×10^3	85	Biretta et al. (1991)	*Einstein* HRI
2.4×10^{17}	4×10^3	160	this work	*ROSAT* HRI
4.8×10^{17}	2×10^5	≤ 700	this work	*ASCA*

In order to examine the nature of the accretion flow in M87, we have attempted to compile the best observational limits on the full band spectrum of the core emission. Our aim is to obtain good observational limits on the core flux over a wide range of frequencies rather than to compile a comprehensive list of all previous observations. For radio through optical, we use the highest spatial resolution data available from the literature in order to minimize the contribution to the flux from the synchrotron emitting jet and the galaxy. However, contributions from the jet and the underlying galaxy are unavoidable and so the derived spectrum should be considered an upper limit to that of the accretion flow at the core of M87. These data are summarized in Table 1.

We have examined the *ROSAT HRI* and ASCA data sets in order to constrain further the nuclear X-ray flux of M87. The *ROSAT* data were retrieved from the public archive situated at the GODDARD SPACE FLIGHT CENTER and result from a 14 200 s exposure performed on 1992-June-7. Figure 1 shows the *ROSAT*

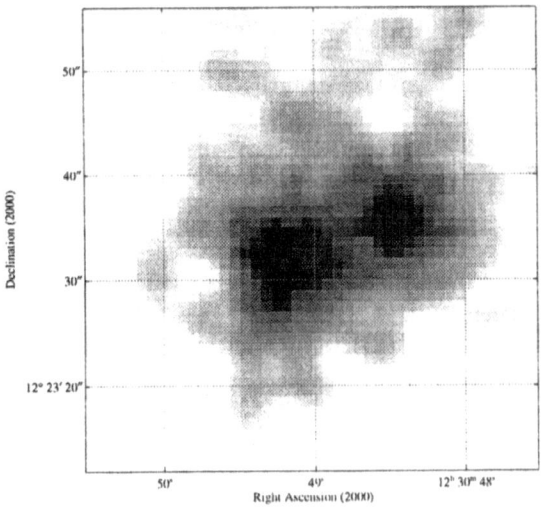

Fig. 1. The core regions of M87/Virgo as imaged in a 14ks exposure with the *ROSAT* HRI. Two distinct sources are seen embedded in general diffuse emission. The easternmost source corresponds to the core of M87 whereas the western source coincides with the brightest knot within the jet (knot-A). The diffuse emission is from the hot interstellar medium.

HRI image of the central regions of M87. Emission from the core and knot-A ($\sim$ 10 arcsecs west of the core) are clearly separated from each other as they were in the *Einstein* HRI image (Biretta, Stern & Harris 1991). The *ROSAT* HRI count rate is 0.093 cts s^{-1} (determined using the XIMAGE software package). Assuming the spectrum to be a power-law with a canonical photon index $\Gamma = 1.7$ (Mushotzky, Done & Pounds 1993) modified by the effects of Galactic absorption (with column density $N_{\mathrm{H}} = 2.5 \times 10^{20}$ cm^{-2}; Stark et al. 1992), this count rate implies a flux density at 1 keV of $F(1\,\mathrm{keV}) = 0.67\,\mu$Jy. This result is fairly insensitive to choosing a different power-law index.

ASCA observed M87 during the PV phase: 12 600 s of good data were obtained in 1993 June. We performed a variety of fits to the spectrum in the central regions, incorporating multiple thermal components, possible excess low-energy absorption, and possible cooling-flow emission. In no case does the addition of a power law component give a noticeable improvement in the fit. Our limit (at 90 per cent confidence) is listed in Table 1.

High resolution VLBI observations probably provide the strongest constraints on the core emission (since they can separate the core emission from knots of jet emission even within the innermost parsec of the source). High resolution optical (HST) and X-ray (*ROSAT* HRI) measurements are likely to be free of galactic contamination but may still possess a significant contribution from the inner jet. Sub-mm and far-IR studies provide uninteresting limits on the core flux: the comparatively poor spatial resolution leads to severe contamination from galactic emission (predominantly thermal emission by dust).

2.2   The 3C 273 data

For comparison with M87, the open squares on Fig. 2 show data for the radio-loud quasar 3C 273. These data are from the compilation of Lichti et al. (1995) and are simultaneous or near-simultaneous. Much of this emission is likely to originate from the jet and be relativistically beamed. However, the fact that we see a big blue bump and optical/UV emission lines from 3C 273 implies that a significant part of the optical/UV emission is unbeamed and likely to originate from the accretion flow.

3   The advection-dominated model

It is well known that sub-Eddington accretion can proceed via a thin accretion disk (see review by Pringle 1981 and references therein). Such disks are characterised by being radiatively efficient so that the energy generated by viscous dissipation is radiated locally. As a consequence, such disks are cold in the sense that the gas temperature is significantly below the virial temperature. However, for sufficiently low mass accretion rates ($\dot{m} < \dot{m}_{\rm crit} \approx 0.3\alpha^2$, where α is the standard disk viscosity parameter) there is another stable mode of accretion (see Narayan & Yi 1995 and references therein). In this second mode, the accretion flow is very tenuous and, hence, a poor radiator. The energy generated by viscous dissipation can no longer be locally radiated – a large fraction of this energy is advected inwards in the accretion flow as thermal energy and, eventually, passes through the event horizon. These are known as advection-dominated disks (ADDs).

For convenience, we rescale the radial co-ordinate and define r by

$$r = \frac{R}{R_{\rm Sch}},\qquad(2)$$

where R is the radial coordinate and $R_{\rm Sch}$ is the Schwarzschild radius of the hole. It is an important feature of ADDs that, in the region $r < 1000$, the ions and electrons do not possess the same temperature. The ions attain essentially the virial temperature. They couple weakly via Coulomb interactions to the electrons which, due to various cooling mechanisms, possess a significantly lower temperature.

By assuming that the system is undergoing advection-dominated accretion, we can predict the radio to X-ray spectrum of the accretion flow. Since the gas is optically-thin, the emission in different regions of the spectrum is determined by synchro-cyclotron, bremsstrahlung and inverse Compton processes. The amount of emission from these different processes and the shape of the spectrum can be determined as a function of the model variables: the viscosity parameter, α, the ratio of magnetic to total pressure, β, the mass of the central black hole, M, and the accretion rate, $\dot{m}$. We take $\alpha = 0.3$ and $\beta = 0.5$ (i.e. magnetic pressure in equipartition with gas pressure). The electron temperature at a given point in the ADD, T_e, can then be determined self-consistently for a given $\dot{m}$ and M by balancing the heating of the electrons by the ions against the various radiative cooling mechanisms. Within $r < 1000$, it is found that $T_e \approx 2 \times 10^9$ K. To determine the observed spectrum, we must integrate the emission over the volume and take account of self-absorption effects. We have taken the inner radius of the disk to correspond with the innermost stable orbit around a Schwarzschild black hole, $r_{in} = 3$, and the outer radius to be $r_{out} = 10^3$. The spectrum is rather insensitive to the choice of r_{out} since most of the radiation originates within r_{out}. Details of the model, which is based on that of Narayan & Yi (1995), can be found in Di Matteo & Fabian (1996).

In Fig. 2 we show the spectrum of the advection-dominated disk for $M = 3 \times 10^9 \, M_\odot$ and $\dot{m} = 10^{-3.5}, 10^{-3.0}$ and $10^{-2.5}$. The peak in the radio band is due to synchro-cyclotron emission by the thermal electrons in the magnetic field of the plasma. The X-ray peak is due to thermal bremsstrahlung emission. The power-law emission extending through the optical band is due to Comptonization of the synchro-cyclotron emission: more detailed calculations show this emission to be comprised of individual peaks corresponding to different orders of Compton scattering. The positions at which the synchrotron and bremsstrahlung peaks occur and their relative heights depend on the parameters of the model. The synchrotron radiation is self-absorbed and gives a black body spectrum, up to a critical frequency, ν_c. Above the peak frequency the spectrum reproduces the exponential decay in the emission expected from thermal plasma (Mahadevan & Narayan 1996). The bremsstrahlung peak occurs at the thermal frequency $\nu \sim k_B T_e / h$.

4   The comparison

Figure 2 demonstrates the comparison between the core data and the advection dominated disk model described above. The accretion rate for all three models shown is comparable (within an order of magnitude) to the Bondi rate. Each of these models represents a physically plausible accretion rate. The two lower model curves ($\dot{m} = 10^{-3.5}, 10^{-3.0}$) are completely acceptable in the sense that they do not exceed any observational bounds. Furthermore, it can be seen that a substantial portion of the VLBI core flux may originate from an advection dominated disk. In particular two of the VLBI data points seem to reproduce almost exactly the slope of self-absorbed cyclo-synchrotron spectrum. We note that radio observations of early-type galaxy cores typically show rising or flat spectra with very similar slope (≈ 0.3) (Slee et al. 1994), which is well accounted for by the spectrum of an ADD (Fig. 2 and Mahadevan 1996).

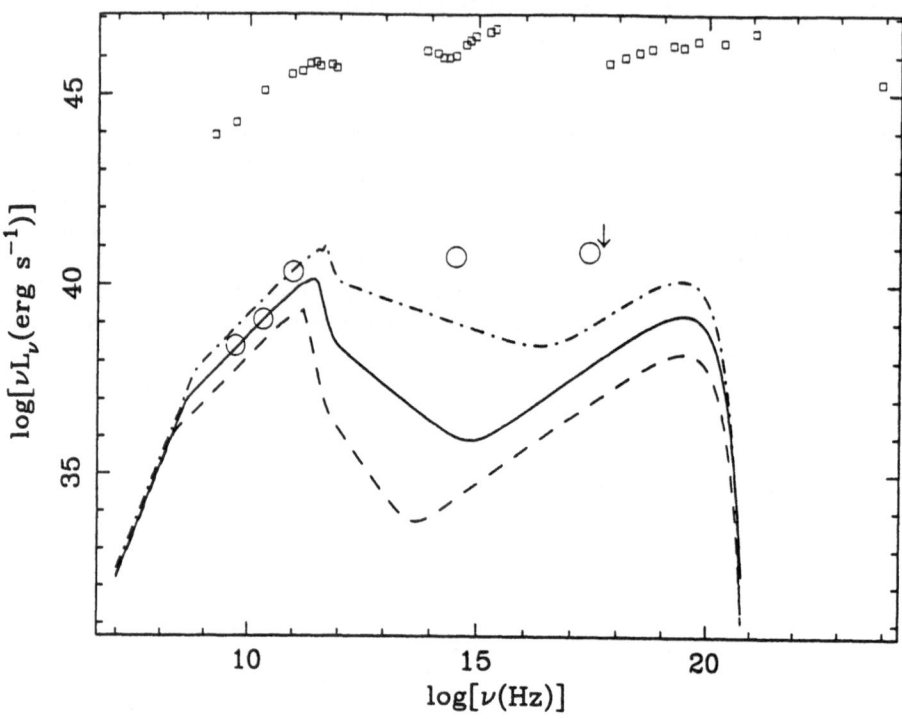

Fig. 2. Spectra of M87 calculated with an advection-dominated flow extending from $r_{min} = 3$ to $r_{max} = 1000$. The parameters are $m = 3 \times 10^9$, $\alpha = 0.3$, $\beta = 0.5$. Three models are shown: (i) $\dot{m} = 10^{-3.5}$-dashed line, (ii) $\dot{m} = 10^{-3}$-solid line, (iii) $\dot{m} = 10^{-2.5}$-dot-dashed line. The circles and squares represent various measurements of the spectrum of M87 and 3C 273, respectively, taken from different the references explained in the text. For M87, a distance of 16 Mpc is assumed, whereas for 3C 273 we have taken a Hubble constant of 50 km s^{-1} Mpc^{-1} and a deceleration parameter of $q_0 = 0.0$. In our model, the spectrum of M87 would lie only 2 decades below that of 3C273 if it had a standard, rather than advection-dominated, disk.

The core as seen in the 100 GHz VLBI data, the HST data and the *ROSAT* HRI X-ray data requires some additional component. The synchrotron jet would be a candidate for this additional component (provided the jet becomes self-absorbed at frequencies $\nu \approx 100\,\mathrm{GHz}$ or less).

Figure 2 shows the contrast between M87 and 3C 273. 3C 273 is observed to be at least 5 orders of magnitude more luminous than M87 at all wavelengths. The big blue bump in the spectrum of this quasar is often interpreted to be thermal emission from a standard thin-disk. Assuming a thin-disk efficiency of $\eta = 0.1$, the inferred accretion rate is $\dot{M} \sim 50\,\mathrm{M_\odot}$. Thus, it is possible that the mass accretion rates in M87 and 3C 273 differ by only 2 orders of magnitude despite the fact that the luminosities differ by 5–6 orders of magnitude.

5 Summary and discussion

We conclude that accretion from the hot gas halo at rates comparable with the Bondi rate is compatible with the low-luminosity of this core provided the final stages of accretion (where most of the gravitational energy is released) involve an advection dominated disk. This is in complete accord with the suggestion of Fabian & Rees (1995). An important test of this model will be the micro-arcsec radio imaging such as is promised by the VLBI Space Observatory Program (VSOP). This will provide the capability to image the core of M87 on scales comparable to the Schwarzschild radius of the black hole. Any advection-dominated disk should be directly revealed as a synchrotron self-absorbed structure with a position angle that is perpendicular to the jet axis.

The consistency of this model with the data allows us to explore some astrophysical implications of ADDs in giant elliptical galaxies. Rees et al. (1982) have argued that electromagnetic extraction from black holes surrounded by ion-supported tori (which share many of the basic physical properties of ADDs) can power relativistic jets in radio galaxies. The jets can be collimated in the inner regions of the ion-supported torus. If we apply the Blandford-Znajek (1977) mechanism to M87, we predict that the amount of energy extracted from a Kerr black hole is of the order of

$$L_{\mathrm{EM}} \approx 10^{43} \left(\frac{a}{m}\right)^2 B_2^2 M_9^2 \,\mathrm{erg\,s^{-1}}, \qquad (3)$$

where $a < m$ is the usual angular momentum of the black hole (in dimensionless units), $M = 10^9 M_9\,\mathrm{M_\odot}$ is the mass of the black hole ($M_9 \approx 3$) and $B = 10^2 B_2$ G is the magnetic field in the vicinity of the hole. From our model (assuming equipartition), we determine B to be

$$B_2 = 1.50 \left(\frac{r}{3}\right)^{-5/4} \left(\frac{\dot{m}}{10^{-3}}\right)^{1/2}. \qquad (4)$$

This determination of L_{EM} is completely consistent with the kinetic luminosity of M87 jet ($L_K \sim 10^{43}\,\mathrm{erg\,s^{-1}}$) obtained by Reynolds et al. (1996).

ADDs may be relevant to the creation of a unified model for radio-loud AGN. We are suggesting that M87, a classic FR-I radio source, possesses an ADD. However, ASCA observations of broad iron Kα fluorescence lines in the two FR-II sources 3C 390.3 (Eracleous, Halpern & Livio 1996) and 3C 109 (Allen et al. 1996) strongly point to the presence of standard cold (thin) accretion disks in these objects. This leads to the speculation that the FR-I/FR-II dichotomy is physically the dichotomy between advection-dominated and standard disks: i.e. FR-I sources possess ADDs whereas FR-II sources possess 'standard' thin accretion disks. In this sense, the accretion disks in FR-II sources and Seyfert nuclei would be intrinsically similar (Seyfert nuclei also display broad iron emission lines that are believed to originate from the central regions of a thin accretion disk.) We note that a very similar unification scheme (in the context of ion-supported tori versus thin accretion disks) has been discussed by Begelman, Blandford & Rees (1984) and Begelman (1985; 1986). The only difference is the increased observational evidence.

Acknowledgement: TDM and CSR acknowledge PPARC and Trinity College, Cambridge, for support. ACF thanks the Royal Society for support. UH thanks the NRC for support. CRC acknowledges partial support from NASA LTSA grant NAGW 2681 through Smithsonian grant SV2-62002 and contract NAS8-38249.

References

Abramowicz M., Chen X., Kato S., Lasota J. P., Regev O., 1995, ApJ, 438, L37
Allen S. W., Fabian A. C., Idesawa E., Inoue H., Kii T., Otani C., 1996, submitted.
Bääth L. B. et al., 1992, A&A, 257, 31
Begelman M. C., Blandford R. D., Rees M. J., 1984, Rev. Mod. Phys., 56, 255
Begelman M. C., 1985, in Astrophysics of active galaxies and quasi-stellar objects, eds Miller J. S., University Science Books, Mill Valley, P411
Begelman M. C., 1986, Nat, 322, 614
Biretta J. A., Stern C. P., Harris D. E., 1991, AJ, 101, 1632
Blandford R. D., Znajek R., 1977, MNRAS, 179, 433
Bondi H., 1952, MNRAS, 112, 195
Di Matteo T., Fabian A. C., 1996, in press
Eracleous M., Halpern J. P., Livio M., 1996, ApJ, 459, 89
Fabian A. C., Canizares C. R., 1988, Nat, 333, 829
Fabian A. C., Crawford C. S., 1990, MNRAS, 247, 439
Fabian A. C., Rees M. J., 1995, MNRAS, 277, L55
Ford H. C. et al. 1995, ApJ, 1994, 435, L27
Hansen C., Skinner G., Eyles C., Wilmore A. 1990, MNRAS, 242, 262
Harms R. J. et al., 1994, ApJ, 435, L35
Lea S., Mushotzky R.,Holt S., 1982, ApJ, 262, 24
Lichti G. G. et al., 1995, A&A, 298, 711
Mahadevan R., Narayan R., 1996, ApJ, 465, 327
Mahadevan R., 1996, in press.
Matsumoto H., Koyama K., Awaki H., Tomida H., Tsuru T., Mushotzky R., Hatsukade I., 1996, PASJ, 48, 201
Mushotzky R. F., Done C., Pounds K. A., 1993, ARAA, 31, 717
Narayan R., Yi I., 1995, ApJ, 452, 710

Narayan R., Yi I., Mahadevan R., 1995, Nat, 374, 623

Pauliny-Toth I. I. K., Preuss E., Witzel A., Graham D., Kellerman K. I., Ronnang B., 1981, AJ, 86, 371

Pringle J. E., 1981, ARAA, 19, 137

Rees M. J., 1982, in Riegler G., Blandford R., eds, The Galactic Center. Am. Inst. Phys., New York, p. 166.

Rees M. J., Begelman M. C., Blandford R. D., Phinney E. S., 1982, Nat., 295, 17

Reid M. J., Biretta J. A., Junor W., Muxlow T. W. B., Spencer R. E., 1989, ApJ, 336, 112

Reynolds C. S., Fabian A. C., Celotti C., Rees M. J., 1996, MNRAS, in press

Sadler E. M., Jenkins C. R., Kotanji C. G., 1989, MNRAS, 240, 591

Slee O. B., Sadler E. M., Reynolds J. E., Ekers R. D., 1994, MNRAS, 269, 928

Spencer R. E., Junor W., 1986, Nat., 321, 753

Stark A. A., Gammie C. F., Wilson R. W., Bally J., Linke R. A., Heiles C., Hurwitz M., 1992, ApJS, 79, 77

Stone J. M., Hawley J. F., Gammie C. F., Balbus S. A., 1996, ApJ, 463, 656

Takano S., Koyama K., 1991, PASJ, 43, 1

Wrobel J. M., Heeschen D. S., 1991, AJ, 101, 148

ROSAT observations of warm absorbers in AGN

S. Komossa, H. Fink

Max-Planck-Institut für extraterrestrische Physik, 85740 Garching, Germany

Abstract:

We study the properties of warm absorbers in several active galaxies (AGN) and present new candidates, using *ROSAT* X-ray observations. Several aspects of the characteristics of warm absorbers are discussed: (i) constraints on the density and location of the ionized material are provided; (ii) the impact of the presence of the warm gas on other observable spectral regions is investigated, particularly with respect to the possibility of a warm-absorber origin of one of the known high-ionization emission-line region in AGN; (iii) the possibility of dust mixed with the warm material is critically assessed; and (iv) the thermal stability of the ionized gas is examined. Based on the properties of known warm absorbers, we then address the question of where else ionized absorbers might play a role in determining the traits of a class of objects or individual peculiar objects. In this context, the possibility to produce the steep soft X-ray spectra of narrow-line Seyfert-1 galaxies by warm absorption is explored, as compared to the alternative of an accretion-produced soft excess. The potentiality of explaining the strong spectral variability in RX J0134-42 and the drastic flux variability in NGC 5905 via warm absorption is scrutinized.

1   Introduction and motivation

Absorption edges have been found in the X-ray spectra of several Seyfert galaxies and are interpreted as the signature of ionized gas along the line of sight to the active nucleus. This material, the so-called 'warm absorber', provides an important new diagnostic of the AGN central region, and hitherto revealed its existence mainly in the soft X-ray spectral region. The nature, physical state, and location of this ionized material, and its relation to other nuclear components, ist still rather unclear. E.g., an outflowing accretion disk wind, or a high-density component of the inner broad line region (BLR) have been suggested.

The presence of an ionized absorber was first discovered in *Einstein* observations of the quasar MR 2251-178 (Halpern 1984). With the availability of high quality soft X-ray spectra from *ROSAT* and *ASCA*, several more warm absorbers were found: they are seen in about 50% of the well studied Seyfert galaxies (e.g. Nandra & Pounds 1992, Turner et al. 1993, Mihara et al. 1994, Weaver et al. 1994, Cappi et al. 1996) as well as in some quasars (e.g. Fiore et al. 1993, Ulrich-Demoulin & Molendi 1996, Schartel et al. 1997). More than one warm absorber imprints its presence on the soft X-ray spectrum of MCG-6-30-15 (Fabian 1996, Otani et al. 1996) and NGC 3516 (Kriss et al. 1996). Evidence for an influence of the ionized material on non-X-ray parts of the spectrum is still rare: Mathur and collaborators (e.g. Mathur et al. 1994) combined UV and X-ray observations of some AGN to show the X-ray and UV absorber to be the *same* component. Emission from NeVIIIλ774, that may originate from a warm absorber, was discovered in several high-redshift quasars (Hamann et al. 1995).

The warm material is thought to be photoionized by emission from the central continuum source in the active nucleus. Its degree of ionization is higher than that of the bulk of the BLR and the gas temperature is typically of order 10^5 K. The properties of hot photoionized gas were studied e.g. in Krolik & Kallman (1984), Netzer (1993), Krolik & Kriss (1995), and Reynolds & Fabian (1995).

Here we analyze the X-ray spectra of several Seyfert galaxies, using *ROSAT* PSPC observations that nicely trace the spectral region in which the warm absorption features are located. In a first part, the properties of the 'safely' identified warm absorbers in the Seyfert galaxies NGC 4051, NGC 3227, and Mrk 1298 are derived, as far as X-ray spectral fits allow. These absorbers span a wide range in ionization parameter. Their properties are then further discussed in light of the known multi-wavelength characteristics of the individual Seyfert galaxies. E.g., the optical spectra show several features that may be directly (high-ionization emission lines, strong reddening), or indirectly (FeII complexes) linked to the presence of the warm absorber. In a second part, we will assess whether some properties of other classes of objects or individuals can be understood in terms of warm absorption. In particular, the existence of objects with *deeper* absorption complexes that recover only beyond the *ROSAT* energy range is expected (unless an as yet unknown fine-tuning mechanism is at work, that regulates the optical depths in the important metal ions). In this line, it is examined whether the presence of a warm absorber is a possible cause of the X-ray spectral steepness in narrow-line Seyfert 1 (NLSy1) galaxies. Furthermore, we scrutinize whether the strong spectral variability in RX J0134-42, and the drastic flux variability in NGC 5905 can be traced back to the influence of a warm absorber.

2 Models

The warm material was modeled using the photoionization code *Cloudy* (Ferland 1993). We assume the absorber (i) to be photoionized by continuum emission of the central pointlike nucleus, (ii) to be homogeneous, (iii) to have solar abundances (Grevesse & Anders 1989), and (iv) in those models that include dust mixed with the warm gas, the dust properties are like those of the Galactic interstellar medium, and the abundances are depleted correspondingly (Ferland 1993).

The ionization state of the warm absorber can be characterized by the hydrogen column density N_w of the ionized material and the ionization parameter U, defined as

$$U = Q/(4\pi r^2 n_H c) \tag{1}$$

where Q is the number rate of incident photons above the Lyman limit, r is the distance between central source and warm absorber, n_H is the hydrogen density (fixed to $10^{9.5}$ cm^{-3} unless noted otherwise) and c the speed of light. Both quantities, N_w and U, are determined from the X-ray spectral fits.

The spectral energy distribution (SED) chosen for the modeling corresponds to the observed multi-wavelength continuum in case of NGC 4051 and two of the NLSy1s, a mean Seyfert continuum with energy index $\alpha_{uv-x} = -1.4$ else. The influence of the non-X-ray parts of the continuum shape will be commented on in Sect. 3.1, which will show the assumption of a mean Seyfert SED to be justified. However, the EUV part of the SED contributes to the numerical value of U. For better comparison of the ionization states of the discussed warm absorbers, we also give the quantity $\tilde{U}$ for each object, which is U normalized to the mean Seyfert continuum with α_{uv-x}=-1.4.

The X-ray data reduction was performed in a standard manner, using the EXSAS software package (Zimmermann et al. 1994).

3   NGC 4051

NGC 4051 is a low-luminosity Seyfert galaxy with a redshift of z=0.0023, classified as Seyfert 1.8 or NLSy1. It is well known for its rapid X-ray variability and has been observed by all major X-ray missions. Evidence for the presence of a warm absorber was discovered by Pounds et al. (1994) in the *ROSAT* survey observation. We have analyzed all archival observations of NGC 4051 (part of those were independently presented by McHardy et al., 1995) and PI data taken in Nov. 1993. Emphazis is put on the latter observation, and the results refer to these data if not stated otherwise (for details on the Nov. 93 observation see Komossa & Fink 1997; for the earlier ones Komossa & Fink 1996).

3.1   Spectral analysis

A single powerlaw gives a poor fit to the X-ray spectrum of NGC 4051 (χ^2_{red} = 3.8) and the resulting slope is rather steep (photon index $\Gamma_x = -2.9$). A warm absorber model describes the spectrum well, with an ionization parameter of $\log U = 0.4$ ($\log \tilde{U} = 0.8$), a large warm column density of $\log N_w = 22.67$, and an intrinsic powerlaw spectrum with index $\Gamma_x = -2.3$. The unabsorbed (0.1-2.4 keV) luminosity is $L_x = 9.5 \times 10^{41}$ erg/s (for a distance of 14 Mpc). The addition of the absorber-intrinsic emission and far-side reflection to the X-ray spectrum, calculated for a covering factor of the warm material of 0.5, only negligibly changes the results ($\log N_w = 22.70$) due to the weakness of these components.

We note, that the intrinsic, i.e. unabsorbed, powerlaw with index $\Gamma_x = -2.3$ is steeper than the Seyfert-1 typical value with $\Gamma_x = -1.9$ and it is also in its steepest observed state in NGC 4051. We have verified that no additional soft excess causes an apparent deviation from the canonical index, although the presence of such a component, with $kT_{bb} \simeq 0.1$ keV, is found in the high-state data.

The observed multi-wavelength SED for NGC 4051 was used for the above analysis, with α_{uv-x}=-0.7. In some further model sequences, the influence of the continuum shape on the results was tested, by (i) the addition of an EUV black body component with variable temperature and normalization (chosen such that there is no contribution of that component to the observed UV or X-ray part of the spectrum), and (ii) the inclusion of a strong IR spectral component as observed by IRAS (although at least part of it is most probably due to emission by cold dust from the surrounding galaxy; e.g. Ward et al. 1987). We find: (i) an additional black body component in the EUV has negligible influence on the X-ray absorption structure; (ii) an additional IR component strongly increases the free-free heating of the gas and the electron temperature rises. The best-fit ionization parameter changes to $\log U = 0.2$.

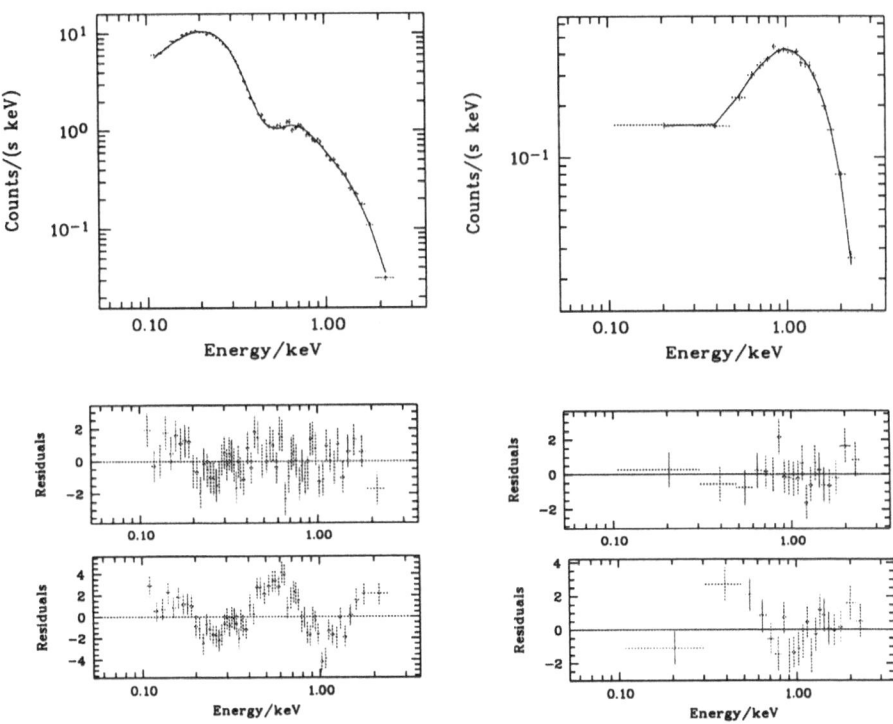

Fig. 1. Left: The upper panel shows the observed X-ray spectrum of NGC 4051 (crosses) and the best-fit warm absorber model (solid line). The next panel displays the fit residuals for this model. For comparison, the residuals resulting from a single powerlaw description of the data are shown in the lower panel (note the different scale in the ordinate). Right: The same for NGC 3227.

3.2   Temporal analysis

We detect strong variability: The largest amplitude is a factor of ~ 30 in count rate during the 2 year period of observations. In the Nov. 1993 observation, NGC 4051 is variable by about a factor of 6 within a day. The amplitude of variability is essentially the same in the low energy ($E \leq 0.5$ keV; dominated by the cold-absorbed powerlaw) and high energy (dominated by ionized-absorption) region of the *ROSAT* band (Fig. 2). To check for variability of the absorption edges in more detail, we have performed spectral fits to individual subsets of the total observation (referred to as 'orbits' hereafter). The best-fit ionization parameter turns out to be essentially constant over the whole observation despite strong changes in intrinsic luminosity. This constrains the, a priori unknown, density and hence location of the warm absorber (see below).

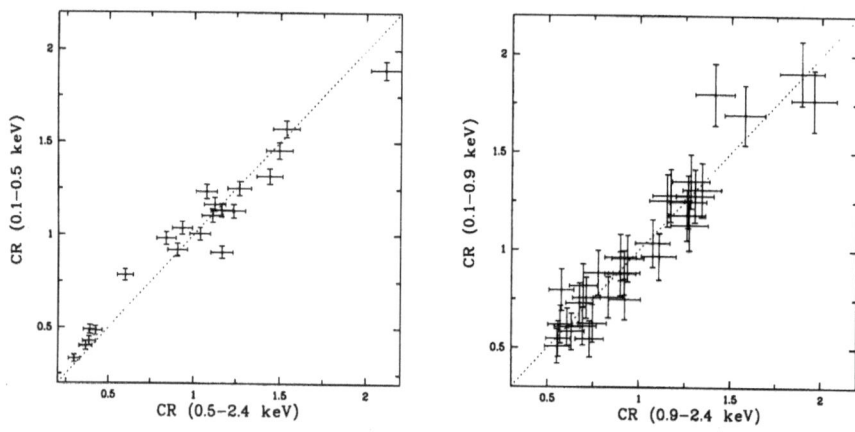

Fig. 2. Count rate CR in the soft *ROSAT* energy band versus count rate in the hard band, each normalized to the mean count rate in the corresponding band. Left: NGC 4051, right: NGC 3227. The dotted line represents a linear correlation.

3.3   Properties of the warm absorber

Density A limit on the density n can be drawn from the spectral (non-)variability, i.e. the constancy of U during a factor of > 4 change in intrinsic luminosity. The reaction timescale t_{rec} of the ionized material is conservatively estimated from the lack of any reaction of the warm material during the long low-state in orbit 7, resulting in a recombination timescale $t_{rec} \gtrsim 2000$ s. The upper limit on the density is given by

$$n_e \approx \frac{1}{t_{rec}} \frac{n_i}{n_{i+1}} \frac{1}{A} \left(\frac{T}{10^4}\right)^X \tag{2}$$

where n_i/n_{i+1} is the ion abundance ratio of the dominant coolant and the last term is the corresponding recombination rate coefficient $\alpha_{i+1,i}^{-1}$ (Shull & Van

Steenberg 1982). We find $n \lesssim 3 \times 10^7 \mathrm{cm}^{-3}$, dismissing a high-density BLR component as possible identification of the warm absorber that dominates the Nov. 93 spectrum. The corresponding thickness of the ionized material is $D \gtrsim 2 \times 10^{15}$ cm.

Location The location of the warm material is poorly constrained from X-ray spectral fits alone. Using $Q = 1.6 \times 10^{52}$ s^{-1} and the upper limit on the density, yields a distance of the absorber from the central power source of $r \gtrsim 3 \times 10^{16}$ cm. Conclusive results for the distance of the BLR in NGC 4051 from reverberation mapping, that would allow a judgement of the position of the warm absorber relative to the BLR, do not yet exist: Rosenblatt et al. (1992) find the continuum to be variable with high amplitude, but no significant change in the Hβ flux.

Warm-absorber intrinsic line emission and covering factor For 100% covering of the warm material, the absorber-intrinsic Hβ emission is only about 1/220 of the observed $L_{\mathrm{H}\beta}$. The strongest optical emission line predicted to arise from the ionized material is [FeXIV]$\lambda 5303$. Its scaled intensity is [FeXIV]$_{\mathrm{wa}}$/Hβ_{obs} ≈ 0.01. This compares to the observed upper limit of [FeXIV]/H$\beta \lesssim 0.1$ (Peterson et al. 1985) and it is consistent with a covering of the warm material of less or equal 100%. Due to the low emissivity of the warm gas in NGC 4051, no strong UV – EUV emission lines are produced (e.g. HeII$\lambda 1640_{\mathrm{wa}}$/H$\beta_{\mathrm{obs}} \leq 0.06$, NeVIII$\lambda 774_{\mathrm{wa}}$/H$\beta_{\mathrm{obs}} \leq 0.2$). Consequently, no known emission-line component in NGC 4051 can be fully identified with the warm absorber. We have verified that this result holds independently of the exact value of density chosen, as well as (IR or EUV) continuum shape.

UV absorption lines Due to the low column densities in the relevant ions, the predicted UV absorption lines are rather weak. The expected equivalent widths for the UV lines CIV$\lambda 1549$, NV$\lambda 1240$ and Lyα are $\log W_\lambda/\lambda \simeq -4.0$, $\log W_\lambda/\lambda \simeq -4.7$, and $\log W_\lambda/\lambda \simeq -3.0$, respectively (for a velocity parameter $b = 60$ km/s; Spitzer 1978).

Influence of dust Dust might be expected to survive in the warm absorber, depending on its distance from the central energy source and the gas-dust interactions. A dusty environment was originally suggested as an explanation for the small broad emission-line widths in NLSy1 galaxies. Warm material with internal dust was proposed to exist in the quasar IRAS 13349+2438 (Brandt et al. 1996). The evaporation distance r_{ev} of dust due to heating by the radiation field is provided by $r_{\mathrm{ev}} \approx \sqrt{L_{46}}$ pc (e.g. Netzer 1990). For NGC 4051 we derive $r_{\mathrm{ev}} \approx 0.02$ pc.

Mixing dust of Galactic ISM properties (including both, graphite and astronomical silicate; Ferland 1993) with the warm gas in NGC 4051 and self-consistently re-calculating the models leads to maximum dust temperatures of 2200 K (graphite) and 3100 K (silicate), above the evaporation temperatures (for a density of $n_{\mathrm{H}} = 5 \times 10^7$ cm^{-3} and U, N_{w} of the former best-fit model). For $n_{\mathrm{H}} \lesssim 10^6$ cm^{-3}, dust can survive throughout the absorber. However, it changes the equilibrium conditions and ionization structure of the gas. For relatively high ionization parameters, dust very effectively competes with the gas in the absorption of photons (e.g. Netzer & Laor 1993).

Firstly, re-running a large number of models, we find no successful fit of the X-ray spectrum. This can be traced back to the relatively higher importance of edges from more lowly ionized species, and particularly a very strong carbon edge (Fig. 3). There are various possibilities to change the properties of dust mixed with the warm material. The one which weakens observable features, like the 2175 Å absorption and the 10μ IR silicon feature, and is UV gray, consists of a modified grain size distribution, with a dominance of larger grains (Laor & Draine 1993). However, again, such models do not fit the observed X-ray spectrum, even if silicate only is assumed to avoid a strong carbon feature and its abundance is depleted by 1/10.

The absence of dust in the warm material would imply either (i) the *history* of the warm gas is such that dust was never able to form, like in an inner-disc driven outflow (e.g. Murray et al. 1995, Witt et al. 1996) or (ii) if dust originally existed in the absorber, the *conditions in* the gas have to be such that dust is destroyed. In the latter case, one obtains an important constraint on the density (location) of the warm gas, which then has to be high enough (near enough) to ensure dust destruction. For the present case (and Galactic-ISM-like dust) only a narrow range in density around $n_H \approx 5 \times 10^7$ cm^{-3} is allowed. For lower densities, dust can survive in at least part of the absorber and higher densities have already been excluded on the basis of variability arguments.

3.4   Evidence for an 'EUV bump'

The warm absorber fit shown in Fig. 1 shows some residual structure between 0.1 and 0.3 keV that can be explained by an additional very soft excess component. Of course, the parameters of such a component are not well constrained from X-ray spectral fits. One with $kT_{bb} = 13$ eV and an integrated absorption-corrected flux (between the Lyman limit and 2.4 keV) of $F = 7 \times 10^{-11}$ erg/cm^2/s fits the data. It contributes about the same amount to the ionizing luminosity as the powerlaw continuum used for the modeling and has the properties of the EUV spectral component for which evidence is as follows:

A lower limit on the number rate Q of ionizing photons in the unobserved EUV-part of the SED can be estimated by a powerlaw interpolation between the flux at 0.1 keV (from X-ray spectral fits) and the Lyman-limit (by extrapolating the observed UV spectrum), which was used for the present modeling. This gives $Q = 1.6 \times 10^{52}$ s^{-1}.

Q can also be deduced from the observed Hβ luminosity. Both quantities are related via $Q = (2.1 - 3.8) \times 10^{12} L_{H\beta}$ (Osterbrock 1989). The mean observed $L_{H\beta} = 7.8 \times 10^{39}$ erg/s (Rosenblatt et al. 1992) yields $Q = (1.6 - 2.5) \times 10^{52}$ s^{-1}. It is interesting to note that this is of the same order as the lower limit determined from the assumption of a single powerlaw EUV-SED, suggesting the existence of an additional EUV component in NGC 4051.

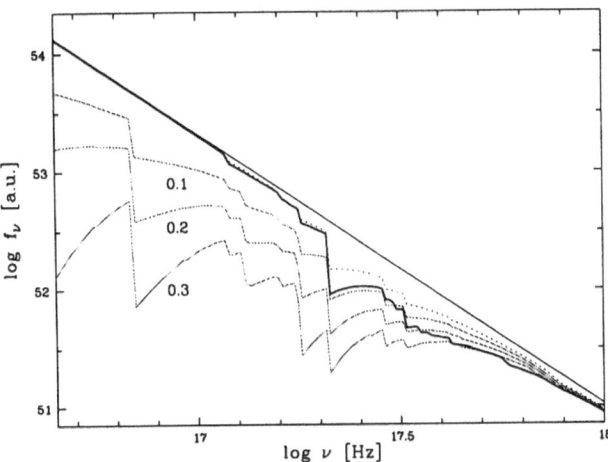

Fig. 3. Change in the X-ray absorption spectrum when dust is added to the warm absorber. The thin straight line marks the intrinsic continuum, the fat line shows the best-fit warm-absorbed spectrum of NGC 4051. The dotted line corresponds to the same model, except for depleted metal abundances. The thin solid lines represent models including dust (and depleted abundances); the depletion factor of dust, relative to the standard Galactic-ISM mixture is marked.

A third approach to Q is via the ionization-parameter sensitive emission-line ratio [OII]λ3727/[OIII]λ5007 (e.g. Schulz & Komossa 1993), and using this method we find $Q \simeq (1.7 - 3.4) \times 10^{52}$ s^{-1}.

An additional EUV component may also explain the observational trend that broad emission lines and observed continuum seem to vary independently in NGC 4051 (e.g. Peterson et al. 1985, Rosenblatt et al. 1992), as expected if the observed optical-UV continuum variability is not fully representative of the EUV regime.

4 NGC 3227

NGC 3227 is a Seyfert 1.5 galaxy at a redshift of z=0.003. Although studied in many spectral regions, most attention has been focussed on the optical wavelength range. NGC 3227 has been the target of several high spectral and spatial resolution studies and BLR mapping campaigns. Ptak et al. (1994) found evidence for a warm absorber in an *ASCA* observation.

4.1 Spectral analysis

For a single powerlaw description of the soft X-ray spectrum we obtain a very *flat* spectrum ($\Gamma_x \simeq -1.2$), and strong systematic residuals remain. A powerlaw with the canonical index of $\Gamma_x = -1.9$, modified by warm absorption, describes the data well. The alternative description, a soft excess on top of a powerlaw, results in an unrealistically flat slope with $\Gamma_x \approx -1.1$, in contradiction to higher-energy observations ($\Gamma_x \simeq -1.9$, George et al. 1990; $\Gamma_x \simeq -1.6$, Ptak et al. 1994).

4.2   Temporal analysis

We find the source to vary by a factor of 3.5 in count rate during the 11 days of the pointed observation. The shortest resolved doubling timescale is 380 minutes. An analysis of the survey data reveals strong variability between survey observation and pointing (which are separated by $\sim$ 3y) with a maximum factor of $\sim$ 15 change in count rate. Unfortunately, the low number of photons accumulated during the survey does not allow a detailed spectral analysis.

Dividing the data in a hard and soft band, according to a photon energy of larger or less than 0.9 keV, we find correlated variability between both bands (Fig. 2), implying that no drastic spectral changes have taken place. The soft band includes most of the warm absorption features (which extend throughout this whole band; cf. Fig. 4) as well as the influence of the cold absorbing column, which consequently cannot be disentangled from that of the ionized material. Spectral fits to individual 'orbits' of the total observation show the ionization parameter U to be constant throughout the whole observation, independent of luminosity. (In this study, the cold and warm column densities were fixed to the values determined for the total observation.)

4.3   Properties of the warm absorber

A *dusty* absorber? We find an ionization parameter of $\log U \approx -1.0$ and a warm column density of $\log N_w \approx 21.5$. The cold column, $N_H = 0.55 \times 10^{21}$ cm^{-2} (corresponding to a reddening of $A_v = 0^m.3$, if accompanied by dust), is larger than the Galactic value ($N_H^{Gal} \simeq 0.22 \times 10^{21}$ cm^{-2}), but not as large as indicated by line reddening. The narrow Hα/Hβ ratio measured by, e.g., Cohen (1983) implies an extinction of $A_v \simeq 1.2$. At another time, Mundell et al. (1995a) report a much higher value, $A_v \simeq 4.5$. The intrinsic Balmer line ratios are expected to be close to the recombination value under NLR conditions. A change in the Hα/Hβ ratio has then to be attributed to extinction by dust. However, dust *intrinsic* to the narrow line clouds is not expected to vary by factors of several within years, suggesting that the extinction is caused by an external absorber along the line of sight. If this dust is accompanied by an amount of gas as typically found in the Galactic interstellar medium, a large cold absorbing column of $N_H \simeq 8.3 \times 10^{21}$ cm^{-2} is expected to show up in the soft X-ray spectrum. (We note that the present X-ray observation and the optical observations by Mundell et al. (1995a) are not simultaneous, but they are only separated by $6\frac{1}{2}$ months.) Such a large neutral column is *not* seen in soft X-rays. Dust mixed with the warm gas could supply the reddening without implying a corresponding cold column.

Fitting a sequence of (Galactic-ISM-like) dusty absorbers provides an excellent description of the soft X-ray spectrum, with $\log U \approx -0.25$ and $\log N_w \approx 21.8$. The cold column is still slightly larger than the Galactic value, consistent with Mundell et al. (1995b), who find evidence for HI absorption towards the continuum-nucleus of NGC 3227.

Another possibility to explain the comparatively low cold column observed in X-rays is that we see scattered light originating from an extended component outside the nucleus. However, in that case the X-ray emission should not be that rapidly variable (e.g. Ptak et al. 1994; our Sect. 4.1). An alternative that cannot be excluded, due to the non-simultaneity of the observations, is that the material responsible for the reddening may have appeared only after the X-ray observation. In this scenario, it may not be associated with the warm absorber at all, but consist of thick blobs of dusty cold material crossing the line of sight. Since this is a possibility, we will include the dust-free warm absorber description of the data in the following discussion.

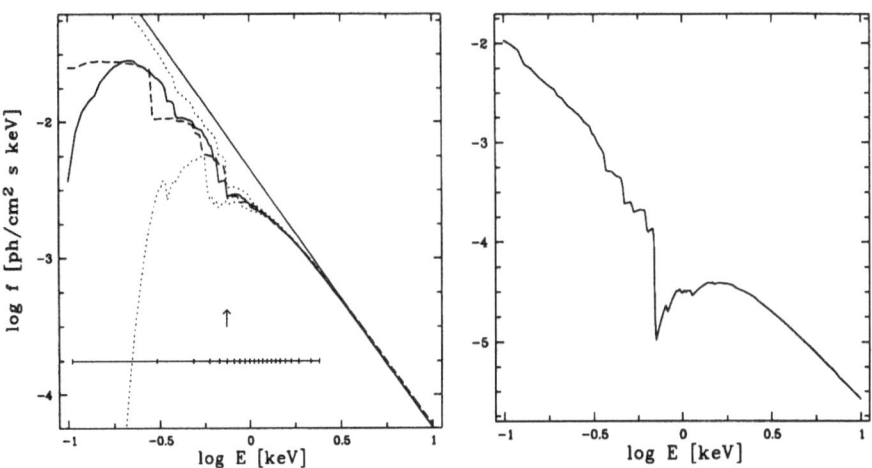

Fig. 4. Left: Warm-absorbed X-ray spectrum of NGC 3227 (solid line: dust-free, dashed: dusty model) between 0.1 and 10 keV, corrected for cold absorption. The straight line corresponds to the unabsorbed powerlaw. The dotted lines show the change in the absorption structure for a factor of 2 (upper curve) and a factor of $\frac{1}{2}$ change in luminosity, and consequently in U. The two curves were shifted to match the normalization of the best-fit model, to allow a better comparison. The horizontal line at the bottom brackets the *ROSAT* sensitivity range, the vertical bars indicate the bin sizes in energy that where used in the spectral fitting. The arrow marks the position of the OVII edge. Right: Warm-absorbed X-ray spectrum of Mrk 1298.

Density and Location (i) For the dust-free best fit ($\log U \simeq -1.0$), the density-scaled distance of the warm absorber is $r \simeq 10^{16}$cm $\times (n_{9.5})^{-0.5}$, which compares to the typical BLR radius of $r \simeq 17$ ld $= 4 \times 10^{16}$ cm, as determined from reverberation mapping (Salamanca et al. 1994, Winge et al. 1995). (ii) For the dusty warm absorber, the gas density has to be less than about 10^7 cm^{-3}, i.e. $r \geq 9 \times 10^{16}$ cm $\times (n_7)^{-0.5}$, to ensure dust survival. Constant ionization parameter throughout the observed low-state in flux (Sect. 4.2) further implies $n \lesssim 2 \times 10^6$ cm^{-3}; or even $n \lesssim 2.5 \times 10^4$ cm^{-3}, based on the assumption of constant U during the total observation (less certain, due to time gaps in the data), and still using t_{rec} as an estimate. The ionized material has to be located at least between BLR and NLR, or even in the outer parts of the NLR, to extinct a non-negligible part of the NLR.

Thermal stability The thermal stability of the warm material is addressed in Fig. 5. Between the low-temperature ($T \sim 10^4$ K) and high-temperature ($T \sim 10^{7-8}$ K) branch of the equilibrium curve, there is an intermediate region of multi-valued behavior of T in dependence of U/T, i.e. pressure (e.g., Guilbert et al. 1983; their Fig. 1). It is this regime, where the warm absorber is expected to be located, with its temperature typically found to lie around $T \sim 10^5$ K. The detailed shape of the equilibrium curve in this regime depends strongly on the shape of the ionizing continuum, and properties of the gas, e.g. metal abundances. An analysis by Reynolds & Fabian (1995) placed the warm absorber in MCG-6-30-15 in a small stable regime within this intermediate region (its position is marked by an arrow in Fig. 5). The ionized material in NGC 3227 is characterized by a comparatively low ionization parameter. Its location in the T versus U/T diagram is shown for the models that provide a successful description of the X-ray spectrum. The shape of the equilibrium curve is clearly modified for the model including dust and correspondingly depleted metal abundances.

The locations of the warm absorbers in other objects of the present study are also plotted. They are found to lie all around the intermediate stable region. Interesting behavior is shown by the warm material in two observations of NGC 4051. A detailed asessment of the thermal stability of ionized absorbers is still difficult due to the strong influence of several parameters on the shape of the equilibrium curves.

Line emission From a sample of objects chosen to asses whether one of the known emission-line regions in active galaxies can be identified with the warm absorber, NGC 3227 represents the low-ionization end of the known warm absorbers. It also exhibits broad wings in Hα that do not follow continuum variations (Salamanca et al. 1994), indicative of a higher than usual ionized component, that may be identified with the warm absorber. (i) Dust-free warm absorber: The absorber-intrinsic luminosity in Hβ is only about 1/30 of the broad observed Hβ emission (Rosenblatt et al. 1994; de-reddened with A_v=0.9). The scaled optical lines are weak, e.g., [FeX]$_{wa}$/H$\beta_{obs} \approx 0.03$. The strongest line in the ultraviolet is CIVλ1549, with CIV$_{wa}$/H$\beta_{obs} \approx 6$. A comparison of the expected line strengths with a typical BLR spectrum is shown in Fig. 6. (ii) In

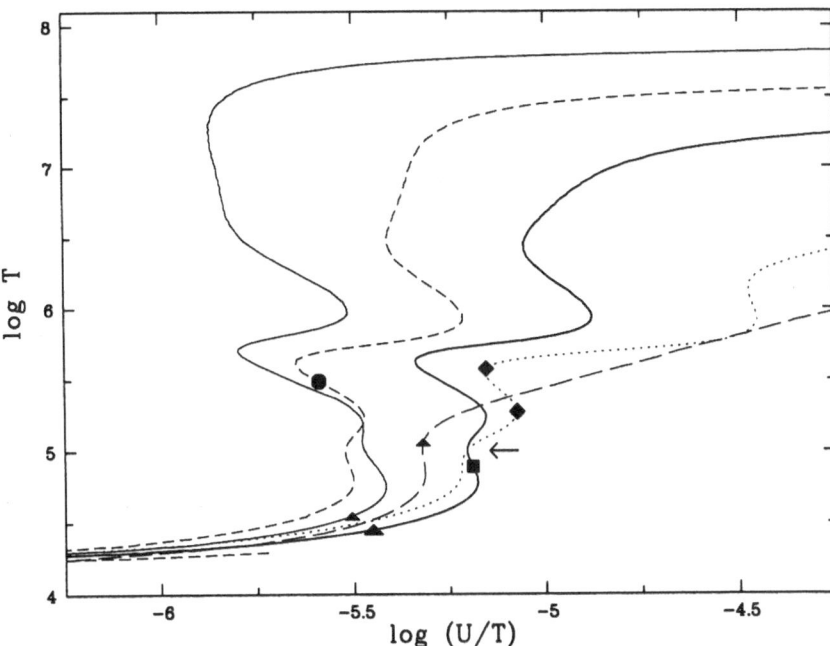

Fig. 5. Equilibrium gas temperature T versus U/T and locations of warm absorbers. The equilibrium curves are shown for different ionizing continua and gas properties, and the positions of the warm absorbers discussed in the present study are marked. The solid lines correspond to SEDs with (i) $\Gamma_x = -1.6$ (left) and (ii) $\Gamma_x = -1.9$ (right), and $\alpha_{uv-x} = -1.4$; long-dashed: the same continuum as (ii), but gas of depleted metal abundances and with dust; short-dashed: ionizing continua employed for the NLSy1 galaxies RX J1239 (left, $\alpha_{uv-x} = -0.75$) and RX J1225 ($\alpha_{uv-x} = -1.9$); dotted: observed SED of NGC 4051 with $\Gamma_x = -2.3$. The symbols mark the positions that we derive for the warm absorbers from X-ray spectral fits; triangles: NGC 3227 (for the dusty and dust-free absorber model); square: Mrk 1298; lozenges: NGC 4051, Nov. 1991 observation (lower symbol) and Nov. 1993 observation; arrow pointed to solid line: MCG-6-30-15 (Reynolds & Fabian 1995); circles: RX J1239 (left) and RX J1225.

case of dust mixed with the ionized material, the overall emissivity is reduced and no strong emission lines are predicted.

UV absorption lines A comparison of UV absorption lines and X-ray absorption properties of active galaxies was performed by Ulrich (1988). Her analysis of an IUE spectrum of NGC 3227 in the range $2000 - 3200$ Å revealed the presence of MgIIλ2798 absorption with an equivalent width of $\log W_\lambda/\lambda \simeq -2.8$. The X-ray warm absorber in NGC 3227 is of comparatively low ionization parameter. Nevertheless, the column density in MgII is very low, with $\log W_\lambda/\lambda \simeq -5.7$ (and considerably lower for the dusty model). Predictions of UV absorption lines that arise from the ionized material are made for the more highly ionized species, e.g. $\log W_\lambda/\lambda \simeq -2.9\ (-3.0)$ for CIVλ1549, and $\log W_\lambda/\lambda \simeq -3.0\ (-3.1)$ for NVλ1240 (for $b = 60$ km/s). The values in brackets refer to the warm absorber model that includes dust.

5 Mrk 1298

Mrk 1298 (PG 1126-041) is a luminous Seyfert 1 galaxy at a redshift of $z = 0.06$. The optical spectrum exhibits strong FeII emission. The presence of a warm absorber in the X-ray spectrum was found by Wang et al. (1996).

5.1 Spectral analysis

Again, the deviations of the X-ray spectrum from a single powerlaw fit are strong ($\chi^2_{red}=4.3$) with a rather steep slope of index $\Gamma_x \simeq -2.6$. Among various spectral models compared with the data (including those consisting of a soft excess on top of a powerlaw), only the warm absorber model provides a successful description. Due to the rather low number of detected photons, the underlying powerlaw index is not well constrained and was fixed to the canonical value of $\Gamma_x = -1.9$.

5.2 Properties of the warm absorber

We find an ionization parameter of $\log U \simeq -0.3$ and a warm column density of $\log N_w \simeq 22.2$. The absorber-intrinsic line emission turns out to be negligible, with $L^{wa}_{H\beta} = 1/740\ L^{obs}_{H\beta}$, except for OVI$\lambda$1035$_{wa}$/H$\beta_{obs} = 0.4$. Wang et al. (1996) mention the existence of strong UV absorption lines in an IUE spectrum. Here we find indeed rather large column densities in C^{3+} and N^{4+}, corresponding to an equivalent width for, e.g., CIV of $\log W_\lambda/\lambda \simeq -2.9$ (for $b = 60$ km/s).

Table 1. Properties of the warm absorbers in the three Seyfert galaxies. For comparison results from a single powerlaw fit are shown.

name	date of obs.	warm absorber					single powerlaw		
		$N_H^{(1)}$	Γ_x	$\log U$	$\log N_w$	χ^2_{red}	$N_H^{(1)}$	Γ_x	χ^2_{red}
NGC 4051	Nov. 1991	$0.13^{(2)}$	-2.2	0.2	22.5	1.1	0.18	-2.8	2.8
	Nov. 1993	0.13	-2.3	0.4	22.7	1.1	0.17	-2.9	3.8
NGC 3227	May 1993	0.55	-1.9	-1.0	21.5	0.8	0.38	-1.2	1.7
		0.55	-1.9	-0.3	21.8	$0.7^{(3)}$			
Mrk 1298	June 1992	$0.44^{(2)}$	-1.9	-0.3	22.2	0.9	$0.44^{(2)}$	-2.6	4.3

$^{(1)}$ in 10^{21} cm^{-2} $^{(2)}$ Galactic value $^{(3)}$ model including dust

6   Narrow Line Seyfert 1 Galaxies

Several extremely X-ray soft Seyferts have been found recently that belong to the class of NLSy1 galaxies (e.g. Puchnarewicz et al. 1992), which are characterized by narrow Balmer lines, and often strong FeII emission (e.g. Goodrich 1989). One interpretation of the X-ray spectral steepness is the dominance of the hot tail of emission from an accretion disk (e.g. Boller et al. 1996). Another one, which will be explored below, is the presence of a warm absorber.

In fact, a natural consequence of the observation of absorption *edges* in the X-ray spectra of Seyfert galaxies, is to also expect objects with deeper absorption, with mainly the 'down-turning' part of the absorption-complex being visible in the *ROSAT* sensitivity region. Good such candidates are ultrasoft AGN, and some are studied in the following.

6.1   RX J1239.3+2431 and RX J1225.7+2055

The 2 NLSy1 galaxies were discovered in *ROSAT* survey observations and exhibit very steep X-ray spectra (formally $\Gamma_x \simeq -3.7$ and -4.3; Greiner et al. 1996) and there is slight evidence for a spectral upturn within the *ROSAT* band. A description of the X-ray spectra in terms of warm absorption is found to be successful, albeit not unique due to the poor photon statistics of the survey data. We find $\log U \approx 0.8$ ($\log \tilde{U} \approx 0.4$) and $\log N_w \approx 23.2$ for RX J1225, and $\log U \approx -0.1$ ($\log \tilde{U} \approx 0.2$) and $\log N_w \approx 22.8$ for RX J1239.

The absorber-intrinsic Hβ emission is rather large and corresponds to about 1/10 and 1/7 of the observed $L_{H\beta}$ for RX J1225 and RX J1239, respectively. Scaling the predicted [FeXIV]λ5303 emission of the warm material in order not to conflict with the observed upper limit ($\lesssim 0.15$) constrains the covering factor of the gas to $\lesssim 1/6$ in RX J1239 and is consistent with 1 in RX J1225. Given the high discovery rate of supersoft AGN among X-ray selected ones (e.g. Greiner et al. 1996), the covering factors indeed have to be high to account for this fact. For both objects, several strong ultraviolet emission lines are predicted, e.g. HeIIλ1640$_{wa}$/H$\beta_{obs} \simeq 1.3$, FeXXIλ1345$_{wa}$/H$\beta_{obs} \simeq 1.1$ in RX J1225. The corresponding emission-line spectrum is illustrated in Fig. 6 assuming the maximal covering consistent with the data.

6.2   RX J0119.6-2821

RX J0119 (1E 0117.2-2837) was discovered as an X-ray source by *Einstein* and is at a redshift of z=0.347 (Stocke et al. 1991). It is serendipituously located in one of the *ROSAT* PSPC pointings.

The X-ray spectrum is very steep. When described by a single powerlaw continuum with Galactic cold column, the photon index is $\Gamma_x \simeq -3.6$ (-4.3, if N_H is a free parameter). The overall quality of the fit is good ($\chi^2_{red} \simeq 1$), but there are slight systematic residuals around the location of absorption edges. Again, a successful alternative description is a warm-absorbed flat powerlaw of

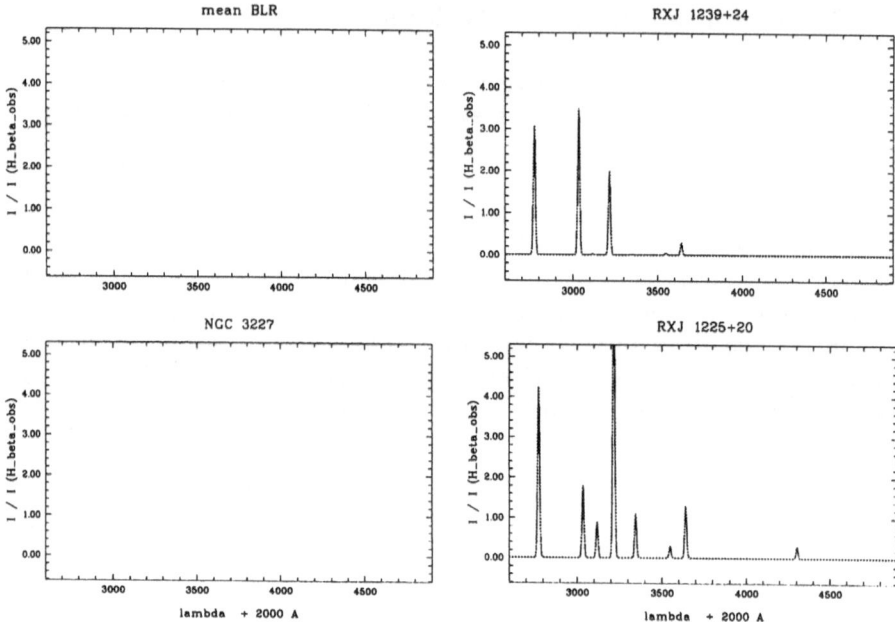

Fig. 6. Emission line spectra in the range 600 – 2900 Å predicted to arise from the warm material in individual objects compared to a mean BLR spectrum, to give an impression on the strengths of these lines and allow a judgement of the detectability. The y-axis gives the intensity of the absorber-intrinsic emission lines relative to the observed broad Hβ emission for the individual objects.

canonical index. We find a very large column density N_w in this case, and the contribution of emission and reflection is no longer negligible; there is also some contribution to Fe Kα. For the pure absorption model, the best-fit values for ionization parameter and warm column density are $\log U \simeq 0.8$, $\log N_w \simeq 23.6$ (N_H is consistent with the Galactic value), with $\chi^2_{red} = 0.74$. Including the contribution of emission and reflection for 50% covering of the warm material gives $\log N_w \simeq 23.8$ ($\chi^2_{red} = 0.65$).

Several strong EUV emission lines arise from the warm material. Some of these are: FeXXIλ2304/Hβ = 10, HeIIλ1640/Hβ = 16, FeXXIλ1354/Hβ = 37, FeXVIIIλ975/Hβ = 16, NeVIIIλ774/Hβ = 9, and, just outside the IUE sensitivity range for the given z, FeXXIIλ846/Hβ = 113. No absorption from CIV and NV is expected to show up. Both elements are more highly ionized.

6.3   RX J0134.3-4258

This object exhibits an ultrasoft spectrum in the *ROSAT* survey data (Dec. 1990; formally $\Gamma_x \simeq -4.4$). Interestingly, the spectrum has changed to flat in a

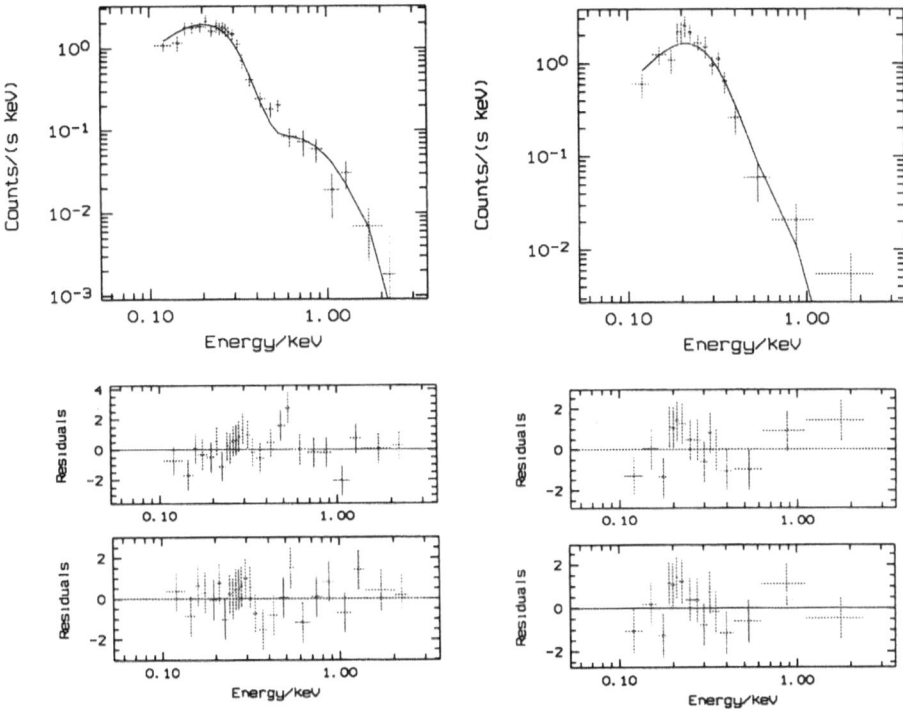

Fig. 7. Left: Powerlaw fit to the X-ray spectrum of RX J0119 (upper panel) and residuals, compared to the residuals for the warm absorber fit (lower panel; note the different scale in the ordinate). Right: Accretion disk fit to the 'outburst' observation of NGC 5905 (upper panel) and residuals, compared to residuals for the warm absorber fit (lower panel).

subsequent pointing (Dec. 1992; $\Gamma_x \simeq -2.2$). This kind of spectral variability is naturally predicted within the framework of warm absorbers, making the object a very good candidate.

We find that a warm-absorbed, intrinsically flat powerlaw can indeed describe the survey observation. Due to the low number of available photons, a range of possible combinations of U and N_w explains the data with comparable success. A large column density N_w (of the order 10^{23} cm^{-2}) is needed to account for the ultrasoft observed spectrum. The most suggestive scenario within the framework of warm absorbers is a change in the *ionization state* of ionized material along the line of sight, caused by *varying irradiation* by a central ionizing source. In the simplest case, lower intrinsic luminosity would be expected, in order to cause the deeper observed absorption in 1990. However, the source is somewhat brighter in the survey observation (Fig. 8). Some variability seems to be usual, though, and the count rate changes by about a factor of 2 during the pointed observation. If one whishes to keep this scenario, one has to assume that the ionization state of the absorber still reflects a preceding (unobserved) low-state in intrinsic flux. Alternatively, gas heated by the central continuum source may have *crossed the line of sight*, producing the steep survey spectrum, and has (nearly) disappeared in the 1992 observation.

It is interesting to note, that the optical spectrum of RX J0134 rises to the blue (Grupe 1996), and a soft excess on top of a powerlaw continuum may represent an alternative explanation of the data.

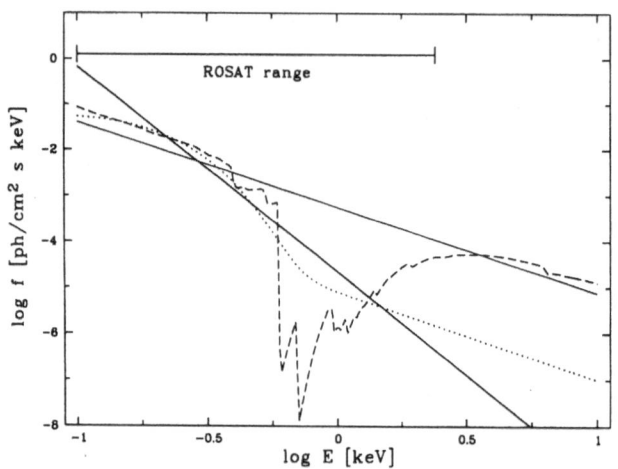

Fig. 8. Comparison of different X-ray spectral fits to the *ROSAT* survey observation of RX J0134 (thick lines) and the pointed observation (thin solid line). Thick solid: single powerlaw with $\Gamma_x = -4.4$; dashed: warm-absorbed flat powerlaw; dotted: powerlaw plus soft excess, parameterized as a black body. All models are corrected for cold absorption.

Table 2. Comparison of different spectral fits to the NLSy1 galaxies: (i) single powerlaw (pl), (ii) accretion disk after Shakura & Sunyaev (1973), and (iii) warm absorber. Γ_x was fixed to -1.9 in (ii) and (iii), except for RX J0134, where $\Gamma_x = -2.2$ (see text).

name	powerlaw[1]			acc. disk + pl[2]			warm absorber[2]		
	$N_H^{(3)}$	Γ_x	χ^2_{red}	$M_{BH}^{(4)}$	$\frac{\dot{M}}{M_{edd}}$	χ^2_{red}	$\log U$	$\log N_w$	χ^2_{red}
RX J1239.3+2431	0.32	−4.3	1.2	0.7	0.4	1.4	−0.1	22.8	1.0
RX J1225.7+2055	0.29	−3.7	1.9	1.0	0.5	2.1	0.8	23.2	1.8
RX J0119.6−2821	0.30	−4.3	0.8	0.6	0.6	0.7	0.8	23.6	0.7
RX J0134.3−4258[5]	0.16	−4.4	0.5	11.4	0.1	0.6	0.5	23.1	0.6

[1] N_H free, if $> N_H^{Gal}$ [2] N_H fixed to N_H^{Gal} [3] in $10^{21} cm^{-2}$ [4] in $10^4 M_\odot$ [5] survey obs.

6.4   Warm absorbers in *all* NLSy1 galaxies ?

There are some NLSy1 galaxies in which the warm absorber is clearly present and causes indeed at least most of the observed spectral steepness, as e.g. in NGC 4051. Mrk 1298 is another example. It exhibits all characteristics of NLSy1s, except that its observed FWHM of Hβ, 2200 km/s, just escapes the criterion of Goodrich (1989). One of the comfortable properties of the warm-absorber interpretation is the presence of enough photons to account for the strong observed FeII emission in NLSy1s, due to the intrinsically flat powerlaw. It does not immediately explain the occasionally observed trend of *stronger* FeII in objects with *steeper* X-ray spectra, but it is interesting to note that Wang et al. (1996) find a trend for stronger FeII to preferentially occur in objects whose X-ray spectra show the presence of absorption edges.

There are, however, other objects where neither a flat powerlaw with soft excess nor a warm absorber can account for most of the spectral steepness, although the presence of such a component cannot be excluded, as we have verified for **1ZwI** and **PHL 1092**.

In any model – warm absorber, accretion-produced soft excess, or intrinsically steep powerlaw spectrum – it is still difficult to elegantly account for all observational trends; as are, besides strong FeII, the tendency for narrower FWHM of broad Hβ for steeper X-ray spectra (e.g. Boller et al., their Fig. 8), and a slight anti-correlation of optical and soft X-ray spectral index (Grupe 1996). Progress is reported in Wandel (1996). Critical comments on the distinction between different models concerning photoionization aspects are given in Komossa & Greiner (1995).

To conclude, a warm absorber was shown to successfully reproduce the X-ray spectra of several NLSy1s and thus may well mimic the presence of a soft excess. High quality spectra are needed to distinguish between several possible models. In fact, although the warm absorber dominates the X-ray spectrum of the well-studied galaxy NGC 4051, an additional soft excess is present in flux high-states, and the underlying powerlaw is steeper than usual (Sect. 3.1). This hints to the complexity of NLSy1 spectra, with probably more than one mechanism at work to cause the X-ray spectral steepness.

7   The X-ray outburst in the HII galaxy NGC 5905

NGC 5905 underwent an X-ray 'outburst' during the *ROSAT* survey observation, with a factor ~ 100 change in count rate. The high-state spectrum is simultaneously very soft and luminous (Bade, Komossa & Dahlem 1996). Its classification (based on an optical pre-outburst spectrum) is HII type. Comparable variability was found in two further objects (IC3599; Brandt et al. 1995, Grupe et al. 1995a, WPVS007; Grupe et al. 1995b). The outburst mechanism is still unclear; tidal disruption of a star by a central black hole has been proposed. In the following, we comment on this and other possibilities and then discuss whether the X-ray outburst spectrum of NGC 5905 can be explained in terms of warm absorption.

SN in dense medium The possibility of 'buried' supernovae (SN) in *dense* molecular gas was studied by Shull (1980) and Wheeler et al. (1980). In this scenario, X-ray emission originates from the shock, produced by the expansion of the SN ejecta into the ambient interstellar gas of high density. Since high luminosities can be reached this way, and the evolutionary time is considerably speeded up, an SN in a dense medium may be an explanation for the observed X-ray outburst in NGC 5905.

Assuming the observed $L_x \approx 3 \times 10^{42}$ erg/s of NGC 5905 to be the peak luminosity, results in a density of the ambient medium of $n_4 \simeq \times 10^6 \mathrm{cm}^{-3}$ (using the estimates from Shull (1980) and Wheeler et al. (1980), and assuming line cooling to dominate with a cooling function of $\Lambda \propto T^{-0.6}$), but is inconsistent with the observed softness of the spectrum: The expected temperature is $T \approx 10^8$ K, compared to observed one of $T \approx 10^6$ K. Additionally, fine-tuning in the column density of the surrounding medium would be needed in order to prevent the SNR from being completely self-absorbed.

Tidal disruption of a star The idea of tidal disruption of stars by a super-massive black hole (SMBH) was originally studied as a possibility to fuel AGN, but dismissed. Later, Rees (1988, 1990) proposed to use individual such events as tracers of SMBHs in nearby galaxies. The debris of the disrupted star is accreted by the BH. This produces a flare, lasting of the order of months, with the peak luminosity in the EUV or soft X-ray spectral region.

The luminosity emitted if the BH is accreting at its Eddington luminosity can be estimated by $L_{edd} \simeq 1.3 \times 10^{38} M/M_\odot$ erg/s. In case of NGC 5905, a BH mass of $\sim 10^4$ $M_\odot$ would be necessary to produce the observed L_x (assuming it to be observed near its peak value). In order to reach this luminosity, a mass supply of about $1/2000$ $M_\odot$/y would be sufficient, assuming $\eta = 0.1$.

Accretion disk instabilities If a massive BH exists in NGC 5905, it has to usually accrete with low accretion rate or radiate with low efficiency, to account for the comparatively low X-ray luminosity of NGC 5905 in 'quiescence'. An accretion disk instability may provide an alternative explanation for the observed X-ray outburst. Thermally unstable slim accretion disks were studied by Honma et al. (1991), who find the disk to exhibit burst-like oscillations for the case of the standard α viscosity description and for certain values of accretion rate.

Using the estimate for the duration of the high-luminosity state (Honma et al. 1991; their eq. 4.8), and assuming the duration of the outburst to be less than 5 months (the time difference between the first two observations of NGC 5905), a central black hole of mass in the range $\sim 10^4 - 10^5 M_\odot$ can account for the observations. The burst-like oscillations are found by Honma et al. only for certain values of the initial accretion rate. A detailed quantitative comparison with the observed outburst in NGC 5905 is difficult, since the behavior of the disk is quite model dependent, and further detailed modeling would be needed.

Warm-absorbed hidden Seyfert Since all previous scenarios are either unlikely (SN in dense medium) or require some sort of fine-tuning, we finally asses the possibility of a warm-absorbed Seyfert nucleus. In this scenario, a hidden

Seyfert resides within the HII-type galaxy, that is slightly variable and usually hidden by a cold column of absorbing gas. The nucleus gets 'visible' during its flux high-states by ionizing the originally cold column of surrounding gas that becomes a *warm* absorber, thereby also accounting for the softness of the outburst spectrum.

We find that the survey spectrum can be well described by warm absorption with $\log U \simeq 0.0$ and $\log N_w \simeq 22.8$. A source-intrinsic change in luminosity by a factor of less than 10 is needed to change the absorption to complete within the *ROSAT* band. Variability of such order is not unusual in low-luminosity AGN; e.g., NGC 4051 has been found to be variable by about a factor of 30 within 2 years (Sect. 3.2). Since there is no evidence for Seyfert activity in the optical spectrum, the nucleus must be hidden. Mixing dust with the warm gas results in an optical extinction of $A_v \approx 34^m$ (assuming a Galactic gas-to-dust ratio) and would hide the Seyfert nucleus completely. The spectrum in the low-state – with a luminosity of $L_x \approx 4 \times 10^{40}$ erg/s and a shape that can be described by a powerlaw with $\Gamma_x \approx 2.4$ – can be accounted for by the usual X-ray emission of the host galaxy (e.g. Fabbiano 1989). However, dust with Galactic ISM properties internal to the warm gas was shown to strongly influence the X-ray absorption structure. An acceptable fit to the X-ray spectrum can only be achieved by tuning the dust properties and the dust depletion factor.

There are several ways to decide upon different scenarios: A hidden Seyfert nucleus should reveal its presence by a permanent hard X-ray spectral component, as observable e.g. with *ASCA*. One may also expect repeated flaring activity in the accretion-disk and warm-absorber scenario, and a *ROSAT* HRI monitoring (PI: N. Bade) is underway.

Acknowledgement: The *ROSAT* project is supported by the German Bundesministerium für Bildung, Wissenschaft, Forschung und Technologie (BMBF/DARA) and the Max-Planck-Society. We thank Gary Ferland for providing *Cloudy*.

References

Bade N., Komossa S., Dahlem M., 1996, A&A 309, L35

Boller T., Brandt W.N., Fink H.H., 1996, A&A 305, 53

Brandt W.N., Pounds K.A., Fink H.H., 1995, MNRAS 273, L47

Brandt W.N., Fabian, A.C., Pounds K.A., 1996, MNRAS 278, 326

Cappi M., Mihara T., Matsuoka M., Hayashida K., Weaver K.A., Otani C., 1996, ApJ 458, 149

Cohen R.D., 1983, ApJ 273, 489

Fabbiano G., 1989, ARA&A 27, 87

Fabian A.C., 1996, MPE Report 263, H.U. Zimmermann, J. Trümper, H. Yorke (eds.), 403

Ferland G.J., 1993, University of Kentucky, Physics Department, Internal Report

Fiore F., Elvis M., Mathur S., Wilkes B.J., McDowell J.C., 1993, ApJ 415, 129

George I.M., Nandra K., Fabian, A.C., 1990, MNRAS 242, 28p

Goodrich R.W., 1989, ApJ 342, 224

Greiner J., Danner R., Bade N., Richter G.A., Kroll P., Komossa S., 1996, A&A 310, 384

Grevesse N., Anders E., 1989, in Cosmic Abundances of Matter, C.J. Waddington (ed.), AIP 183

Grupe D., Beuermann K., Mannheim K., Bade N., Thomas H.-C., de Martino D., Schwope A., 1995a, A&A 299, L5

Grupe D., Beuermann K., Mannheim K., Thomas H.-C., Fink H.H., de Martino D., 1995b, A&A 300, L21

Grupe D., 1996, Ph.D. thesis, Universität Göttingen

Guilbert P.W., Fabian A.C., McCray R., 1983, ApJ 266, 466

Halpern J.P., 1984, ApJ 281, 90

Hamann F., Zuo L., Tytler D., 1995, ApJ 444, L69

Honma F., Matsumoto R., Kato S., 1991, PASJ 43, 147

Komossa S., Greiner J., 1995, AG abstract series 11, 217

Komossa S., Fink H., 1996, AG abstract series 12, 227

Komossa S., Fink H., 1997, A&A, in press

Kriss G.A., Krolik J.H., Otani C., Espey B.R., Turner T.J., Kii T., Tsvetanov Z., Takahashi T., Davidson A.F., Tashiro H., Zheng W., Murakami S., Petre R., Mihara T., 1996, ApJ 467, 629

Krolik J.H., Kallman T.R., 1984, ApJ 286, 366

Krolik J.H., Kriss G.A., 1995, ApJ 447, 512

Laor A., Draine B.T., 1993, ApJ 402, 441

Mathur S., Wilkes B., Elvis M., Fiore F., 1994, ApJ 434, 493

McHardy I.M., Green A.R., Done C., Puchnarewicz E.M., Mason K.O., Branduardi-Raymont G., Jones M.H., 1995, MNRAS 273, 549

Mihara T., Matsuoka M., Mushotzky R., Kunieda H., Otani C., Miyamoto S., Yamauchi M., 1994, PASJ 46, L137

Mundell C., Holloway A.J., Pedlar A., Meaburn J., Kukula M.J., Axon D.J., 1995a, MNRAS 275,67

Mundell C., Pedlar A., Axon D.J., Meaburn J., Unger S.W., 1995b, MNRAS 277, 641

Murray N., Chiang J., Grossman S.A., Voit G.M., 1995, ApJ 451, 498

Nandra K., Pounds K.A., 1992, Nature 359, 215

Netzer H., 1990, in Saas-Fee Lecture Notes 20, T.J.-L. Courvoisier and M. Mayor (eds.)

Netzer H., 1993, ApJ 411, 594

Netzer H., Laor A., 1993, ApJ 404, L51

Otani C., Kii T., Reynolds C. S., et al., 1996, PASJ 48, 211

Peterson B.M., Crenshaw D.M., Meyers K.A., 1985, ApJ 298, 283

Pounds K.A., Nandra K., Fink H.H., Makino F., 1994, MNRAS 267, 193

Ptak A., Yaqoob T., Serlemitsos P.J., Mushotzky R., Otani C., 1994, ApJL 436, L31

Puchnarewicz E.M., Mason K.O., Cordova F.A., Kartje J., Branduardi-Raymont G., Mittaz P.D., Murdin P.G., Allington-Smith J., 1992, MNRAS 256, 589

Rees M.J., 1988, Nature 333, 523

Rees M.J., 1990, Science 247, 817

Reynolds C.S., Fabian, A.C., 1995, MNRAS 273, 1167

Rosenblatt E.I., Malkan M.A., Sargent W.L.W., Readhead A.C.S., 1992, ApJS 81, 59

Rosenblatt E.I., Malkan M.A., Sargent W.L.W., Readhead A.C.S., 1994, ApJS 93, 73

Salamanca I., Alloin D., Baribaud T., et al., 1994, A&A 282, 742

Schartel N., Komossa S., Brinkmann W., Fink, H.H., Trümper, J., Wamsteker, W., 1997, A&A, in press

Schulz H., Komossa S., 1993, A&A 278, 29

Shakura N.I., Sunyaev R.A., 1973, A&A 24,337

Shull J.M., 1980, ApJ 237, 769

Shull J.M., Van Steenberg M., 1982, ApJS 48, 95 & ApJS 49, 351

Spitzer L., 1978, Physical Processes in the Interstellar Medium

Stocke J.T., Morris S.L., Gioia I.M., Maccacaro T., Schild R., Wolter A., Fleming T.A., Henry J.P., 1991, ApJS 76, 813

Turner T.J., Nandra K., George I.M., Fabian A.C., Pounds K.A., 1993, ApJ 419, 127

Ulrich M.-H., 1988, MNRAS 230, 121

Ulrich-Demoulin M.-H., Molendi S., 1996, ApJ 457, 77

Wandel A., 1996, in X-Ray Imaging and Spectroscopy of Cosmic Hot Plasmas, Abstracts, 73

Wang T., Brinkmann W., Bergeron J., 1996, A&A 309, 81

Ward M., Elvis M., Fabbiano G., Carleton N. P., Willner S. P., Lawrence A., 1987, ApJ 315, 74

Weaver K.A., Yaqoob T., Holt S.S., Mushotzky R.F., Matsuoka M., Yamauchi M., 1994, ApJ 436, L27

Wheeler J.C., Mazurek T.J., Sivaramakrishnan A., 1980, ApJ 237, 781

Witt H.J., Czerny B., Zycki P.T., 1996, MNRAS, in press

Winge C., Peterson B., Horne K., Pogge R.W., Pastoriza M., Storchi-Bergmann T., 1995, ApJ 445, 680

Zimmermann H.U., Becker W., Belloni T., Döbereiner S., Izzo C., Kahabka P., Schwentker O., 1994, MPE report 257

Can soft X–ray spectra of AGN be taken as emission from accretion disks?

R. Staubert[1], T. Dörrer[1], P. Friedrich[2], H. Brunner[2], C. Müller[1]

[1] Institut für Astronomie und Astrophysik, Abt. Astronomie, Waldhäuserstraße 64, D-72076 Tübingen, Germany

[2] Astrophysikalisches Institut Potsdam, An der Sternwarte16, D-14482 Potsdam, Germany

Abstract: Soft X-ray spectra of many Active Galactic Nuclei (AGN) show structure which suggest excess emission at low energies, mostly below 1 keV. This component may be the high energy tail of the so called blue bump which in turn may be due to the integrated emission from an accretion disk around the central black hole. ROSAT has observed a large number of AGN with unprecedented sensitivity in the energy range 0.1–2.4 keV.

Here we present results of our spectral analysis of two different samples of AGN:

1) QSO/Seyfert-I from the ROSAT All Sky Survey and
2) radio–quiet quasars from ROSAT Pointed Observations.

The ROSAT data are combined with UV Data from IUE and hard X-ray data from various hard X-ray missions. We give results on individual objects as well as on statistical properties of the samples.

The ROSAT spectra of AGN in both samples are found to generally have power law spectra steeper than the canonical value of 0.7, establishing a steep soft X–ray component. The soft X–ray component – together with UV and hard X-ray data – is then described in terms of thermal emission from a thin α-accretion disk, including Comptonization and relativistic corrections. We have developed disk models of increasing complexity. In fitting these models to the observational data constraints on the physical parameters of the systems - the mass of the black hole, the accretion rate, the viscosity and the inclination angle - are derived.

Our general conclusion is that emission from an accretion disk can in many cases account for the observed spectral features, such as the soft X-ray excess as the high energy tail of the big blue bump.

1 Sample of QSO and Seyfert-I galaxies from the ROSAT All Sky Survey (Sample 1)

1.1 Selection of the sample

- QSO and Seyfert-I from the Veron–Cetty & Veron (1993) catalogue
- IUE low dispersion spectra available
- ROSAT All Sky Survey detection

- Significant detection in each of the following energy bands:
  0.07–0.4 keV, 0.4–1.0 keV, 1.0–2.4 keV.

From the 420 IUE–selected AGN finally 89 remained. Since the ROSAT spectral range is restricted to soft X–rays additional data were needed to determine the hard X–ray power law spectrum. Those data were taken from the literature: from *Einstein* (Kruper et al. 1990, Masnou et al. 1992), from EXOSAT (Wilkes & Elvis 1987, Turner & Pounds 1989, Comastri et al. 1992), and from GINGA (Williams et al. 1992).

1.2   Model 1: Vertically averaged

We adopt the standard geometrically thin α–accretion disk, based on the paper of Shakura & Sunyaev (1973) and Novikov & Thorne (1973).

Model Description (see Dörrer 1991):

- Standard geometrically thin α–accretion disk
- Viscosity proportional to total pressure (gas + radiation)
- Vertically averaged structure (e. g. Lightman 1974, Maraschi & Molendi 1990) including relativistic corrections
- Eddington approximation for radiative transfer with saturated Comptonization included (Czerny & Elvis 1987)
- General relativistic effects incorporated by calculation of the Cunningham transfer function (Cunningham 1975) for any set of parameters (Speith et al. 1995)

Model Parameters:

- Mass M of the central black hole (in $10^8 M_\odot$)
- Accretion rate $\dot{M}$ (in units of $\dot{M}_{Edd}$)
- Viscosity parameter α_{vis}
- Inclination angle Θ of observer with respect to disk axis

1.3   Power Law Fits

A fundamental result of the analysis of ROSAT AGN spectra is that they are generally steeper than in the hard X–ray range. In our sample we find a mean spectral index $\alpha_{ROSAT} = 1.3$ whereas the slopes of the hard power laws (from Einstein, EXOSAT, and GINGA) in our sample have a mean spectral index of $\alpha_{hard} = 0.7$ (Fig. 1), which is in agreement with the canonical power law spectrum (Turner & Pounds 1989). Regarding Seyfert–I galaxies and QSOs of our sample separately we find a difference of about 0.2 in their mean spectral indices α_{ROSAT}. This steepening can be expressed in terms of an excess countrate which we define as the excess count rates in the spectral bands from 0.07–0.4 keV and from 0.4–1.0 keV (Tab. 1 and 2) over the extrapolated hard power law spectrum seen in the 1.0–2.4 keV band. These excess countrates confirm the soft X–ray excess found in the EXOSAT ME+LE data by Turner & Pounds (1989).

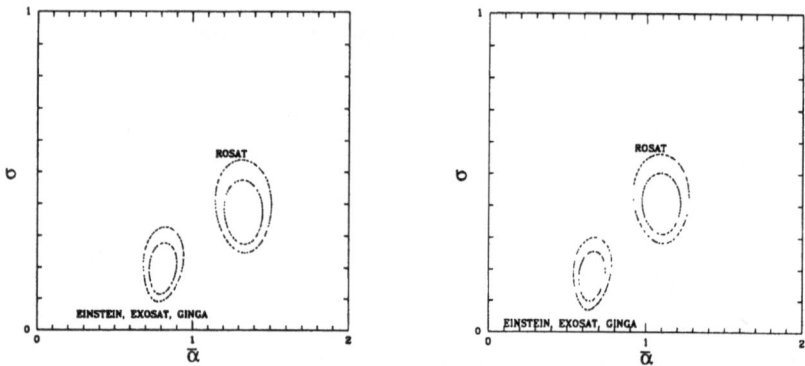

Fig. 1. Intrinsic distribution of the ROSAT spectral indices α_{ROSAT} of the sample in comparison to that of the hard power law spectral indices for QSOs (left) and Seyfert-I galaxies (right); σ is the width of the Gaussian distribution.

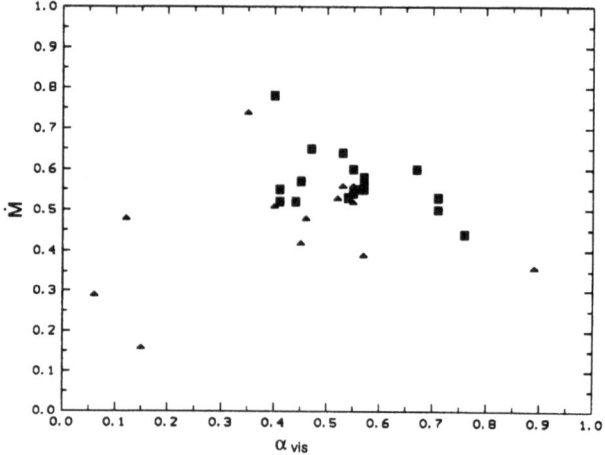

Fig. 2. Distribution of the best–fit parameters α_{vis} and $\dot{M}$ with $\chi^2 \leq 1.5$; the squares represent QSOs, the triangles Seyfert-I galaxies.

1.4   Accretion Disk Fits (Model 1)

Assuming the soft X–ray excess is caused by the hard end of the accretion disk spectrum we fit the complete disk model together with a hard X–ray power law spectrum to the combined X–ray and UV data and thus get constraints on the accretion disk model parameters. Because this method demands a high signal to noise ratio, the models can only be fitted to the brightest AGN for which such data are available. The fitted models include the following components:

1. the accretion disk model described above with four free parameters (M, $\dot{M}$, α_{vis}, $\cos\Theta$),

Table 1. QSOs from the ROSAT All Sky Survey: ROSAT spectral indices, N_H and accretion disk fits

AGN name	gal. N_H	ROSAT spec ind. α	rel. excess 0.07–0.4 keV	0.4–1.0 keV	from accretion disk fit M [10^8 M$_\odot$]	$\dot{M}$	α_{vis}	$\cos\Theta$	χ^2(d.o.f.)
III ZW 2	6.1	0.59±0.30	-0.07	-0.10					
I ZW 1	5.1	1.57±0.12	0.44	0.41	1.9	0.53	0.71	0.13	1.8 (18)
PG 0052+251	4.5	1.42±0.17	0.33	-0 09					
TON S210	1.7	1.46±0.17	0.74	0.50					
FAIRALL 9	2.7	1.10±0.05	0.32	0.07	1.0	0.54	0.55	0.44	1.0 (42)
3C 48.0	4.4	1.59±0.12	0.50	-0.01	14.9	0.60	0.67	0.13	2.4 (9)
PKS 0405-123	3.9	1.25±0.18	0.40	0.20					
PKS 0454-220	3.1	0.90±0.27	0.09	0.12					
PKS 0537-441	4.3	1.69±0.26	0.66	0.16	97.5	0.64	0.65	0.25	0.2 (5)
PKS 0637-752	10.9	2.54±0.26	0.74	-0.04	95.6	0.60	0.55	0.50	1.3 (9)
VII ZW 118	4.8	1.61±0.10	0.83	0.55	2.0	0.56	0.57	0.48	0.7 (11)
1E 0754+3928	5.0	1.81±0.41	0.83	0.62					
PG 0804+761	3.1	1.19±0.08	0.21	0.16	8.4	0.53	0.54	0.22	1.2 (22)
PKS 0837-120	5.9	0 70±0 38	-0.12	0.18					
S5 0836+71	2 9	0.35±0.15	-0.73	-0.19					
PG 0844+349	3.4	1.51±0.13	0.74	0.42					
B2 0923+392	1.7	0.95±0.18	0.20	0.07					
PG 0953+414	1.4	1.56±0.09	0 83	0.47	—	—	—	—	
4C 41.21	1.2	1.25±0.17	0.48	0.25					
B2 1028+313	2.0	0.99±0.14	0.32	0.06	5.8	0.50	0.71	0.22	1.4 (11)
PKS 1049-090	3.2	1.51±0.19	0.72	0.28	—	—	—	—	
3C 249.1	2.9	1.14±0.14	-0.26	-0.06					
PG 1116+215	1.4	1.43±0.10	0.50	0.21	9.4	0.52	0.41	0.61	2.8 (16)
3C 263.0	0.8	1.41±0.10	0.76	0.59					
LB 2136	2.0	0.85±0.19	0.54	0.51					
4C 29.45	1.6	0.66±0.32	0.33	0.41	42.1	0.16	0.87	0.56	1.2 (5)
GQ COM	1.7	1.01±0.10	0.32	0.19	4.6	0.57	0.45	0.21	2.8 (11)
PG 1211+143	2.8	1.82±0.07	0.71	0.32	3.5	0.65	0.47	0.26	1.4 (26)
PKS 1217+023	2.0	1.33±0 17	0.66	0.15	4.1	0.52	0.58	0.40	0.3 (3)
3C 273	1.8	1.14±0.03	0.62	0.23	48.7	0.44	0.76	0.66	0.3 (27)
PKS 1302-102	3.2	1.32±0.17	0.54	0.26	65.6	0.54	0.42	0.06	0.7 (8)
PG 1307+085	2.2	1.34±0.12	0.54	0.11					
1E 1352+1820	1.8	1.44±0.16	0.88	0 81	3.1	0.52	0.52	0.49	0.6 (7)
PG 1416-129	7.2	1.30±0.34	0.71	0.35	2.6	0.51	0.74	0.44	0.6 (9)
MKN 1383	2.6	1.37±0.08	0.78	0.39					
MKN 478	1.0	1.92±0.06	0.89	0.59	1.4	0.78	0.40	0.52	1.5 (26)
PG 1444+407	1.1	1.85±0.15	0.79	0.27	3.6	0.55	0.56	0.99	1.6 (8)
B2 1512+370	1.4	1.10±0.16	0.69	0.55	14.7	0.55	0.41	0.26	1.0 (11)
MKN 876	2.7	1.34±0.05	0.30	0.13	1.8	0.58	0.57	0.93	1.2 (41)
TON 256	3.8	0.91±0.14	0.48	0.21	1.2	0.64	0.53	0.39	0.9 (10)
MKN 877	4.4	0.94±0.29	-0.45	-0.21	1.1	0.57	0.50	0.37	0.3 (2)
3C 345.0	0.7	0.87±0.11	0.39	0.42	20.5	0.52	0.44	0.43	2.0 (10)
PG 1718+481	2.2	0.79±0.21	0.10	0.27	98.1	0.90	0.07	1.00	2.4 (8)
B2 1721+343	3.1	1.20±0.06	0.59	0.32	8.8	0.55	0.57	0.59	1.2 (28)
1831+731	5.6	1.26±0.37	0.40	-0 27	5.1	0.17	0.62	0.75	0 8 (9)
ESO 141-G55	5.9	1.08±0.13	0.57	0.31					
MKN 509	4.9	1.63±0.04	0.60	0.24					
II ZW 136	4.2	2.02±0.10	0.48	0.05					
MKN 926	3.5	0.80±0.08	0.21	0.23					
4C 09.72	4.2	1.20±0.24	0.50	0.37					
PKS 2349-014	3.6	1.18±0.13	0.64	0.48					

2. the hard power law spectrum with a given slope and the normalization as a free parameter,

3. a second power law spectrum with a fixed slope of $\alpha = 1.2$, a cutoff at $\nu = 10^{16}$ Hz (≈ 42 ev) and the normalization as a free parameter, describing the IR to optical/UV continuum (see below),

4. the hydrogen column density N_H which was fixed to the galactic column density.

Table 2. Seyfert-I galaxies from the ROSAT All Sky Survey: ROSAT spectral indices, N_H and accretion disk fits

AGN name	gal. N_H	ROSAT spec. ind. α	rel. excess 0.07–0 4 keV	rel. excess 0.4–1.0 keV	from accretion disk fit M [10^8 $M_\odot$]	$\dot{M}$	$\alpha_{vis.}$	$\cos\Theta$	χ^2(d.o.f.)
MKN 1148	4.3	0.56±0.21	-1.16	0.11	0.78	0.29	0.06	0.27	2.5 (6)
MKN 352	5.5	1.65±0.11	0.49	0.09	0.43	0.48	0.46	0.05	0.8 (14)
MKN 1152	1.7	0.64±0.22	0.31	0.09	0.61	0.36	0.89	0.15	1.4 (9)
II ZW 1	3.5	0.69±0.28	0.31	0.07	0.56	0.34	0.80	0.42	4.0 (2)
MKN 359	4.8	1.40±0.09	0 69	0.53	0.39	0.42	0.45	0.07	2.3 (11)
MKN 1018	2.5	1.00±0.13	0.52	0.32	0.78	0.33	0.90	0.23	2.3 (5)
MKN 590	3.1	1.25±0.05	0.52	0.14					
NGC 985	3.1	1.42±0.09	0.70	0 19					
NGC 1566	1.8	1.09±0.10	0.11	-0.17					
AKN 120	9 6	1.27±0.20	-0.03	0.21	3.8	0.52	0.55	0.26	0.9 (38)
PKS 0518–458	4.0	-0.17±0.28	0.18	0.40	0.35	0.16	0.15	0.15	2.7 (10)
MKN 374	6.6	2.03±0.19	0.86	0.49	0.35	0.74	0.35	0.51	0.3 (10)
MKN 79	5.4	0.78±0.09	0.13	0.29					
MKN 10	4.7	1.31±0.13	0.71	0.51					
MKN 705	4.0	1.58±0.08	0.69	0.39					
MKN 734	2.7	1.92±0.15	0.81	0.32	0.91	0.43	0.69	0.24	0.6 (9)
NGC 3783	8.4	1.35±0.20	0.50	0.42					
NGC 3998	1.3	1.44±0.11	-0.39	-0.06	<0.1	—	—	—	—
NGC 4051	1.2	1.52±0.05	0.70	0.48	<0.1	—	—	—	—
NGC 4151	1.8	0.61±0.11	—	—	—	—	—	—	—
MKN 205	2.7	0.79±0.08	0.14	0.16					
TON 1542	2.6	1.56±0.11	0.76	0.53	0.89	0.56	0.53	0.83	2.0 (13)
NGC 4593	2.0	1.33±0.09	0.71	0.46	<0.1	—	—	—	—
MCG 6-30-15	4.1	0.85±0.07	0.16	0.23	<0.1	—	—	—	—
NGC 5548	1.8	1.15±0.03	0.62	0.26	0.30	0.53	0.52	0.10	2.3 (64)
PG 1448+273	2.6	1.46±0.07	0.75	0.60	2.04	0.51	0.40	0.67	3.6 (14)
MKN 841	2.2	1.30±0.01	0.80	0 37	0.75	0.39	0.57	0.48	1.2 (12)
MKN 290	2.3	1.29±0.09	0.65	0.21	0.54	0.48	0.12	0.11	1.0 (13)
E 1556+274	3.6	-0.57±0.21	-1.24	-0.17	—	—	—	—	—
H 1613+06	4.8	1.35±0.16	0.55	0.08					
MKN 506	3.2	1.16±0.05	0.79	0.45	0.41	0.56	0.55	0.54	1.0 (38)
3C 382.0	6.8	1.10±0.08	0.10	0.18	0.94	0.54	0.55	0.74	1.2 (44)
3C 390.3	4.3	-0.11±0.19	-1.43	0.08					
NGC 7213	3.1	1.38±0.04	0.50	0.03					
AKN 564	5.8	1.76±0.05	0.77	0.47	—	—	—	—	—
NGC 7469	4.8	1.22±0.07	0.65	0 41					
PG 2304+042	5.5	0.89±0.22	0.11	-0.08					
NGC 7603	4.3	0.77±0.21	0.49	0.09	0.82	0.51	0.50	0.18	1.9 (9)

The results which are summarized in Tab. 1 and Tab. 2 show that the observed soft excess of QSOs and Seyfert–I galaxies can be explained with a limited range of parameters α_{vis} and $\dot{M}$ by our accretion disk model. Fig. 2 contains all AGN with χ^2 values $\lesssim 1.5$; the three Seyfort I galaxies outside that range ($\alpha_{vis} < 0.2$) are probably intrinsically absorbed.

2   Sample of radio–quiet quasars from the ROSAT pointed observation phase (Sample 2)

2.1   Selection of the sample

- Quasars from the Veron–Cetty & Veron (1993) and Hewitt & Burbidge (1993) catalogues.
- Radio to optical spectral index α_{ro} flatter than 0.3
- At the time of analysis (summer 1995) UV spectra were available and the pointed ROSAT PSPC data had entered the public domain

- Object located within 44 minutes of arc from the optical axis of the ROSAT observation
- At least 200 PSPC source counts had been measured.

The final sample defined in this way which was investigated in our further analysis consists of 31 objects.

2.2 Model 2: Self–consistent vertical structure

A detailed description of this model is given in Dörrer et al. (1996). The main differences with respect to model 1 are listed below.

- Self–consistent description of the vertical structure and radiation field of the disk around a Kerr black hole by solution of hydrostatic equilibrium, radiative transfer, energy balance, and equation of state
- Viscosity entirely due to turbulence
- Multiple Compton scattering treated in Fokker–Planck approximation using the Kompaneets operator

2.3 Power Law Fits

To determine the steepening of the spectra in the soft X–ray range we have fitted power law spectra to the ROSAT countrates for all sample members. The resulting spectral indices α_{fit} and α_{fix} for free and fixed N_H, respectively, is shown in Fig. 3. Note that – in agreement with sample 1 – the spectra are significantly steeper than the canonical value $\alpha_{hard} = 0.7$ which is indicated as a vertical line.

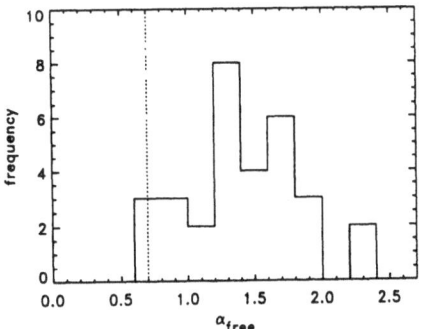 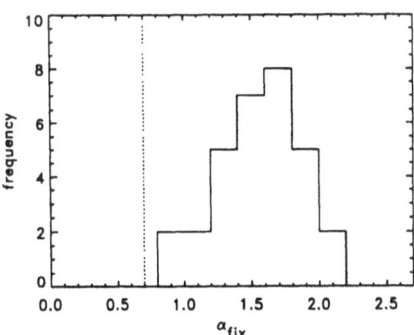

Fig. 3. Histogram of the fitted ROSAT spectral indices for a free absorbing column density N_H (left) and N_H fixed to the galactic value (right).

2.4   Accretion Disk Fits (Model 2)

The accretion disk model 2 and additional components as decribed in 1.4 have
been fitted to the radio–quiet quasars of our sample 2 leading to similar results
as obtained for model 1 and sample 1. To illustrate the overall fit function, one
example for these spectral fits is given in Fig. 4. The resulting best–fit parame-

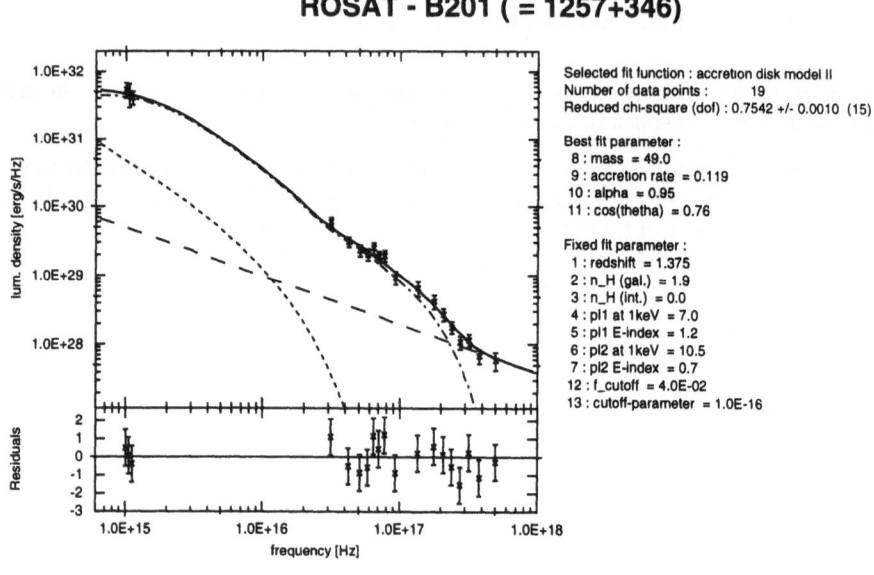

Fig. 4. Example of a spectral fit as described in the text; N_H in $10^{20}/cm^2$,
normalization of power laws (pl1 and pl2) at $1keV$ in $10^{27}erg/sec/Hz$

ters, their 1σ errors, the hard power law spectral index, and the corresponding
χ^2–values are shown in Tab. 3. Cases, where the upper or lower errors lie beyond the limit of our calculated grid (in most cases because the influence of M
and Θ on the normalization of the spectrum is similar), are denoted by a minus
sign. Note, that the fitted accretion rates $\dot{M}$ for all objects are below 0.3, thus
fullfilling the requirement for the geometrically thin disk approximation. On the
other hand , the viscosity parameter is relatively high, in most sources between
0.5 and 1.0. For an α– disk, α should not greatly exceed unity, according to its
definition.

Fig. 5 shows the 68%, 90%, and 99% confidence contours of the α and $\dot{M}$
distributions: In our sample 2 we find a mean accretion rate of $< \dot{M} >= 0.13$
within a relatively narrow parameter range, whereas the viscosity parameter
($< \alpha_{\mathrm{vis}} >= 0.76$) is spread over a wider range.

Table 3. Accretion disk fits to spectral data of radio-quiet quasars (Sample 2 / Model 2). Listed are the hard power law index α_{hard} used, the best–fit parameters (M, $\dot{M}$, α_{vis}, $cos\Theta$) with 1σ uncertainties and the reduced χ^2–values.

name	α_{hard}	M [$10^8 M_\odot$]	$-\sigma_M$	$+\sigma_M$	$\dot{M}$	$-\sigma_{\dot{M}}$	$+\sigma_M$	α_{vis}	$-\sigma_{\alpha_{\text{vis}}}$	$+\sigma_{\alpha_{\text{vis}}}$	$cos\,\Theta$	$-\sigma$	$+\sigma$	χ^2
0026+129	0.91	10.01	7.68	25.56	0.1034	0.0015	0.0038	0.904	0.120	-	0.20	-	-	0.34
0052+251	0.95	10.05	6.92	37.95	0.1092	0.0016	0.0067	0.793	0.036	0.098	0.30	-	-	0.73
0119−286	0.70	4.70	2.97	7.51	0.1287	0.0081	0.0120	0.801	0.017	0.033	0.29	0.22	0.56	2.36
0157+001	0.70	2.91	0.58	34.57	0.1064	-	0.0012	0.886	0.107	0.046	0.94	-	-	0.67
0205+024	0.70	3.43	1.31	11.78	0.1331	0.0111	0.1064	0.807	0.018	0.033	0.54	0.47	0.32	1.47
0804+761	1.04	2.40	0.56	27.98	0.1089	-	0.0027	0.870	0.082	0.052	0.91	-	-	0.66
0914−621	0.70	3.78	1.84	31.96	0.1471	0.0369	0.1422	0.593	0.090	0.145	0.64	-	-	1.22
1029−140	0.70	12.50	4.98	4.88	0.1147	0.0019	0.0018	0.812	0.010	0.015	0.27	0.11	0.25	3.53
1049−005	0 70	19.06	14.95	75.82	0.1112	0 0089	0 1285	0.601	-	0 280	0.27	-	-	1 07
1100−264	0.70	92.01	27.62	-	0.2028	-	-	0.755	-	-	1.00	0.39	-	1 20
1116+215	1.00	18.54	11.55	6.17	0.1073	0.0011	0.0013	0 838	0.016	0 018	0 27	0.08	0.36	3.44
1202+281	0.70	6.99	-	1.54	0.1066	0 0017	0.0034	0.920	0.057	0.049	0.12	-	-	1.07
1216+069	0.70	6.12	1.43	-	0.1123	0.0065	0.0325	0.978	0.144	-	0.96	-	-	0.88
1247+267	0.70	89 97	44.48	-	0.2450	0.0945	-	0.471	0.253	0.485	0.74	0.35	-	1 05
1257+346	0.70	49.00	21.28	-	0.1185	0.0087	0.0218	0.947	-	- -	0.76	0.47	-	0.75
1307+085	1.08	2.52	-	-	0.1103	0.0051	0.0001	0 784	0.119	0.046	1.00	-	-	1 66
1309+355	0.70	3.21	2.15	15 09	0 2052	0.0044	0.0064	0.350	0.045	0.042	0.25	0.25	-	1.47
1334+246	0.70	0.73	-	5.91	0 1374	0.0117	0.0629	0.689	0.050	0.090	0.41	-	-	0.91
1352+183	0.70	2.99	1.57	17 56	0.1100	-	0 0043	0.786	0.044	0 091	0.44	-	-	0.75
1407+265	0 70	92.61	61 49	-	0.2861	0.0166	0.0190	0.546	0.039	0.027	0.11	-	0.42	1.04
1415+451	0.70	0 88	0.51	1.87	0.1121	0.0024	0.0106	0.650	0.049	0.061	0.50	0.43	-	2.07
1416−129	0.51	2.10	-	5.45	0.1138	0.0019	0.0048	0.987	0.032	-	0.50	0.42	0.25	0.76
1444+407	0.70	4.97	2.03	-	0.1135	0.0066	-	0.903	0.062	0.073	0.66	-	-	1.20
1521+101	0.70	42.57	13.20	-	0.2141	0 1043	0.1890	0.483	0.162	-	1.00	0.72	-	1.10
1543+489	0.70	9.30	3.60	89.68	0.1081	0.0043	0.0046	0.925	0.060	-	0.72	-	-	0.92
1613+658	0.70	1.95	0.28	22.62	0.1210	0.0111	-	0.863	0.073	0.033	0.91	-	-	0.50
1617+175	0.70	4.93	3.46	26.07	0.1410	-	0.0734	0.304	-	0.141	0.28	-	-	1.57
1630+377	0.70	26.39	5.95	-	0.2136	0.0067	0.0096	0.585	0.178	0.061	1.00	0.83	-	1.19
1700+642	0.70	58.69	24.11	-	0.2003	0.0991	-	0.794	-	-	0.84	0.54	-	1.35
1821+643	0.89	20.92	8.61	-	0.1990	0.0681	0.0029	0.562	0.035	-	0.60	0.56	-	3.97
2251−178	0.53	2.51	1.17	2.55	0.1227	0.0041	0.0044	0.816	0.011	0.010	0.35	0.21	0.43	2.76

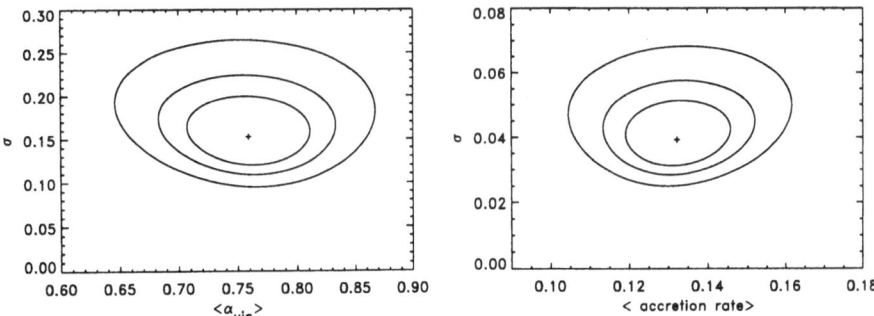

Fig. 5. 68%, 90%, and 99% confidence contours of α_{vis} (left) and $\dot{M}$ (right)

3 Summary

The soft X–ray excess in two samples of AGN has been established and quantified, not only for individual objects but also as a property of the complete samples. We find that this soft excess can be explained with thermal emission from an accretion disk. Note that our calculations are for a bare disk without the ad-hoc addition of a hot corona.

The fitted accretion rates using our model 2 generally are lower than the corresponding values from model 1 and are consistent with the thin disk approximation. This is mainly because model 2 also takes into account the temperature gradient in the vertical direction of the disk. This means, that the local spectra differ from the blackbody even in the optically thick case, leading to harder spectra. An additional result of our model is that accretion rates – at least of radio quiet QSOs – higher than 30% of the Eddington accretion rate are not observed. The viscosity parameters required on the other hand are relatively high, in most sources between 0.4 and 1.0. Full scale publications about the analysis of both samples are in preparation (Friedrich et al. 1996, Brunner et al. 1996).

References

Brunner, H., Müller, C., Friedrich, P., Dörrer, T., Staubert, R. 1996, in preparation

Comastri, A., Setti, G., Zamorani, G., Elvis, M., Giommi, P., Wilkes, B. J., McDowell, J. C., 1992, Astrophys. J. **384**, 62

Cunningham, C. T., 1975, Astrophys. J., **202**, 788

Czerny, B., and Elvis, M., 1987, Astrophys. J., **321**, 305

Dörrer, T. 1991, Diploma thesis University of Tübingen

Dörrer, T., Riffert, H., Staubert, R., Ruder, H., 1996, Astron. Astrophys., **311**, 69

Friedrich, P., Dörrer, T., Brunner, H., Staubert, R. 1996, in preparation

Hewitt, A., and Burbidge, G., 1993, Astrophys. J. Suppl., **87**, 451

Kruper J. S., Urry, C. M., Canizares, C. R., 1990, Astrophys. J. Suppl., **74**, 347

Lightman, A. P., 1974, Astrophys. J., **194**, 419

Maraschi, L., and Molendi, S., 1990, Astrophys. J., **353**, 452

Masnou, J. L., Wilkes, B. J., Elvis, M., McDowell, M. C., Arnaud, K. A, 1992, Astron. Astrophys., **253**, 35

Novikov, I.D., and Thorne, K.S., 1973, in *Black Holes*, ed. C. DeWitt and B. DeWitt (New York: Gordon & Breach), 343

Shakura, N. I., and Sunyaev, R. A., 1973, Astron. Astrophys., **24**, 337

Speith, R., Riffert, H., Ruder, H., 1995, Comp. Phys. Comm. **88**, 109

Turner, T. J., Pounds, K. A., 1989, Monthly Notices Roy. Astron. Soc., **240**, 833

Veron–Cetty, M.P., and Veron, P., 1993, ESO, Sci. Rep., 13, 1

Wilkes, B. J, Elvis, M., 1987, Astrophys. J., **323**, 243

Williams O. R., Turner, M. J. L., Stewart, G. C., Saxton, R. D., Ohashi, T., Makishima, K., Kii, T., Inoue, H., Makino, F., Hayashida, K., Koyama, K., 1992, Astrophys. J., **389**, 157

The Galactic Center – a Laboratory for AGN

W.J. Duschl[1,2,3]

[1] Institut für Theoretische Astrophysik, Universität Heidelberg, Tiergartenstr. 15, D-69121 Heidelberg, Germany
[2] Max-Planck-Institut für Radioastronomie, Auf dem Hügel 69, D-53121 Bonn, Germany
[3] E-Mail: wjd@ita.uni-heidelberg.de

Abstract: We discuss the center of the Milky Way as a prototype of galactic centers, and argue that there are no important generic differences between the centers of normal and active galaxies. While the Galactic Center currently does not harbor an active nucleus, there are strong indications of an upcoming phase of much higher activity. The observed physical status seems to imply that currently accretion is occuring in an advection-dominated mode.

1 Introduction

It is generally accepted that the activity observed in the nuclei of a fair fraction of all galaxies can be attributed either to accretion into a massive black hole, or to an ongoing starburst, or to a combination of the two processes. Moreover, there is mounting evidence that in most, if not all cases there is no generic difference between an active galaxy and a normal one. It rather seems to be only a matter of the respective *current rate* of starburst and accretion activity going on in the central region of a galaxy.

In this paper, we will concentrate on the center of the Milky Way. We will discuss the evidence for accretion going on on different geometrical scales (Sects. 2 and 3) and the implications for current and possible future activity in the Galactic Center. Finally, we will compare the Galactic Center with the centers of other galaxies, normal and active ones (Sect. 4).

This paper is in part based on a recent review with the same title (Mezger, Duschl and Zylka 1996). For a discussion of many aspects of the physics of the Galactic Center which are not addressed here, and for a more complete list of references, we refer the reader to that review paper.

2 The large scale picture

2.1 Characteristics of the Galactic Disk and the Central Region: An Overview

Seen from outside, our Galaxy would be described most likely as a barred spiral, probably similar to NGC 4303. In the Hubble classification scheme the Milky Way fits between Sbc and SBbc (de Vaucouleurs 1970). Considering that model calculations (e.g., von Linden et al. 1997) suggest repeated changes between barred and non-barred spiral structures to be a common phenomenon in galaxies with large ratios of disk to halo mass, our Galaxy may currently be in a transition state between a barred and a non-barred structure.

The Galactic Disk (GD) with a stellar mass of $\sim 10^{11}\,M_\odot$ extends to galactic radii $R \sim 14\,\mathrm{kpc}$. The Galactic Disk is supposed to be stabilized by a spherical halo of mass $M_\mathrm{halo} \geq M_\mathrm{GD}$. A thin layer of Interstellar Matter (ISM) is located in the plane of symmetry of the Galactic Disk. HII regions, as tracers of massive star formation, indicate a two-armed spiral structure within $3\,\mathrm{kpc} \leq R \leq 12\,\mathrm{kpc}$ (Georgelin and Georgelin 1976; Fig. 1).

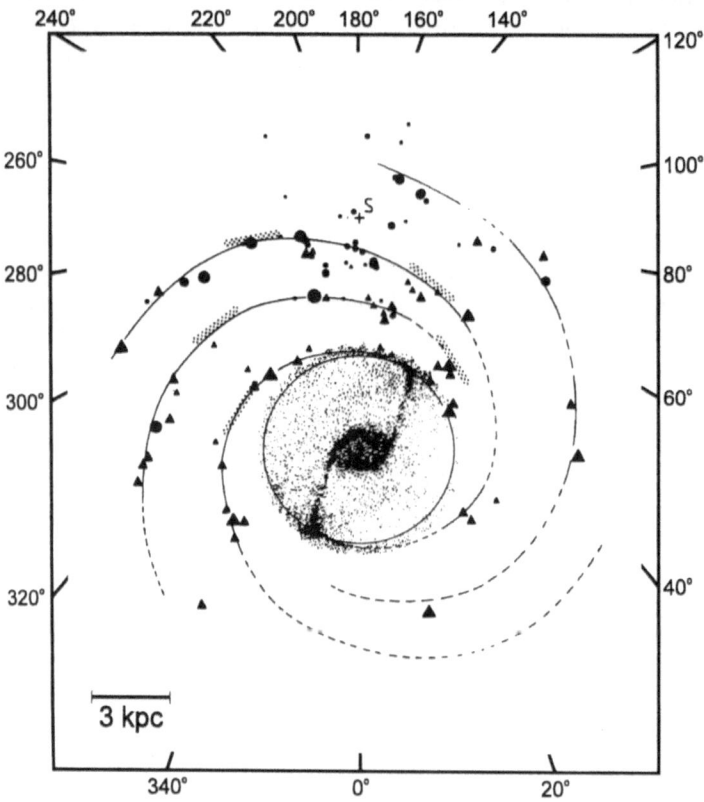

Fig. 1. The morphology of our Galaxy. The spiral structure ($R > 3.5\,\mathrm{kpc}$; Georgelin and Georgelin 1976) is based on observations of Giant HII regions, the central bar structure (Gerhard 1996) is based on both radio and NIR observations and model fitting. The above representation of the structure of the Galaxy has been adapted by Mezger et al. (1996).

In the Central Region, $R \leq 3\,\mathrm{kpc}$, the stars form an ellipsoidal or box-shaped bulge. The NIR surface density due to the increasing volume density of stars rises sharply at $l = \pm 10°$ corresponding to $R \sim 1.5\,\mathrm{kpc}$. The dynamics of the molecular gas and the surface brightness distribution of stars in the NIR suggest that gas and stars have a bar structure which is tilted relative to the line-of-sight by $\sim 16°$, with the near side of the bar lying at positive galactic longitudes. While the stellar bar may extend as far out as to the Inner Lindblad Resonance (ILR) of the spiral structure ($R_{\mathrm{ILR}} \sim 3\,\mathrm{kpc}$) there is a zone of avoidance for the ISM in the range $1.5\,\mathrm{kpc} \leq R \leq 3\,\mathrm{kpc}$ which can be explained by the fact that the bar structure does not permit stable orbits in this range of galactic radii (see the recent review by Gerhard [1996] and references therein). Fig. 1 shows the morphology of the ISM in our Galaxy as derived mainly from radio observations. In the following we refer to the regions $R \leq 0.3\,\mathrm{kpc}$ as *Nuclear Bulge*, to $0.3\,\mathrm{kpc} \leq R \leq 3\,\mathrm{kpc}$ as *Galactic Bulge*, and to the whole region $R \leq 3\,\mathrm{kpc}$ as *Central Region*.

2.2   The Large-Scale Kinematics of the Interstellar Gas

The kinematics and dynamics of the gas between galactocentric radii $\sim 100\,\mathrm{pc}$ and a few pc have been recently reviewed by Morris and Serabyn (1996). Here we address only the most important aspects. The distribution and kinematics of the gas in the central $\sim 300\,\mathrm{pc}$ radius are consistent neither with axial symmetry nor with uniform circular rotation (Bania 1977, Liszt and Burton 1978, Morris et al. 1983, Heiligman 1987, Bally et al. 1988, von Linden et al. 1993a,b, Biermann et al. 1993, Jackson et al. 1996). Approximately 3/4 of the dense molecular gas is located at positive longitudes, and about the same, but not completely identical fraction shows positive radial velocities (Fig. 2). In about 1/3 of the gas large radial and vertical motions are present (Bally et al. 1988).

On the basis of molecular cloud kinematics, it is possible to divide the gas into the following components

- the *3 kpc Arm* ($v \sim -53\,\mathrm{km\,s^{-1}}$) and the *-30 km s⁻¹ Arm*
- *Galactic Center disk clouds*: Their emission follows roughly a diagonal line in the $l-v$–distribution with vanishing radial velocities at $l \sim 0°$; This reflects to a first approximation the rotation curve in the inner Galaxy (Bally et al. 1988, Sofue 1995).
- a *trapezoidal* or *elliptical envelope* enclosing the Galactic Center disk clouds in the $l-v$ diagrams and extending between $(l, v) \sim (1°\!.7, 200\,\mathrm{km\,s^{-1}})$ and $(-1°\!.0, -200\,\mathrm{km\,s^{-1}})$ (Robinson and McGee 1970, Bania 1977, Bally et al. 1988).

As far as the mass estimates are concerned, one has to note that they depend crucially on the H_2-to-CO conversion factor. Sofue (1995) has estimated a total molecular gas mass for $|l| \leq 1°$ of $4.6\,10^7\,M_\odot$ using the new H_2-to-CO conversion factor by Arimoto et al (1996). This mass is only about one third of the previous mass estimates based on older conversion factors. Sodroski et al. (1995) even

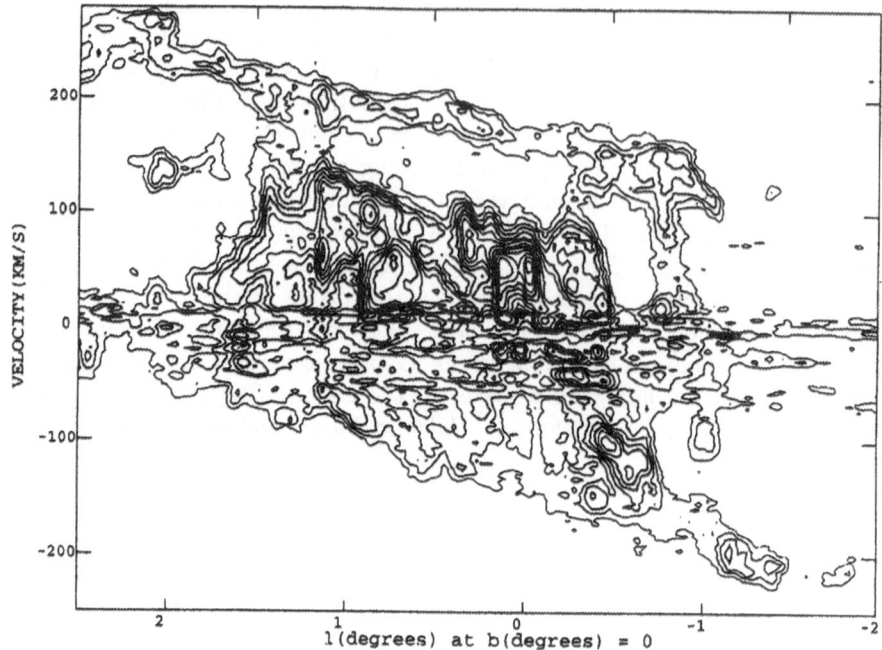

Fig. 2. Position-velocity diagram (^{12}CO contours; after Uchida et al. in prep.); the gas outside of the GC manifest itself in the velocity interval $-20 < v/(\text{km/s}) < 20$ and has very narrow line widths

derive conversion factors for the vicinity of the GC which are by factors $\sim 3 - 10$ lower compared to the solar vicinity.

Due to our location in the disk of the Galaxy, the true geometric arrangement of these clouds and structures remains ambiguous.

von Linden et al. (1993a) and Biermann et al. (1993) proposed a model in which a non-negligible radial velocity component is overlaid on the orbital motion of the molecular gas. This model ansatz allows one to determine the geometrical arrangement of the material as well as the global properties of the flow. They find a remarkably large radial mass flow rate in the range between ~ 10 and ~ 150 pc of $\sim 10^{-2}\,M_{\odot}/\text{yr}$. These values are very similar to mass flow rates derived independently for the Circum-Nuclear Disk, a ring-like structure extending between ~ 1.8 and ~ 6 pc from the Galactic Center (Güsten et al. 1987, Jackson et al. 1993). This large mass flow rate will eventually reach the innermost regions of the Galaxy.

3   Sgr A* and its immediate vicinity

Sgr A* is the radio source at the dynamical center of the Galaxy.

As was first noted by Woltjer (1959) and - nearly three decades later - was reiterated by Lo (1986) the Milky Way seems to be most closely related to Seyfert galaxies, which are spirals with nuclei brighter than those of normal galaxies. According to the presence or lack of broad lines and a strong continuum in the visible and UV wavelength range one subdivides Seyfert galaxies into Seyfert 1 and 2 galaxies with a smooth transition between the two classes. Intermediate types are accordingly explained by intermediate inclination angles of the plane of the torus with the line-of-sight. There also seems to be a continuous transition from Seyfert 1 galaxies to Quasars. Both show very similar line and continuum spectra but differ by the lower nuclear luminosity of the Seyfert galaxies.

Once more it should be reiterated that the Galactic Center is at present not in an active state. However, it will become clear from the observations reviewed here that nearly all elements of the massive black hole/acretion disk model are actually present in the Center of our Galaxy thus giving support to the hypothesis of Lynden-Bell and Rees (1971) that most – if not all – normal galaxies have massive but dormant black holes at their centers. It then is the proximity of the Galactic Center[1] which makes it a unique object for the investigation of the phenomena related to the activity in the centers of galaxies: Only for the Galactic Center does the angular resolution of present-day mm-VLBI come close to resolving the Schwarzschild radius of the central black hole of a few $10^6 \, M_\odot$, so that the emission intrinsic to the black hole can be separated from that of the immediate environment (see below).

3.1   Evidence for super-massive black holes in centers of galaxies

Model calculations demonstrate that galaxies tend to form massive central objects of $\geq 10^6 \odot$ within $\sim 10^{10}$ years (see, e.g., Duschl 1988a,b). The observational evidence for the presence of supermassive compact masses in many "normal", i.e., non-active galactic centers has been recently reviewed by Kormendy and Richstone (1995). Based on the velocity dispersion of stars, these authors derive for six galaxies of different spectral type central dark (i.e. non-stellar) masses between 10^6 and $10^9 \, M_\odot$ and for several galaxies upper limits for their central mass which fall in this mass range. Only one giant galaxy, M33, is suspected to have a less massive dark central object of $\leq 5 \, 10^4 \, M_\odot$. Kormendy and Richstone show that all objects for which the central dark mass could be determined (including the low mass case M33) follow a tight relation between this central mass and the luminosity of the galaxy's galactic bulge.

Specifically, the following recent observations strongly support the existence of supermassive compact objects in the centers of three galaxies of low nuclear activity:

- *M87* is a giant elliptical galaxy with a well investigated radio and optical jet. HST observations show a disk-like distribution of the nuclear gas and

---

[1] 8.5 kpc vs. 700 kpc to the nearest normal galaxy, M31, and 25 Mpc (for $H_0 = 50 \, \mathrm{km \, s^{-1} \, Mpc^{-1}}$) to the closest Seyfert 2 galaxy, NGC 1068; $1''$ corresponds to $4 \, 10^{-2}$ pc at 8.5 kpc, to 3.4 pc at 700 kpc, and to 120 pc at 25 Mpc

dust (Ford et al. 1994) with an orientation perpendicular to the jet. Radial velocity measurements indicate Keplerian motions of this material between ~ 150 and $16\,\text{pc}$ (no velocities have been measured for material closer to the nucleus) corresponding to an enclosed mass of $3\,10^9\,\text{M}_\odot$. Moreover, the width of emission lines from the very center, $\sim 1700\,\text{km}\,\text{s}^{-1}$, agrees with a compact mass of this order.

- Even more compelling is the spiral galaxy *NGC 4258*, a weak AGN of the liner type (Low-Ionization Nuclear Emission-line Region). Based on VLBA observations of H_2O masers located at distances $\sim 0.18 - 0.29\,\text{pc}$ from the center, Miyoshi et al. (1995) derived an enclosed mass of $4.1\,10^7\,\text{M}_\odot$ within $0.18\,\text{pc}$.

- Based on high resolution imaging spectroscopy (FWHM $= 0\rlap{.}''47$), Bender et al. (1996) deduce a black hole of mass $(3.0 \pm 0.5)\,10^6\,\text{M}_\odot$ in the center of M32, the companion of the Andromeda Nebula.

The dynamical evidence for the presence of compact supermassive objects in the centers of some galaxies is strong but it provides no information about the physical nature of this mass. Only for the center of our Galaxy does one know both the mass of $\sim 2 - 3\,10^6\,\text{M}_\odot$ contained within $R \leq 0.1\,\text{pc}$ of Sgr A* (see Sect. 3.3) and the unique intrinsic spectrum of this compact synchrotron source with a size of $10^1 - 10^2\,R_S$ (see Figs. 3 and 4). This strengthens the case for the presence of a starving black hole in the Galactic Center.

3.2   Sgr A*: Spectrum, time variation and morphology

The radio/IR spectrum The most recent spectrum of Sgr A* shown in Fig. 3 is from Beckert et al. (1996) who have added to the Zylka et al. (1995) spectrum three flux densities at 0.408, 0.96 and 1.66 GHz, derived nearly twenty years ago by Davies et al. (1976). Also added are two more upper limits obtained at $\lambda 350\,\mu\text{m}$ by Serabyn and Lis (1994) and at $\lambda 30\,\mu\text{m}$ by Telesco et al. (1996). The gap in the spectrum between $\lambda 450$ and $30\,\mu\text{m}$ allows - as an upper limit - the presence of $\sim 3\,\text{M}_\odot$ of ISM in a $10''$-beam, with a dust temperature of $\sim 50\,\text{K}$. Such a low dust temperature, however, would be hard to reconcile with the high intensity of the radiation field in the Galactic Center where dust assumes temperatures of $\sim 200 - 300\,\text{K}$ (Zylka et al. 1995).

Here, we follow the spectral analysis by Beckert et al. (1996), keeping in mind that other synchrotron emission models have been suggested (see Sect. 3.4). The observed radio/IR characteristics of Sgr A* are given in Table 1. The spectrum of Sgr A* has a high-frequency cut-off $\nu_c \sim 2 - 4\,10^3\,\text{GHz}$ and a low-frequency turnover $\nu_t \sim 1\,\text{GHz}$. In between the time-averaged spectrum increases $S_\nu \propto \nu^{1/3}$ (Duschl and Lesch 1994). The integrated radio luminosity is $L_{\text{radio,IR}}\,(1 - 10^4\,\text{GHz}) \sim 3\,10^2\,\text{L}_\odot$. One possible explanation of this spectrum - in Table 2 referred to as model A - is optically thin synchrotron emission from quasi-monoenergetic electrons (Duschl and Lesch 1994; Zylka et al. 1995) combined with free-free absorption due to the HII region Sgr A West within which

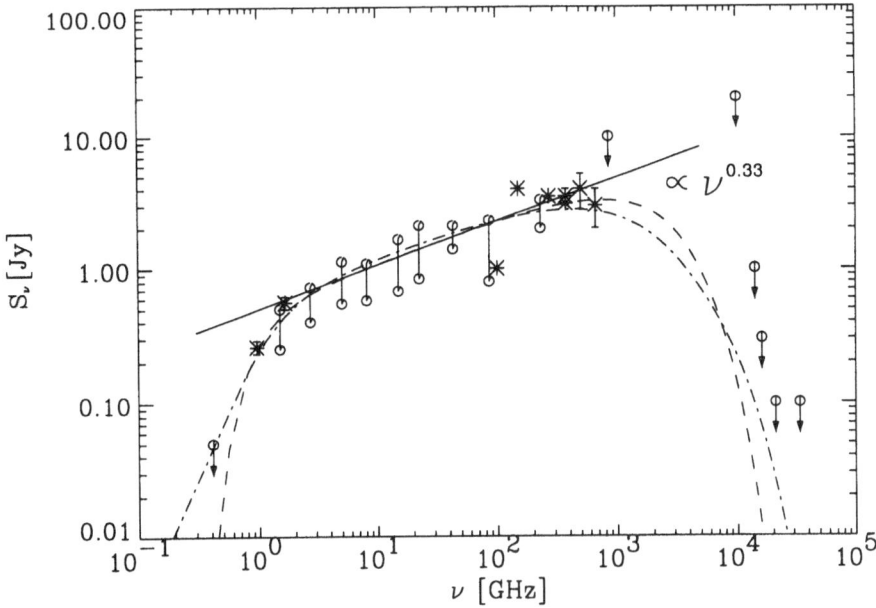

Fig. 3. A comparison of the observed radio spectrum of Sgr A* with best-fit model spectra (Beckert et al. 1996). Note that bars with symbols at both ends denote the variability range and not error bars of individual observations; additional star symbols mean individual flux density measurements. The low-frequency turnover is due to either free-free absorption (dashed curve) or synchrotron self-absorption (dashed-dotted curve). For model parameters see Table 2.

Sgr A* is located. Model parameters for the synchrotron source are an electron density $n_{e,rel} \sim 3\,10^4\,\mathrm{cm}^{-3}$ and a magnetic field strength $B \sim 11\,\mathrm{G}$. The low-frequency turnover would be caused by an absorbing thermal plasma, with an emission measure of $E_t \sim 10^6\,\mathrm{pc\,cm}^{-6}$ for an electron temperature $T_e \sim 6\,000\,\mathrm{K}$ (Fig. 3, dashed curve).

How does this model concur with observations? Sgr A West consists of an extended component and the so-called *Minispiral*, with Sgr A* located at the northern edge of the Central Bar. If Sgr A* is located at the center of Sgr A West, the emission measure provided by one half of the extended component, $0.5\,E_{ext} \sim 10^6\,\mathrm{pc\,cm}^{-6}$, could account for the free-free absorption. The Bar, however, with an estimated emission measure of $9\,10^6 - 4\,10^7\,\mathrm{pc\,cm}^{-6}$ at the position of Sgr A* would have to be located behind the source.

In the second interpretation – in Table 2 referred to as model B – the low-frequency turnover in the spectrum is explained by synchrotron self-absorption in Sgr A*. The required magnetic field strength of $\sim 11\,\mathrm{G}$ is the same as in the above case and is in accordance with independent estimates of the equipartition

Table 1. Observed characteristics of Sgr A*

l_{II}	$-00°03'20\overset{''}{.}72$
b_{II}	$-00°00'02\overset{''}{.}9$
$\alpha(1950)$	$17^{\mathrm{h}}42^{\mathrm{m}}29\overset{\mathrm{m}}{.}31$
$\delta(1950)$	$-28°59'18\overset{''}{.}38$
$\alpha(2000)$	$17^{\mathrm{h}}45^{\mathrm{m}}39\overset{\mathrm{m}}{.}97$
$\delta(2000)$	$-29°00'34\overset{''}{.}88$
Distance R_0	$8.5\,\mathrm{kpc}$
Size	$\leq 2.5 - 4\,10^{13}\,\mathrm{cm}$
Mass	$2 - 3\,10^6\,\mathrm{M_\odot}$
ν_c	$2 - 4\,10^3\,\mathrm{GHz}$
ν_t	$\sim 1\,\mathrm{GHz}$
$S_\nu(\lambda 2.2\,\mu\mathrm{m};\text{ dereddened})$	$\leq 9\,\mathrm{mJy}$
$S_{1-4\,\mathrm{keV}}$ (dereddened)	$\sim 2 - 4\,10^{35}\,\mathrm{erg\,s^{-1}}$
$L_{\mathrm{radio},1-10^4\,\mathrm{GHz}}$	$\sim 3\,10^2\,\mathrm{L_\odot}$
$L_{\mathrm{visual/UV}}$	$\leq 5\,10^4\,\mathrm{L_\odot}$
$L_{\mathrm{X-ray},1-10\,\mathrm{keV}}$	$\leq 2\,10^2\,\mathrm{L_\odot}$
Average extinction $\langle A_{\mathrm{V}}\rangle$ over central pc	31^{m}

magnetic field strength in the inner regions of an accretion disk accreting into a black hole of a few $10^6\,\mathrm{M_\odot}$ with a mass flow rate of $10^{-7\cdots-6}\,\mathrm{M_\odot/yr}$. The source diameter of $\sim 2.4\,10^{13}$ cm required by this model is $\sim 30 - 20\,R_S$ with $R_S \sim 6-9\,10^{11}$ cm the Schwarzschild radius of a black hole of mass $2-3\,10^6\,\mathrm{M_\odot}$. Within the observational errors and the uncertainties of the model parameters, this predicted source size is compatible with source sizes of $\leq 4\,10^{13}$ cm obtained from mm VLBI observations of Sgr A* (see below). Recent 1-mm-VLBI observations of Sgr A* even strengthen this argument (Krichbaum et al., priv. comm.)

It should be noted that both alternatives have already been considered by Davies et al. (1976). However, thanks to the now much better determined spectrum of Sgr A* and source parameters of Sgr A West, Beckert et al. (1996) arrive at firm conclusions regarding the structure of Sgr A West and Sgr A*. Both effects, i.e. free-free absorption by Sgr A West and synchrotron self-absorption in the compact source Sgr A* must affect the Sgr A* spectrum. With the observations available to date one cannot yet decide which is the dominating effect causing the observed low-frequency turnover ν_t. Array observations at or below $\nu_t \leq 1.5\,\mathrm{GHz}$ yielding the accurate shape of the spectrum at frequencies $\nu \leq \nu_t$ may eventually resolve the question.

Table 2. Best fit parameters of the quasi-monoenergetic synchrotron radiation model with the low-frequency turnover due to free-free absorption (A) or synchrotron-selfabsorption (B)

Model		A	B
Mean electron energy	$\langle E \rangle$	147 MeV	114 MeV
electron energy index	$d \log N / d \log E$	-2.0	-2.0
Width of electron distribution	E_{max}/E_{min}	3	3
Magnetic field strength	B	11 G	11 G
Source radius	R	$1.9\,10^{13}$ cm	$1.2\,10^{13}$ cm
Density of relativistic electrons	$n_{e,rel}$	$6.1\,10^3$ cm^{-3}	$2.7\,10^4$ cm^{-3}
Frequency of free-free absorption	$\nu(\tau_{ff} = 1)$	0.8 GHz	< 0.8 GHz

The NIR through X-ray spectrum Eckart et al. (1992) using high-resolution NIR imaging, detected at $\lambda 2.2\,\mu$m an emission ridge of size $\sim 1''$ with a dereddened flux density of 0.06 Jy. Assuming optically thick free-free emission from a source with disk-shaped brightness distribution of electron temperature $T_e \geq 2\,10^4$ K yields a luminosity of

$$\frac{L}{L_\odot} \sim 1.2\,10^5 \left(\frac{T_e}{3\,10^4\,\text{K}} \right)^3 . \qquad (1)$$

With effective temperatures in the range $T_{eff} \sim 2-4\,10^4$ K this yields luminosities of $L_* \sim 3.6 - 29\,10^4\,L_\odot$. More recently, however, with the remarkable K-band resolution of $\sim 0''15$ Eckart et al. (1995) resolved the extended source into a cluster of six (probably stellar) sources of which only one can coincide with Sgr A*. Hence, the above estimates have to be decreased by $\sim 1/6$, yielding upper limits of the optical/UV luminosities for Sgr A* of $\leq 5\,10^4\,L_\odot$. Eckart et al. (1995) refer to the radio and IR source as Sgr A*(R) and Sgr A*(IR), respectively. The relative accuracy of radio and NIR frames then available did not allow them to decide if Sgr A*(R) does actually coincide with one of the NIR sources. Recently, however, Menten et al. (1996) achieved a breakthrough. They detected several SiO and H$_2$O maser sources within the central parsec which arise from the innermost parts of the circumstellar envelopes of (super-)giant stars whose position relative to Sgr A* could be determined with milli-arcsec (mas) accuracy using VLBI. Comparing these radio images with high resolution NIR images the radio relative to the NIR reference frame could be determined with an accuracy of $\pm 0''02$. They find that none of of the NIR sources coincides with the compact radio source Sgr A*. From this a conservative upper limit of < 9 mJy for the dereddened K-band flux density of Sgr A*(R) is obtained. Eckart et al. conclude that Sgr A* (R) is probably dark; its M/L ratio at present has a lower limit of $\sim 100\,M_\odot/L_\odot$ and Sgr A* (IR) represents a small local clustering of moderately luminous stars near or at the position of Sgr A* (R). With one exception the polarization of these NIR stars is similar to other sources in their vicinity.

Several groups report the detection of X-ray emission from Sgr A*, most recently Predehl and Trümper (1994) with ROSAT in the energy range 1.2 - 2.5 keV, and Koyama (1994), Tanaka (priv.comm.) and Koyama et al. (1996a,b) with ASCA in the energy range $\sim 1 - 10$ keV (for a review of previous X-ray observations of the Galactic Center see, e.g., Skinner 1993 and Predehl et al. 1994). The point source detected by ROSAT with an angular resolution of 25″ coincides with the radio position of Sgr A* to within 10″, i.e., the positional accuracy of ROSAT observations, and has a (dereddened) flux density of $S_{1-4\,\mathrm{keV}} \sim 2 - 4\,10^{35}\,\mathrm{erg\,s^{-1}}$. Previous detections at higher energies (3 - 30 keV) with coded mask telescopes on Spacelab-2 and ART-P but with considerably lower positional accuracy were reported by Skinner et al. (1987) and Pavlinsky et al. (1994).

Here we follow the arguments by Beckert et al. (1996). The ART-P source exhibits a non-thermal spectrum. If the ROSAT flux density is corrected for an X-ray absorption corresponding to the standard extinction $A_\mathrm{V} \sim 31^\mathrm{m} (\cong N_\mathrm{H} \sim 5.4\,10^{22}\,\mathrm{cm^{-2}})$ between Sun and Galactic Center its intensity lies ~ 2 to 3 orders of magnitude below the extrapolated ART-P spectrum. To bring the ROSAT flux densities in agreement with the ART-P spectrum requires an additional absorption correction corresponding to a column density of insterstellar hydrogen $N_\mathrm{H} \sim 1 - 1.5\,10^{23}\,\mathrm{cm^{-2}}$ and hence to a visual extinction of $A_\mathrm{V} \sim 56^\mathrm{m} - 83^\mathrm{m}$. This could be explained if the ROSAT source were located deep inside the Sgr A East Core GMC against which Sgr A* and Sgr A West are seen in projection. In this case, however, the ROSAT source could not be identical with Sgr A*, which our observations place at the center of the HII region Sgr A West and in front of the *Bar* of the ionized *Minispiral*. Sgr A West, on the other hand, is located in front of the extended synchrotron source Sgr A East which, in turn, is located in front of the Sgr A East Core GMC (Zylka et al. 1995).

Alternatively, Predehl and Trümper suggest that the ROSAT source in fact coincides with Sgr A* but is subject to a very local or intrinsic absorption. NIR observations of the central 0.5 pc do not indicate localized dust absorption on scales of $\sim 1''$ (Krabbe et al. 1995). To comply with these observations, the X-ray absorption would have to be provided by an ionized shell surrounding Sgr A*, where most of the central dust has evaporated. Beckert et al. (1996) show that this is not consistent with observations.

Above (see Eq. 1) we derived an upper limit for the luminosity of Sgr A* assuming black-body emission with a Planck curve fitted through the $\lambda 2.2\,\mu\mathrm{m}$ measurement. The upper limit for the soft X-ray emission obtained by Predehl and Trümper allows a more sophisticated estimate. If both the NIR upper limit with $S_{2.2\,\mu\mathrm{m}} \sim 0.01\,\mathrm{Jy}$ estimated from the observations by Eckart et al. (1992) and the soft X-ray source with $S_{1-4\,\mathrm{keV}} \sim 2 - 4\,10^{35}\,\mathrm{erg\,s^{-1}}$ observed by Predehl and Trümper (1994) close to the position of Sgr A*(R) are taken as upper limits to the emission by an accretion disk surrounding the central black hole, one can fit the spectrum of such an accretion disk[2] (e.g., Frank et al. 1992). It is found that the spectrum (see Fig. 6) is constrained by the NIR flux density and the mass of the black hole (BH) for which we adopt $M_{BH} \sim 2.5 \, 10^6 \, M_\odot$ but would be undetectable in the soft X-ray domain. The corresponding upper limit to the NIR through X-ray luminosity of Sgr A* is a few $10^2 \, L_\odot$.

Time variations at mm/submm wavelengths The variability of Sgr A* in the cm/dm wavelength range is well established (for a summary of the observations, see Mezger, Duschl and Zylka 1996). The variability of the Sgr A* emission at mm/submm wavelengths was recently investigated by Zylka et al. (1995) at $\lambda 1300 \, \mu$m and $800 \, \mu$m on time scales of $\sim 1 - 3$ yrs. No variations larger than $\sim 10 - 20\%$ at $1300 \, \mu$m and 40% at $800 \, \mu$m were observed. Gwinn et al. (1991) investigated the short term variability of Sgr A* at the same wavelengths with the CSO on time spans between 0.1 s and 24 h, also with negative results.

Zylka et al. (1995) report on more recent investigations of the variability on time scales of a few days . Again, within the observational uncertainties no variability at these wavelengths has been detected, although the $\lambda 800 \, \mu$m data may indicate a systematic variation on time scales of a few years.

At a mass of a few $10^6 \, M_\odot$ and a source size of a few 10^{13} cm, all relevant timescales a very short. The dynamical time scale, for instance, is of the order of an hour. The relevant thermodynamical time scales are also comparatively short, of the order of hours to few days (see, e.g., Duschl and Lesch 1994). This implies that the flux densities of all currently available spectra of Sgr A* that cover a broad frequency range are *not* coeval. To discern between the different models (see Sect. 3.4) needs, however, genuine simultaneous spectra.

Morphology, recent results Over a wide frequency range the apparent size of Sgr A* decreases $\propto \lambda^2$ (see Fig. 4). That indicates that one does not resolve the true source size of Sgr A* but rather observes the unresolved image enlarged by interstellar scattering. At $\lambda 1.35$ cm, for a long time the short-wave limit of VLBI observations, Marcaide et al. (1992) measured an apparent source size of 2 mas ($\cong 2.5 \, 10^{14}$ cm). Serabyn et al. (1992) observed Sgr A* at $\lambda 1.3$ mm with the OVRO mm-wave interferometer and a synthesized beam of $1\rlap{.}''9 \times 4\rlap{.}''3$ and found a source of size $< 2''$ and flux density $S_{1.3\,mm} \sim 2.4$ Jy. A second source of ~ 0.7 Jy located $\sim 7''$ to the north appears to be a restoration artifact (Scoville, pers.comm.). Krichbaum et al. (1993) succeeded for the first time to detect Sgr A* at $\lambda 7$ mm using VLBI. Their image indicates a possible elongation of Sgr A* at a position angle $-25°$. Backer (1994) confirmed the detection of Sgr A* at $\lambda 7$ mm with VLBI observations made three months later. They found a flux density of 2.1 Jy contained in a (0.7 ± 0.07) mas circular Gaussian source. The detection of Sgr A* with VLBI at $\lambda 3.5$ mm was first reported by Krichbaum et al. (1994a) and subsequently confirmed in another experiment by Rogers et al.

---

[2] We have taken a steady standard accretion disk which extends from 3 to $1000 \, R_S$ assuming that it radiates locally like a black body.

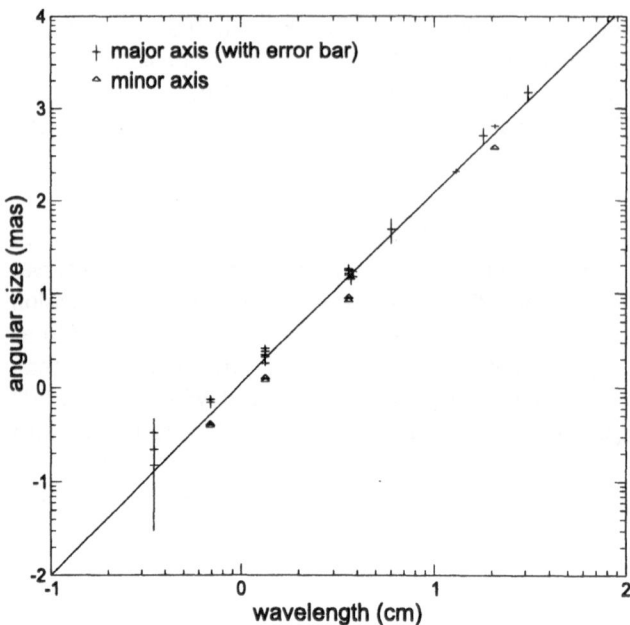

Fig. 4. Observed source sizes of Sgr A* as a function of the wavelength: Major (crosses with error bars) and minor (triangles) axes; the best fit is $\propto \lambda^{2.04\pm0.01}$. For $\lambda < 7\,\mathrm{mm}$ the observed sizes could be above the extrapolation of the data at $\lambda \geq 1\,\mathrm{cm}$.

(1994). With some remaining uncertainty in the determination of the source size, both groups consistently find a flux density of 1.5 Jy contained in a Gaussian source of $0.2 - 0.3\,\mathrm{mas}$ size and a brightness temperature of a few $10^9\,\mathrm{K}$..

Fig. 4 (after Krichbaum, priv. comm.) summarizes the present state of high-resolution imaging of Sgr A* which exhibits at all wavelengths an elliptical shape (Krichbaum et al. 1994a, 1994b, 1994c); it shows the major and minor axes. The best fit is $\propto \lambda^{2.04\pm0.01}$. It appears that at $\lambda < 7\,\mathrm{mm}$ the measured sizes of $\sim 0.2/0.3\,\mathrm{mas}$ fall above the prediction based on an extrapolation of the data at $\lambda \geq 1\,\mathrm{cm}$. This could indicate that at $\lambda \leq 3\,\mathrm{mm}$ one begins to resolve Sgr A*. Very recent 1.3 mm VLBI observations support this point (Krichbaum et al., priv. comm.). The corresponding linear sizes at $\lambda 3.5\,\mathrm{mm}$ are $2.5 - 4\,10^{13}\,\mathrm{cm}$. The position angle of the major axis is roughly constant at $\sim 90°$ over the whole observed wavelength range.

Apart from the λ^2 dependence of the apparent source size of Sgr A* there are independent arguments that the source size for $\lambda \geq 1\,\mathrm{cm}$ is due to interstellar scattering: van Langenfelde and Diamond (1991) showed that VLBI observations of the 1.6 GHz OH maser emission from OH/IR stars in the immediate vicinity of Sgr A* yield the same apparent size as obtained for Sgr A* at that frequency. Frail et al. (1994a) obtained for OH/IR stars close to Sgr A* scattering disks with the same elongation as Sgr A* indicating that this morphology is not an intrinsic characteristic of the Sgr A* compact synchrotron emission source but is rather due to anisotropic electron density variations in the scattering screen. Applying this investigation to other OH maser sources, van Langenfelde et al. (1992) found a region of pronounced scattering centered on Sgr A*. Although this symmetry by itself is a strong indication of a physical correlation between the scattering medium and the Galactic Center, van Langenfelde et al. could nonetheless not rule out a chance coincidence with scattering screen along the line-of-sight to but unrelated with the Galactic Center. Lazio et al. (1996) find a paucity of AGN within $\sim 1°$ of Sgr A* which they attribute to a strong broadening of these sources, indicating that the scattering region is actually local to the GC. Moreover, Wielebinski and Kramer (priv. comm.) extended previous pulsar surveys to shorter wavelengths ($\lambda 6$ cm) and found that the observed number of pulsars in the inner few 10^2 pc is considerably smaller than expected from extrapolation from larger galactocentric radii. This could also be due to large interstellar scattering, but a different stellar population, or environmental effects counterproductive to the formation of pulsars (e.g., high density of the ambient medium) cannot yet be excluded either.

3.3   The kinematics of gas and stars in the Central Cavity and the mass of Sgr A*

Within $R \leq 1.7$ pc most of the gas is ionized and therefore radio recombination lines and the [NeII] $\lambda 12.8\,\mu$m line are used as probes of the kinematics of the gas. For a recent review on recombination line observations of Sgr A West mostly made with the VLA, see Roelfsema and Goss (1992), for observations of the [NeII] line see Lacy (1994). It soon became evident that for $R \geq 1.5$ pc the gravitational potential is dominated by the (nearly isothermal) central star cluster, while for radii $R \leq 1.5$ pc the gravitational potential is dominated by a central compact mass of a few $10^6\,M_\odot$ (see, e.g., Genzel and Townes 1987, Mezger and Wink 1986, Serabyn et al. 1988, Herbst et al. 1993 and Lacy 1994). Roberts and Goss (1993) mapped the H92α line emission from Sgr A West with an angular resolution of $\sim 1''$. They identify three major kinematic features: The Western Arc, the Northern Arm, and the Bar. The Western Arc is explained as part of a ring of radius 1 pc in circular rotation about a point mass near Sgr A*, with a position angle of the major axis of $\sim 22°$ and an inclination angle with respect to the Galactic plane of $\sim 34°$. With a rotational velocity of ~ 105 km s^{-1} the enclosed mass is found to be $\sim 3.5\,10^6\,M_\odot$.

The kinematics of the ionized gas surrounding the minicavity has been recently investigated by mapping of the $\lambda 12.8\,\mu$m [NeII] line with an angular and spectral resolution of $\sim 2''$ and ~ 30 km s^{-1}, respectively (Lacy et al. 1991), and by mapping of the H92α recombination line emission using the VLA with an angular resolution of $\Theta_\alpha \times \Theta_\delta \sim 0.''17 \times 1.''18$ and a spectral resolution corresponding to 14 km s^{-1} (Roberts et al. 1996). Modelling of the [NeII] line data yields a central mass of $(2 \pm 0.5)\,10^6\,M_\odot$. The velocity field of the H92α data is modeled by gas in a hyperbolic orbit about a point mass at the position of Sgr

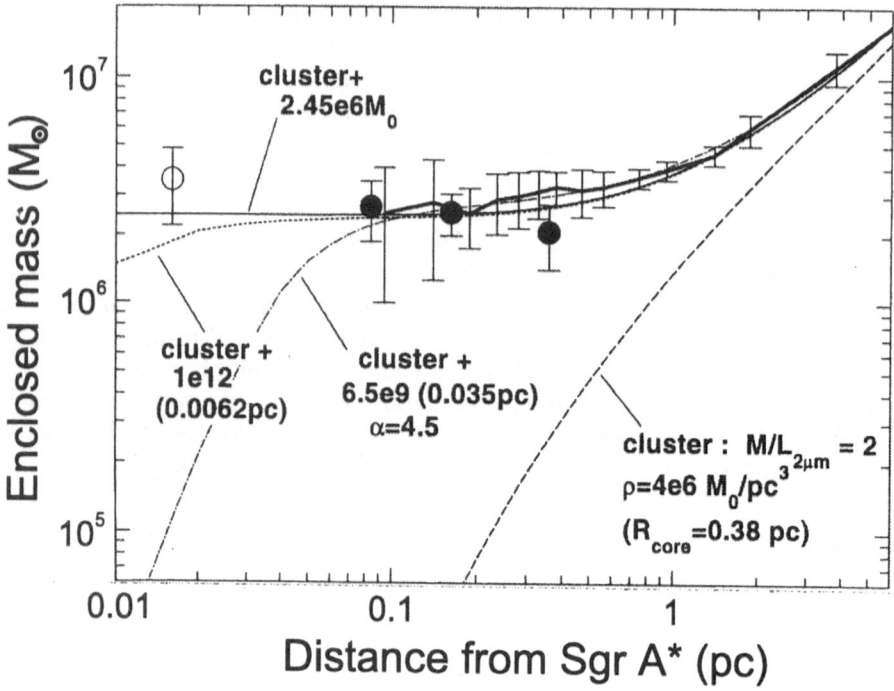

Fig. 5. The enclosed mass $M(R)$ as a function of distance R from Sgr A* (Eckart and Genzel 1996a,b) as derived from radial and proper motions of stars. The heavy full line connects the observations (given with their respective 1σ errors). The thick dashed curve represents the mass distribution of the stellar cluster for a core radius $R_{core} = 0.38$ pc, a central density $\rho = 4\,10^6\,M_\odot/\text{pc}^{-3}$, and $M/L_{2\,\mu m} = 2$ (in solar units). The thin full line shows the sum of this NIR cluster and a central point mass of $2.45\,10^6\,M_\odot$. The other two curve are the sums of the NIR cluster plus an additional dark cluster of core radius 0.035 pc, central density $6.5\,10^9\,M_\odot/\text{pc}^{-3}$ and density for large radii $\propto R^\alpha$ (dash-dotted curve), and for 0.0062 pc and $1\,10^{12}\,M_\odot/\text{pc}^{-3}$ (thin dashed curve), respectively. Note that in this figure, the distance between Sun and Galactic Center is assumed to be 8.0 kpc.

A* with a mass of $\sim (3\pm0.5)\,10^6\,M_\odot$. Herbst et al. (1993) mapped the $\lambda\,2.17\,\mu m$ Brγ line in the Galactic Center. They derive a total mass of gas in the Northern arm and Central Bar of $\sim 5\,M_\odot$. Model fits for the dynamics of the gas require $\sim 4\,10^6\,M_\odot$ within $R \sim 0.17$ pc.

More recently, following the pioneering work by Sellgren et al. (1987) and Rieke and Rieke (1988), stellar velocity dispersions were used to extend the determination of the mass $M(R)$ enclosed within the radius R to distances as close as 0.1 pc. Strengths and weaknesses of this method are thoroughly discussed in Haller et al. (1996a) who also derived a compact mass of $\sim 2\,10^6\,M_\odot$. Genzel et al. (1996) determined radial velocities and velocity dispersions of ~ 25 early-type stars and of ~ 200 red giants and supergiants within the central 2 pc. Both tracers agree that a dark mass of $\sim 2.5 - 3.2\,10^6\,M_\odot$ is located at the dynamical center of our Galaxy, a result which is in excellent agreement with mass estimates based on the kinematics of the ionized gas of the minispiral. Measurements of the proper motion of Sgr A* can yield an independant estimate of its mass. Backer (1994), using VLBI, derived a lower mass limit of $M(\text{Sgr A*}) \sim 200 - 2\,000\,M_\odot$ which – while not setting a very stringent limit – indicates that Sgr A* in all probability is not a stellar object.

Eckart and Genzel (1996a,b) also undertook a new approach, using proper motions of ~ 50 stars in the radial range $R \sim 0.004 - 0.4\,\text{pc}$ as mass tracers. Proper motions and radial velocity dispersions are in very good agreement suggesting a central dark mass of $2 - 3\,10^6\,M_\odot$. Enclosed masses $M(R)$ are shown in Fig. 5 as a function of distance R from Sgr A* (Eckart and Genzel, 1996a,b).

For large radii ($R > 200\,\text{pc}$), $M(R)$ is obtained from H $\lambda\,21$ cm line observations, for $5 \leq R/\text{pc} \leq 100$ from OH/IR stars, and for $R < 5\,\text{pc}$ from terminal velocities of the ionized gas, velocity dispersion of stars and their radial and proper motions. The best fit to these data requires a central compact mass of $\sim 2.5\,10^6\,M_\odot$. This all leads to the conclusion that with a high degree of probability there is a dark compact mass of $2 - 3\,10^6\,M_\odot$ located at or very close to the dynamical center of the Galaxy.

It has to be stressed that the observed relation $M(R)$ indicates the presence of a compact mass of a few million $M_\odot$ within a radius of $R \leq 0.1\,\text{pc}$, but that this result does neither prove that this is actually the mass of Sgr A* nor shows that there exists a massive black hole at the center of our Galaxy. Alternative interpretations have been discussed in Haller et al. (1996a). According to Genzel et al. (1996) a cluster of White Dwarfs or neutron stars are ruled out by the new results; hence the only alternative left to the presence of a single massive black hole of a few $10^6\,M_\odot$ at the center of the Galaxy would be a compact cluster of $\geq 10 - 20\,M_\odot$ black holes adding up to the same total mass. Saha et al. (1996) find that the ratio M/L_K varies from ≤ 1 for $R \geq 0.8\,\text{pc}$ to > 2 at $R \sim 0.35\,\text{pc}$ and conclude that this behaviour is due to an increasing concentration of stellar remnants towards the Galactic Center. Rieke (priv.comm.), on the other hand, points out that – for a better understanding of the kinematics – dynamical simulations of globular clusters for stars with $\sim 0.2 - 0.3\,M_\odot$ are needed which are more representative for the population of old stars in the Galactic Center. If, however, the innermost proper motions derived by Eckart and Genzel (1996a,b; $\sim 600\,\text{km s}^{-1}$ at $R \sim 0.01\,\text{pc}$) are confirmed, the only solution is a massive black hole.

3.4   Models of Sgr A*

By its very nature, an underfed BH is difficult to observe. In addition, the physical conditions in the central parsec – i.e., the volume densities of low- and high-mass stars and the ensuing intensity of the interstellar radiation field with all its consequences for free-free and dust emission – are rather extreme. If the

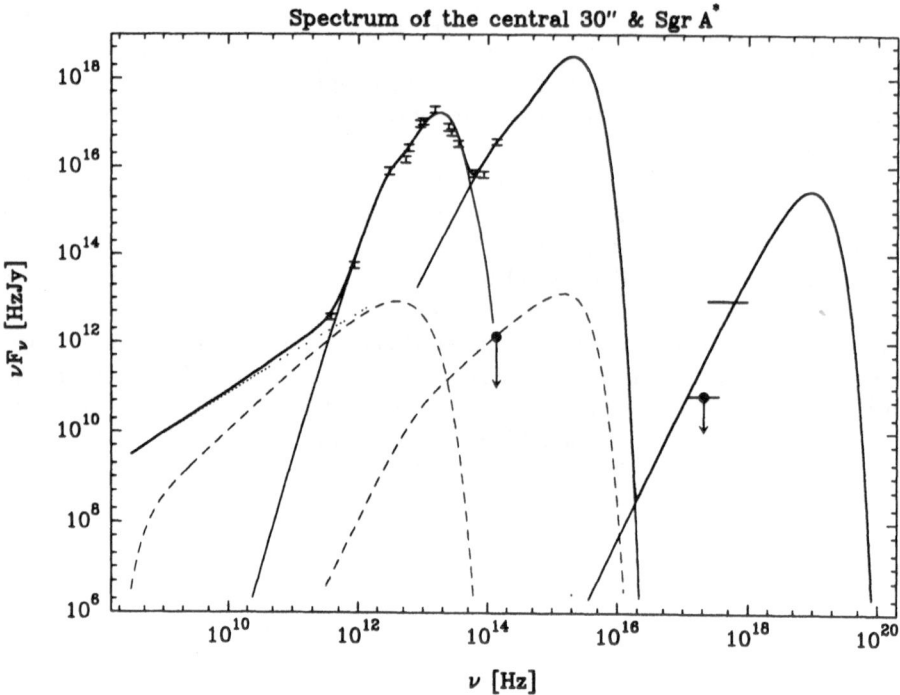

Fig. 6. The radio through X-ray spectrum of the central $30''(\hat{=}1.2\,\mathrm{pc})$ as seen from the Galactic Poles (heavy curve). Free-free emission dominates the spectrum for $\nu < 2\,10^{11}\,\mathrm{Hz}$, dust emission for $2\,10^{11} < \nu/\mathrm{Hz} < 3\,10^{13}$, stellar radiation for $3\,10^{13} < \nu/\mathrm{Hz} < 2\,10^{16}$ and hot plasma for $\nu > 2\,10^{16}\,\mathrm{Hz}$. Note that the stellar flux densities relate to the central parsec only. The full dots denote upper limits in the NIR and X-ray bands for the point source Sgr A*. The dashed curves give the observed spectrum of Sgr A*(radio/IR) and the computed model spectrum of an accretion disk compatible with the NIR upper limit and a black hole mass of $\sim 2.5\,10^6\,\mathrm{M}_\odot$ (see text).

angular resolution of the observing instruments is insufficient and the activity of the nucleus is low, emission from its vicinity can easily mask the emission proper of the black hole/accretion disk configuration. To demonstrate this, we compare in Fig. 6 the radio through X-ray spectrum of the central parsec of our Galaxy (solid heavy curve) with the corresponding spectrum of Sgr A* (dashed curve). Both spectra are shown as would be seen from the Galactic Poles (i.e., without extinction by the ISM characterized by a visual extinction of $A_{\mathrm{V}} \sim 31^{\mathrm{m}}$ between Sun and Galactic Center). Dashed curves show – presented in the form of νS_ν in units of Jy×Hz – the radio/IR spectrum of Sgr A* (from Fig. 3) and an accretion disk spectrum fit to the NIR integrated flux densities[3] (see Sect.

---

[3] The disk spectrum is required to be compatible with both upper limits of the flux densities in the NIR and the soft X-ray regime. But for a given mass of $2 - 3\,10^6\,\mathrm{M}_\odot$ of the accreting black hole, the constraint by the NIR flux density is so much stronger

3.2). The spectrum of the central parsec consists of

1. free-free emission. The flux density $S_{15\,GHz} \sim 8\,Jy$ has been obtained by integration of the radio image over the central $30''$ and subtraction of Sgr A*;
2. dust emission;
3. optical/UV emission from hot and cool stars extrapolated from their K-band flux densities;
4. soft X-ray emission obtained by integration of the ROSAT map (horizontal line; Predehl, priv. comm.). The spectrum fitted to this ROSAT point is that of a 10 keV thermal plasma.

The spectrum shown in Fig. 6 as a solid curve would be the spectrum of the center of the Milky Way seen by an observer with an angular resolution of $\sim 0''5$ if his home planet were located, for instance, in the galaxy M31: A spectrum mimicking a weak Seyfert 1 nucleus of a few $10^8\,L_\odot$. The synchrotron spectrum of Sgr A* proper would be completely masked by background emission due to the central cluster of hot and luminous stars. This should serve as a caveat for interpreting nuclear spectra. Radio/IR spectra similar to that of Sgr A* (Fig. 3) may be characteristic for black hole/accretuion disk configurations, but in most weak AGN could not be separated – due to lack of angular resolution – from the intense background emission caused by direct and reprocessed radiation from stars and a central black hole/accretion disk source.

The Sgr A* spectrum has been discussed and interpreted in terms of synchrotron radiation of quasi-monoenergetic relativistic electrons in Sect. 3.2. Here we review some other models which were suggested recently.

Narayan et al. (1995) start from the black hole/accretion disk configuration but use a different class of accretion disk models, in which most of the liberated potential energy is not radiated locally as in the standard models for accretion disks but rather advected into the black hole. Their best model fit to the spectrum is obtained for a $7\,10^5\,M_\odot$ black hole accreting $1.2\,10^{-5}\alpha\,M_\odot/yr$. α is the usual parameter which measures the efficiency of transport of angular momentum and mass in the disk and is limited to the range $\alpha = 0\ldots 1$. In their model less than 0.1% of the viscously dissipated energy is actually radiated while more than 99.9% is advected into the BH without leaving any directly observational traces.

Falcke et al. (1993) argue that three sources contribute to the spectrum of Sgr A*: (i) an accretion disk with an accretion rate of $10^{-7} \geq \dot{M}/(M_\odot/yr) \geq 10^{-8.5}$ around a black hole of mass $10^6\,M_\odot$ is responsible for the NIR part, (ii) a jet for the radio frequency part and (iii) the jet nozzle for the submm part. These authors parametrize the energy balance between the accretion disk and the hypothezised jet and find that – in the framework of this model – the total jet luminosity and the radiated accretion disk luminosity are about equal. Falcke and Heinrich (1994) have analyzed standard accretion disks taking into account

that the ROSAT limit yields no additional constraint

relativistic effects. They find that at the above accretion rates, a limit-cycle type instability should occur and give rise to luminosity variations on time scales between a few and several thousand years.

Both Mastichiadis and Ozernoy (1994) and Melia (1994) argue for Bondi-type wind accretion into a black hole, which is supplied by the cluster of massive stars referred to as IRS16, but differ considerably in their model parameters. Based on observed X- and γ-ray flux densities attributed to Sgr A*, Mastichiadis and Ozernoy argue for a black hole of mass $\ll 6\,10^3\,M_\odot$ (for a discussion of other arguments against a supermassive black hole, see Ozernoy 1992). Melia, from modeling the spectrum between 10^8 and 10^{20} Hz, finds a most likely black hole mass of $(2 \pm 1)\,10^6\,M_\odot$ and an accretion rate of $\sim 10^{-4}\,M_\odot\,yr^{-1}$.

Most of the models agree that there is a supermassive black hole in the Galactic Center. The differences in the determined accretion rates are mainly due to the lack of spectral information between the NIR and the soft X-ray regime, which makes it impossible to firmly determine the total luminosity of Sgr A*. Moreover, the newly developed models of advection-dominated accretion disks render it doubtful whether the luminosity of a black hole actually depends in a linear way on an accretion disk's mass flow rate at all.

4   The Galactic Center, a laboratory for AGN?

Here we summarize the observed characteristics of the GC (Sect. 4.1) and compare them (in Sect. 4.2) with observations and models of AGN. In Sect. 4.4 special attention is given to a comparison of the Galactic Center with the central regions of M81 and NGC 1068. In Sect. 4.5 we atempt to answer the above question: "The Galactic Center: A laboratory for AGN?".

4.1   A summary of observations

We begin with a summary of the most important observational results which characterize the physical state of the central region:

The Galactic Bulge $(3 \geq R/\text{kpc} \geq 0.3)$: Contains a stellar mass of $\sim 10^{10}\,M_\odot$ and appears to form a transition zone between halo and Nuclear Bulge; contains stars of all ages from $\sim 1 - 10\,\text{Gyr}$ with an abundance varying from $Z/Z_\odot \sim 0.3 - 10$, and – at least for $R \leq 2\,\text{kpc}$ – an indication of an abundance gradient. Stars and gas appear to form a bar-like structure which can explain the "zone of avoidance" of ISM in the range $3 \geq R/\text{kpc} \leq 1.5$. The gas-to-stellar mass ratio is very low, $M_H/M_* \ll 1\%$, and there is no indication of present-day star formation.

Galactic Bar: In the range between a few $10^2\,\text{pc}$ and $\sim 2\,\text{kpc}$ there seems to be a weak bar present. This bar may be responsible for the transport of matter towards the inner few hundred parsec of the Galaxy and thus for the supply of mass there.

The Nuclear Bulge $(R \leq 0.3\,\text{kpc})$: Superimposed on an old halo-type stellar population is a younger $(10^7 - 10^8\,\text{yr})$ generation of high and medium mass stars whose number-ratio relative to the old population increases towards the center. Its total stellar mass amounts to $\sim 4\,10^9\,M_\odot$. The hot stars of the central cluster have a core radius of $\sim 0.2\,\mathrm{pc}$; within the central parsec their stellar luminosity is $\sim 8\,10^7\,L_\odot$.

The *ISM* is concentrated in a narrow ($h \sim 30-50\,\mathrm{pc}$) layer of predominantly molecular gas ($M_{H_2} \sim 0.5 - 1\,10^8\,M_\odot$, $Z/Z_\odot \sim 2$), about half of which forms very compact giant molecular clouds. Modelling of the cloud kinematics yields a radial inflow rate of $\dot{M} \sim 10^{-2}\,M_\odot\,\mathrm{yr}^{-1}$ between $R \sim 150$ and $10\,\mathrm{pc}$. The thin layer of ISM is pervaded by the very narrow ($\sim 20\,\mathrm{pc}$) layer of thermal $7\,000\,\mathrm{K}$ plasma provided by extended low density HII regions. Ratios L_{IR}/M_{H_2} and N_{Lyc}/M_{H_2} are comparable with the Galactic Disk. But with few exceptions, there are no indications of very active present-day high-mass star formation. Two mild starbursts $\sim 10^7$ and $\sim 10^8\,\mathrm{yr}$ ago seem to fit the observations better.

Magnetic fields in the nuclear bulge are strong ($\sim 2\,\mathrm{mG}$). The field lines are oriented parallel to the galactic plane inside giant molecular clouds and perpendicular to the plane in the intercloud medium.

Circum-Nuclear Disk (CND; $1.7 \leq R/\mathrm{pc} \leq 7$): Contains $\sim 10^4\,M_\odot$ of highly clumped ISM which rotates around the Galactic Center. Its inner edge is well defined. In the NIR the CND is seen in absorption against the central star cluster of the nuclear bulge. The CND appears to be inclined by $\sim 25° - 35°$ with respect to the galactic plane and the south-eastern segment of the ring is in front of the central stellar cluster. Estimated radial mass flow rates through the CND towards the Galactic Center are a few $10^{-2}\,M_\odot\,\mathrm{yr}^{-1}$.

Central Cavity and Minispiral ($R \leq 1.7\,\mathrm{pc}$): At $R \sim 1.7\,\mathrm{pc}$ volume and column densities drop by one to two orders of magnitude but there is a smooth transition in radial velocities from the inner edge of the CND to the ionized gas of the low-excitation ($T_{eff} \sim 3\,10^4\,\mathrm{K}$) HII region Sgr A West which fills the central cavity with a total mass of $M_{HII} \sim 260\,M_\odot$. About 40 % of the free-free flux density emerges from a spiral-like structure referred to as Minispiral which extends from the CND to the Galactic Center. Magnetic fields – which follow the arms of the minispiral – are found to be $\geq 2\,\mathrm{mG}$. Northern and Eastern Arm appear to sandwich a feature of $\sim 300\,M_\odot$ of atomic gas.

The *Stellar Population* of the central parsec consists of *(i)* ~ 24 massive and hot stars which form a cluster of core radius $\sim 0.17\,\mathrm{pc}$ and which account for most of the luminosity of $\sim 10^8\,L_\odot$; *(ii)* more than 200 K and M supergiants of intermediate mass; and *(iii)* several million low-mass, low-luminosity ($M_*/L_* \sim 3-4\,L_\odot/M_\odot$) main sequence stars. The different stellar ages of the more massive stars suggest small recurrent star formation bursts 10^7 and $10^8\,\mathrm{yr}$ ago.

Sgr A* is located close to or at the dynamical center of the Galaxy. Upper limits for its size and mass are $\leq 2.5-4\,10^{13}\,\mathrm{cm}$ and $\sim 2-3\,10^6\,M_\odot$, respectively. It is embedded in the Nuclear and Galactic Bulge. Its radio/IR spectrum is best explained by optically thin synchrotron radiation emitted by relativistic electrons with a quasi-monoenergetic distribution. The luminosity contained in the radio/IR part of the spectrum is $\sim 300\,L_\odot$. $L_{opt/UV} \leq 500\,L_\odot$ is an upper limit for the optical and UV luminosity of Sgr A* if a standard accretion disk spectrum is fitted to the upper limit of the K-band flux density and a black hole mass of $\sim 2 - 3\,10^6\,M_\odot$ is adopted. The X-ray luminosity of Sgr A* is less than a few $10^2\,L_\odot$.

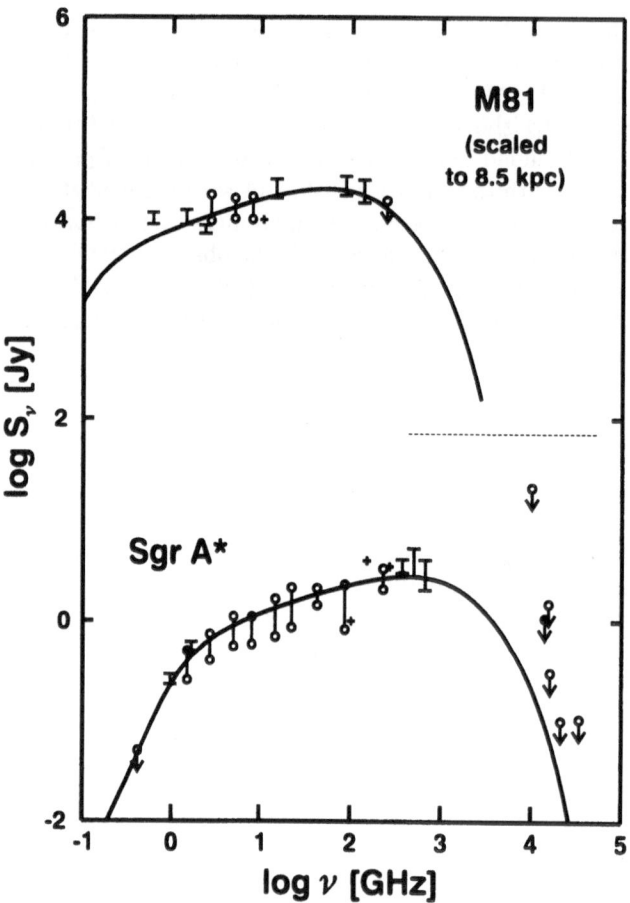

Fig. 7. A comparison of the radio/IR spectrum of Sgr A* (below the broken line) with the core spectrum of M81 (Reuter and Lesch 1996) reduced to the distance of the GC (above the broken line). The full lines are model spectra of synchrotron radiation of quasi-monoenergetic relativistic electrons (see Table 3). In both cases the spectrum is self-absorbed for the lowest frequencies, and optically thin for the higher ones.

4.2 The Galactic Center as compared to AGN

- *Black Hole:* Close to or at the dynamical center of our Galaxy there is a dark compact mass of $2.5 - 3\,10^6\,M_\odot$ which, in all likelihood, is a black

Table 3. Comparison of the model parameters which fit synchrotron spectra of quasi-monoenergetic distributions of relativistic electrons to the observed spectra of M81 and Sgr A*. In both cases the spectrum is self-absorbed for the lowest frequencies, and optically thin for the higher ones.

	M81	Sgr A*	
Magnetic field	0.5	11	G
Source radius	$7\,10^{15}$	$1.2\,10^{13}$	cm
Electron energy	100	114	MeV

hole. This mass is considerably less than that inferred for black holes in the most active galaxies, but well in the range of dark masses detected in the centers of Seyfert and normal galaxies. In our Galaxy the black hole manifests itself in the radio/MIR region as the compact synchrotron source Sgr A*; there exist only upper limits for the flux densities in the NIR−X-ray regime. The radio/IR luminosity of $300\,L_\odot$ is comparable to the upper limits obtained for the optical/UV and X-ray luminosities. If this luminosity is provided by a standard accretion disk around a black hole, it corresponds to an accretion rate of $\dot{M} \sim 10^{-8}\,M_\odot\,\mathrm{yr}^{-1}$. Remember that the Eddington limit for a black hole of $3\,10^6\,M_\odot$ is $\sim 10^{11}\,L_\odot$, the corresponding accretion rate is $\sim 10^{-2}\,M_\odot\,\mathrm{yr}^{-1}$ – with the exact value depending on the efficiency of the accretion.

– *Obscuring torus:* The circum-nuclear disk is a ring of gas and dust which encloses the Central Cavity ($R \sim 1.7\,\mathrm{pc}$). The central cavity is filled with ionized and neutral atomic gas ($M_{HII} + M_H \sim 500 - 600\,M_\odot$). The ionized gas forms spiral-like streamers (minispiral) which are coupled to strong ($2 - 4\,\mathrm{mG}$) magnetic fields. These streamers could transport matter towards the center of the central cavity. The ionization of the gas in the central cavity and the total luminosity $L(R \leq 1\,\mathrm{pc}) \sim 10^8\,L_\odot$ is due to a cluster of highly evolved massive stars. In many respects the CND is similar to the obscuring tori that are held responsible for differences in the observed characteristics of the various AGN classes. The formation of such tori appears to be a direct consequence of the presence of massive central black holes.

– *Mass flow:* While the present-day accretion into the black hole seems to be rather low, there is plenty of gas and dust supply available in the inner $200\,\mathrm{pc}$. For the molecular disk as well as for the CND radial mass inflow rates of the order of $\sim 10^{-2}\,M_\odot\,\mathrm{yr}^{-1}$ are derived. This material can easily be replenished by the action of a bar at larger radii. In a stationary state, this difference between the mass inflow rate into the central cavity and the accretion rate into the black hole would create a real problem since none of the observed matter consuming processes (formation of stars, winds, jets, etc.) can account for a consumption rate of more than a few tens of percent of the available material. Hence, in future, when the material now on its way to the Galactic Center will have reached the inner parsec and the vicinity of the black hole, even another burst of star formation will not keep the black hole from accreting at a much higher rate than currently is the case. The only conclusion left is that the mass flow rate varies in cycles so that the current low accretion rate is due to a genuine phase of low matter supply in the black hole's immediate vicinity. This view is supported by the fact that only $10^{7...8}$ yr ago the mass consumption rate during the recent event of a star burst in the inner parsec could have been as high as $10^{-2} M_\odot yr^{-1}$ if $\sim 3\,10^3 M_\odot$ of ISM were transformed into the observed massive stars during a time of $\sim 3\,10^5$ yr. Since star formation is not a very efficient process for consuming gas and dust, the accretion rate into the black hole, at that time, must have been accordingly much higher than today. If one assumes that the duty cycle of activity in our Galactic Center is similar to what one inferres on statistical grounds for AGN (a few percent), the average accretion rate would be $\sim 10^{-4} M_\odot yr^{-1}$. If such an average accretion rate were characteristic for the entire Hubble time ($\sim 10^{10}$ yr), the lower limit for the mass of the black hole would be $\sim 10^6 M_\odot$, in good agreement with the actually observed value of its mass.

– *The spectrum of Sgr A* and of the central parsec:* The observed and (for the optical/UV band) inferred spectrum of Sgr A* is shown in Fig. 6 together with the observed and dereddened spectrum of the central parsec. The radio/MIR spectrum of Sgr A* can be fitted well by synchrotron radiation from quasi-monoenergetic relativistic electrons. Falcke et al. (1993) suggested that the released gravitational energy of the accretion disk/black hole central energy always divides up in similar fractions of thermal energy and non-thermal energy, i.e. the acceleration of electrons to relativistic speed which subsequently – accelerated in magnetic fields – emit synchrotron radiation. The comparison of the spectrum derived for Sgr A* with that of the central parsec (Fig. 6) supports this view: The luminosity of the observed non-thermal synchrotron component and that of the inferred thermal disk component in the spectrum are approximately equal. We hypothesize that the intrinsic spectra of black hole/accretuion disk engines are similar regardless of the energy output. If this were actually the case the difference between "thermal" and "non-thermal" AGN spectra would not relate to intrinsic source characteristics but rather to environmental effects such as the reprocessing of optical/UV light and obscuration by dust. A comparison of the intrinsic spectrum of Sgr A* with the spectrum of the central parsec supports this view: The latter spectrum is only vaguely related to the Sgr A* spectrum but rather resembles that of a radio-quiet AGN.

– And finally ...

4.3   The Accretion Disk in the Galactic Center

Unfortunately, there is no direct evidence for the presence of a standard accretion disk in the Galactic Center which should have a size of $\geq 10^3$ Schwarzschild radii R_S (i.e. $\geq 10^{14\cdots15}$ cm); neither a compact H_{II} region, nor emission from hot dust nor a counterpart to Sgr A* (radio) in the NIR are observed. Only the compact synchrotron source has a size of $\sim 10\,R_S$ which could be compatible with the inner edge of an accretion disk or ring. The low value of the mass accretion rate of a standard disk $\sim 10^{-8}\,M_\odot\,\mathrm{yr}^{-1}$ of the black hole in the center of the Milky Way falls in the range of stellar accretion rates rather than of those of AGN. One should keep in mind, however, that there is no reasin whatsoever that forbids other than standard accretion disks. In advection-dominated accretion disks a higher accretion rate does not manifest itself by a correspondingly higher luminosity. Because of this the application of advection-dominated models to the situation in the Galactic Center is of great impoertance. It could easuily turn out that there we have the only laboratory for such disk around massive black holes that is directly accessble – to some degree at least.

4.4   The case of M81 and NGC 1068

One may argue that the above discussed difference between the true spectrum of Sgr A* and that of the inner parsec is a consequence of the low luminosity of Sgr A*(R). How can we be sure that with an increased accretion rate the luminosity of the synchrotron emission also increases? The catalogue of AGN spectra (Steppe et al. 1992) shows quite a few compact core sources with inverted spectra. Recently Reuter and Lesch (1996) using the IRAM 30-m-MRT have extended such an inverted spectrum of M81 observed with VLBI at $\lambda \geq 2\,\mathrm{cm}$ to wavelengths as short as $\lambda 1.2\,\mathrm{mm}$ and find it to be qualitatively very similar to that of Sgr A*. But its luminosity of $\sim 3\,10^6\,L_\odot$ is $\sim 10^4$ that of Sgr A*. Both spectra are shown together in Fig. 7, but with the flux densities of M81 reduced to the distance of the Galactic Center. In the frequency range between $< 1\,\mathrm{GHz}$ and $\sim 30\,\mathrm{GHz}$, the flux density scales $S_\nu(\mathrm{M81}) \propto \nu^{1/3}$ as does Sgr A*. The new mm-observations show that the maximum flux density is attained at $\sim 100\,\mathrm{GHz}$ and then begins to decrease. Beckert (priv. comm.) has calculated a theoretical spectrum for M81* using the same model as for Sgr A* (see Sect. 3.2 and Table 2). The electron energies are practically the same in both cases; the magnetic field is lower by a factor of ~ 20 in M81, but the major difference seems to be the volume of the emitting source in M81 that is larger by $\sim 2\,10^8$ compared to Sgr A*. This latter difference is the main reason for the much higher luminosity. The resulting spectrum is compared with the observed flux densities in Fig. 7, the characteristic parameters of M81 and Sgr A* are summarized in Table 3.

M81 is at a distance of $\sim 3\,\mathrm{Mpc}$ which means that an angular distance of $1''$ corresponds to $\sim 15\,\mathrm{pc}$. If the distribution of dust in the central $300\,\mathrm{pc}$ of M81 were similar to that in the nuclear bulge of our Galaxy, Reuter and Lesch would have had no chance to separate the compact synchrotron source from the dust and free-free emission background using the 30-m-telescope with an angular resolution of only $\sim 10''$ to $30''$. It appears to be the absence of dense molecular gas in the nuclear bulge of M81 which makes the compact synchrotron source at the center of M81 such an easily detectable object.

In the previous section, we suggested that a radio spectrum $S_\nu \propto \nu^{1/3}$ is typical for the core emission of AGN. This view is supported by Slee et al. (1994), who find compact radio continuum cores in $\sim 70\%$ of radio-emitting elliptical and S0 galaxies which have a flat or inverted spectrum with a median spectral index of 0.3. This could be interpreted that a radio/IR spectrum similar to that of Sgr A* is always associated with a black hole/accretion disk fueled AGN.

Observations of the Seyfert 2 galaxy NGC 1068 point in the same direction. With MERLIN and an angular resolution of ~ 60 mas (~ 4 pc at a distance of 15 Mpc) Muxlow et al. (1996) detected a compact core with an inverted spectrum ($S_\nu \propto \nu^{0.31}$) between 5 and 22 GHz located close to the starting point of a flaring jet. This compact core is surrounded by four more sources with steep radio spectra which have masked the inverted spectrum in earlier observations of lower angular resolution. Wittkowski et al. (1997) observed with the Russian 6 m telescope the center of NGC 1068 in the IR with a speckle technique and showed that in the IR K-band, the nuclear flux comes from a region of size of a few parsec. Interstingly enough, it seems as if in this source the range of the nuclear spectrum with a $S_\nu \propto \nu^{1/3}$ dependency extends all the way into the NIR spectral range. They derive source parameters along the line discussed in the present paper, and find that the major difference to its less active relatives seems to be a considerably higher electron energy.

One is tempted to speculate that the higher efficiency of the acceleration mechanism for the electrons may be the true difference between an active galactic center and a normal one.

4.5   Conclusions – The Galactic Center: A Laboratory for AGN

The observations and theoretical investigations compiled and discussed in this review confirm our view that the Center of our Galaxy belongs to the class of mildly active Seyfert Nuclei which account for $\sim 10\%$ of all spiral galaxies. It therefore can serve as a true laboratory for AGN as alluded to in the title of this paper. The black holes in quasars and radio galaxies are more massive, compared with the mass of the black hole in our Galaxy, by up to 4 orders of magnitude and we do not yet know how the characteristics of AGN scale with black hole mass. In using the center of the Milky Way as a laboratory of AGN one should always keep in mind that currently it appears to be in a phase of rather low activity. But at the same time one must not forget, that both, past and future phases of much higher activity, are indicated by observations of its current status.

Acknowledgement: This work was in part supported by the *Deutsche Forschungsgemeinschaft* through Sonderforscghungsbereich 328 (*Evolution of galaxies* at the University of Heidelberg).

References

Arimoto N., Sofue Y., Tsujimoto T., 1996, PASJ 48, 275

Backer D.C., 1994, in: Genzel and Harris (1994), 403

Bally J., Stark A.A., Wilson R.W., Henkel C., 1988, ApJ 324, 223

Bania T.M., 1977, ApJ 216, 381

Beckert T., Duschl W.J., Mezger P.G., Zylka R., 1996, A&A 307, 450

Bender R., Kormendy J., Dehnen W., 1996, ApJ 464, L123

Biermann P.L., Duschl W.J., von Linden S., 1993, A&A 275, 153

Burton W.B., Liszt H.S., 1978, ApJ 225, 815

Davies R.D., Walsh D., Booth R.S., 1976, MNRAS 177, 319

de Vaucouleurs G., 1970, IAU-Symp. 38, 18

Duschl W.J., 1988a, A&A 194, 33

Duschl W.J., 1988b, A&A 194, 43

Duschl W.J., Lesch H., 1994, A&A 286, 431

Eckart A., Genzel R., 1996a, Nature 383, 415

Eckart A., Genzel R., 1996b, MNRAS (in press)

Eckart A., Genzel R., Krabbe A. et al., 1992, Nature 355, 526

Eckart A., Genzel R., Hofmann R., Sams B.J., Tacconi-Garman L.E., 1995, ApJ 445, L23

Falcke H., Heinrich O.M., 1994, A&A 292, 430

Falcke H., Mannheim K., Biermann P.L., 1993, A&A 278, L1

Ford H.C., Harms R.J., Tsvetanov Z.I. et al., 1994, ApJ 435, L27

Frail D.A., Diamond P.J., Cordes J.M., van Langenfelde H.J., 1994a, ApJ 427, L43

Frank J., King A., Raine D., 1992, Accretion Power in Astrophysics

Genzel R., Townes C.H., 1987, ARAA 25, 377

Genzel R., Thatte N., Krabbe A., Kroker H., Tacconi-Garman L.E., 1996, ApJ (submitted)

Georgelin Y.M., Georgelin Y.P., 1976, A&A 49, 57

Gerhard O.E., 1996, IAU-Symp. 169, 79

Güsten R., Genzel R., Wright M.C.H. et al., 1987, ApJ 318, 124

Gwinn C.R., Danen R.M., Middleditch J., Ozernoy L.M., Tran K.Th., 1991, ApJ 381, L43

Haller J.M., Rieke M.J., Rieke G.H., Tamblyn P., Close L., Melia F., 1996a, ApJ 456, 194

Heiligman G.M., 1987, ApJ 314, 747

Herbst T.M., Beckwith S.V.W., Forrest W.J., Pipher J.L., 1993, AJ 105, 956

Jackson J.M., Geis N., Genzel R. et al., 1993, ApJ 402, 173

Jackson J.M., Heyer M.H., Paglione T.A., Bolatto A.D., 1996, ApJ 456, L91

Kormendy J., Richstone D., 1995, ARAA 33, 581

Koyama K., 1994, in: New horizons of X-ray astronomy – First results from ASCA (F. Makino, T. Ohashi, eds.), 181

Koyama K., Maeda Y., Sonobe T., Tanaka Y., 1996a, in: Röntgenstrahlung from the Universe (H.U. Zimmermann, J.E. Trümper, H. Yorke, eds.), MPE Report 263, 315

Koyama K., Maeda Y., Sonobe T. et al., 1996b, PASJ 48, 249

Krabbe A., Genzel R., Eckart A. et al., 1995, ApJ 447, L95

Krichbaum T.P., Zensus J.A., Witzel A. et al., 1993, A&A 274, L37

Krichbaum T.P., Schalinski C.J., Witzel A. et al., 1994a, in: Genzel and Harris (1994), 411

Krichbaum T.P., Standke K.J., Graham D.A. et al., 1994b, IAU-Symp. 159, 187

Krichbaum T.P., Witzel A., Standke K.J. et al., 1994c, in: Compact Extragalactic Radio Sources (J.A. Zensus, K.I. Kellermann, eds.), 39

Lacy J.H., 1994, in: Genzel and Harris (1994), 165

Lacy J.H., Achtermann J.M., Serabyn E., 1991, ApJ 380, L71

Lazio T.J.W., Cordes J.M., 1996, in: Gredel (1996),

Lo K.-Y., 1986, PASP 98, 179

Lynden-Bell D., Rees M.J., 1971, MNRAS 152, 461

Marcaide J.M., Alberdi A., Bartel N. et al., 1992, A&A 258, 295

Mastichiadis A., Ozernoy L.M., 1994, ApJ 426, 599

Melia F., 1994, ApJ 426, 577

Menten K.M., Reid M.J., Eckart A., Genzel R., 1996, ApJ (in press)

Mezger P.G., Wink J.E., 1986, A&A 157, 252

Mezger P.G., Duschl W.J., Zylka R., Beckert T., 1996, in: Amazing Light (R.Y. Chiao, ed.), 485

Mezger P.G., Duschl W.J., Zylka R., 1996, AAR 7, 289

Miyoshi M., Moran J., Herrnstein J. et al., 1995, Nature 373, 127

Morris M., Serabyn E., 1996, ARAA 34, 645

Morris M., Davidson J.A., Werner M.W. et al., 1993, BAAS 25, 851

Narayan R., Yi I., Mahadevan R., 1995, Nature 374, 623

Ozernoy L., 1992, in: Testing the AGN paradigm (S.S. Holt, S.G. Neff, C.M. Urry, eds.), 40

Pavlinski M.N., Grebenev S.A., Sunyaev R.A., 1994, ApJ 425, 110

Predehl P., Trümper J., 1994, A&A 290, L29

Predehl P., Genzel R., Trümper J., Zinnecker H., 1994, in: Genzel and Harris (1994), 21

Reuter H.-P., Lesch H., 1996, A&A 310, L5

Rieke G.H., Rieke M.J., 1988, ApJ 330, L33

Roberts D.A., Goss W.M., 1993, ApJS 86, 133

Roberts D.A., Yusef-Zadeh F., Goss W.M., 1996, ApJ 459, 627

Robinson B.J., McGee R.X., 1970, Austr. J. Phys. 23, 405

Roelfsema P.R., Goss W.M., 1992, A&AR 4, 161

Rogers A.E.E., Doeleman S., Wright M.C.H. et al., 1994, ApJ 434, L59

Saha P., Bicknell G.V., McGregor P.J., 1996, ApJ 467, 636

Sellgren K., Hall D.N.B., Kleinmann S.G., Scoville N.Z., 1987, ApJ 317, 881

Serabyn E., Lis D.C., 1994, quoted in: Mezger (1994)

Serabyn E., Lacy J.H., Townes C.H., Bharat R., 1988, ApJ 326, 171

Serabyn E., Carlstrom J.E., Scolville N.Z., 1992, ApJ 401, L87

Skinner G.K., 1993, A&AS 97, 149

Skinner G.K., Willmore A.P., Eyles C.J. et al., 1987, Nature 330, 544

Slee O.B., Sadler E.M., Reynolds J.E., Ekers R.D., 1994, MNRAS 269, 928

Sodroski T.J., Odegard N., Dwek E. et al., 1995, ApJ 452, 262

Sofue Y., 1995, PASJ 47, 527

Steppe H., Liechti S., Mauersberger R. et al., 1992, A&AS 96, 441

Telesco C.M., Davidson J.A., Werner M.W., 1996, ApJ 456, 541

van Langenfelde H.J., Diamond P.J., 1991, MNRAS 249, 7p

van Langenfelde H.J., Frail D.A., Cordes J.M., Diamond P.J., 1992, ApJ 396, 686

von Linden S., Lesch H., Combes F., 1996, in: Formation and Evolution of Galaxies (W.J. Duschl, N. Arimoto, eds.) (in press)

von Linden S., Duschl W.J., Biermann P.L., 1993a, A&A 269, 169

von Linden S., Biermann P.L., Duschl W.J., Lesch H., Schmutzler T., 1993b, A&A 280, 468

Wittkowski M., Balega Y., Beckert T., Duschl W.J., Weigelt G., 1997, A&A (submitted)

Woltjer L., 1959, ApJ 130, 38

Zylka R., Mezger P.G., Ward-Thompson D., Duschl W.J., Lesch H., 1995, A&A 297, 83

A bright X–ray source inside a molecular cloud

R. Sunyaev[1,2], E. Churazov[1,2]

[1] Max-Planck-Institut für Astrophysik, Karl-Schwarzschild-Str. 1, 85740 Garching bei Munchen, Germany
[2] Space Research Institute, Profsouznaya 84/32, 117810 Moscow, Russia

Abstract:
  We consider how the observations of an X–ray source embedded into a molecular cloud could be used for diagnostic of the cloud parameters and time history of the source flux, using data on the neutral iron 6.4 keV fluorescent line. We concentrate on the particular cases of the source 1E1740.7–2942 and the Galactic Center region. For a steady source an equivalent width of the 6.4 keV line is a convenient indicator of the amount of gas in the cloud. For a variable source the shape and flux of the line could provide information on the activity of the source in the past (few light crossing times of the cloud). Observations of detailed structure of the recoil profile of the line (due to scattering by electrons bound in neutral atoms and molecules) could provide information on the helium abundance in the scattering matter. Appearance of a new generation of X–ray spectrometers, combining high energy resolution and large effective area, will make observations of the 6.4 keV line a very powerful tool for diagnostic of neutral scattering media.
  More detailed discussion is given in Churazov, Gilfanov, Sunyaev, 1996; Sunyaev & Churazov, 1996; Churazov & Sunyaev, 1997; Churazov, Sunyaev, Vainshtein, 1997.

1 1E1740.7–2942, constraints on the depth of the cloud

1E1740.7–2942, located $\sim 50'$ from the dynamic center of our Galaxy, was discovered by the Einstein/IPC (Hertz & Grindlay 1984) in soft X–rays (< 4 keV). Observations at higher energies (Skinner et al., 1991; Cook et al., 1991) indicated unusual hardness of the source spectrum. This result was further confirmed by long term monitoring of the GC region by GRANAT/SIGMA (e.g. Sunyaev et al., 1991), which demonstrated that most of the time 1E1740.7–2942 gives dominant contribution to the hard X–ray flux of the few degrees region around GC.
  Observations at millimeter wavelength revealed the dense ($\sim 10^5$ cm^{-3}) molecular cloud in the direction of 1E1740.7–2942 (Bally & Leventhal 1991, Mirabel et al. 1991). This discovery raised the hypothesis that the unique properties of this object are due to the presence of a dense gas near the compact object. In particular, Bondi-Hoyle accretion onto a single black hole (Bally & Leventhal 1991, Mirabel et al. 1991) or onto a black hole in a binary system (Chen et al., 1994) has been discussed. The presence of dense and cold gas near the compact object has also other important consequences — e.g. as a possible site for cooling

and annihilation of hot positrons, possibly produced by 1E1740.7–2942 during hard states (e.g. Ramaty et al., 1992).

Since the Thomson optical depth of the cloud (as derived from observations of molecular lines at millimeter wavelength) is large enough, $\tau_T \sim 0.2$, the scattered component of the same relative intensity could be present (Sunyaev et al., 1991). Detailed verification of the variability of the source flux observed by GRANAT allowed one to constrain averaged depth of the gas surrounding the source to the value $\tau_T \leq 0.1$ (Churazov et al., 1993; Pavlinsky et al., 1994). Additional constraints can be obtained studying another feature, associated with reprocessed X-ray emission in the neutral gas — 6.4 keV iron K-line, using ASCA observations of 1E1740.7–2942 during the PV phase.

Reprocessing of the X-ray emission in the neutral gas, surrounding the steady compact source should lead to the appearance of a scattered component with the intensity proportional to the averaged Thomson optical depth of the cloud: $F_{scat} = F_0 \times \tau_T$, where F_0 is the flux of the compact source. The expected flux in the 6.4 keV line is defined by the source flux above 7.1 keV, the averaged radial Thomson depth of the cloud, an iron abundance, the absorption crossection at the iron K-edge ($\sigma_K \sim 3.5 \times 10^{-20}$ cm^2) and the fluorescent yield ($W \sim 0.3$). One can easily show that for the solar abundance of the iron expected equivalent width of the 6.4 keV in the scattered component is ~ 1000 eV (see e.g. Fabian, 1977; Vainshtein & Sunyaev, 1980), weakly depending on the Thomson depth of the cloud. In the observed spectrum (steady source in the center of the cloud) the equivalent width should be a factor of $\sim \tau$ lower. This allows one to estimate the radial Thomson depth of the cloud from the observed value of the 6.4 keV line equivalent width. It is important to mention that 1E1740.7–2942 source is strongly variable on the 1 year time scale (Sunyaev et al., 1991). Therefore if the source is inside the cloud (having linear size of the order of 1 pc), the equivalent width of the K_α line may be maximal during the low state of the source.

Shown in Fig. 1 are the results of Monte-Carlo simulations of the emerging spectrum for the steady source (having intrinsic power law spectrum with photon index $\Gamma = 1.8$) in the center of the uniform spherical cloud with a radial column density of $18 \times 10^{22} cm^{-2}$. The latter value approximately corresponds to the low energy absorption in the source spectrum. In this case one can expect an equivalent width of the 6.4 keV line $\sim 115 eV$. For comparison the right panel in Fig. 1 shows the observed SIS0 spectrum and best fit power law model plus the 6.4 keV line with an equivalent width of $115 eV$. One can see that the actual flux in the 6.4 keV line is much lower. The observed equivalent width of the 6.4 keV line (~ 11 eV) indicates that the radial column density (averaged over all directions) of the gas around the source does not strongly exceed $N_H \sim 2 \times 10^{22} cm^{-2}$. Taking into account the ASCA mirrors angular resolution (3' half power diameter) and area, used for the spectrum extraction (see Churazov, Gilfanov, Sunyaev, 1996), the linear size of the region to which the above limit should be applied is about of several parsecs at the GC distance. The above value of the averaged N_H is a factor of 5 to 15 lower than the typical values of the line of sight column density in the direction of the source derived from

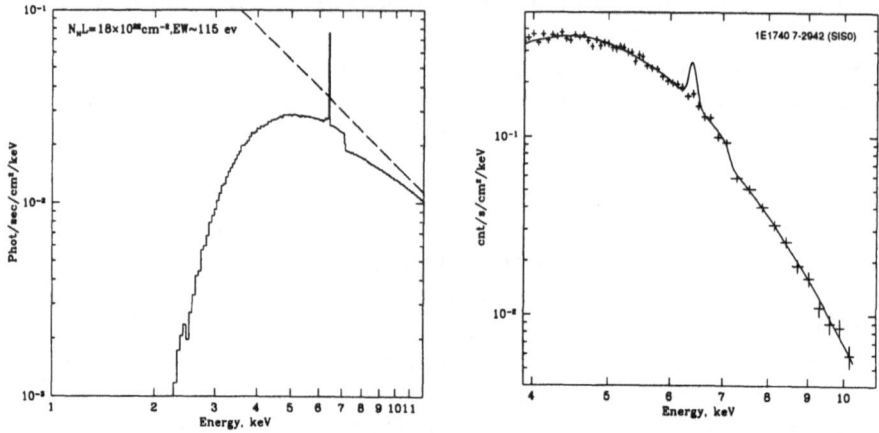

Fig. 1. Left: Monte-Carlo simulation of the expected spectrum for the source in the center of the uniform cloud with the radial column density $18 \times 10^{22} cm^{-2}$. **Right:** Observed SIS0 spectrum. Solid line shows best fit power law plus 6.4 keV line with equivalent width 115 eV. From Churazov, Gilfanov & Sunyaev, 1996.

other data, suggesting that the bulk of the absorbing gas is not located in the immediate vicinity of the source. Taking into account the possible contributions to the 6.4 keV line flux from the large scale diffuse component and the reflection from the accretion disk, the actual value of the line flux originating from the molecular gas surrounding the source may be even lower. From above one can conclude that, if most of the line of sight absorption is due to the dense cloud then the cloud must be located in front of the compact source.

More detailed discussion is given in Churazov, Gilfanov, Sunyaev, 1996.

2   Galactic Center region and variable source

Observations of the Galactic Center region with the GRANAT/ART–P telescope revealed a diffuse component above ~ 8 keV elongated parallel to the Galactic plane and resembling the distribution of the molecular gas clouds (Sunyaev et al., 1993, Markevitch et al., 1993). It was suggested that this component is due to the Thomson scattering by a dense molecular gas of the X-rays from nearby compact sources. It was shown that the detailed iron K_α line observations may prove this interpretation of the GRANAT/ART–P data. Further evidence in support of this assumption came from ASCA observations of this region, which revealed strong 6.4 keV line flux from the general direction of the largest molecular complexes (e.g. Koyama, 1994). The Sgr B2 complex was found to be especially bright in the 6.4 keV line, at the level of a few $10^{-4} phot/s/cm^2$ (Koyama et al., 1996). Such a high flux in the fluorescent line requires a powerful source of primary X-ray emission which gives rise to the reprocessed radiation. Since the light crossing time of the Galactic Center region is as long as several hundred years, observed reprocessed radiation may be associated with the source which is dim at present, but was very bright some hundred years ago. Such a scenario has been suggested by Sunyaev et al. (1993) and supported by Koyama (1994), the Galactic Center (i.e. Sgr A*) itself being the primary candidate. Given the distance from the Sgr B2 complex to the GC ($\sim 40'$ or about 100 pc in projection) Koyama et al. (1996) estimated the luminosity of the putative GC source in excess of $10^{39} ergs/s$ some hundred years ago. Although high enough this value is still much below the Eddington limit for the $10^6 M_\odot$ black hole and even a rather short flare at the Eddington level could provide the required flux (Sunyaev et al., 1993). A less energetic object is required if one assumes that the primary source of continuum emission is located in the vicinity of the Sgr B2 complex (or even inside it). The situation when the compact source is embedded into molecular cloud was considered also for the source 1E1740.7–2942 near the Galactic Center, since observations at millimeter wavelength revealed giant ($N_H \sim 210^{23}$ cm^{-2}) molecular cloud in the direction of this object (Bally & Leventhal 1991, Mirabel et al. 1991).

Note that light crossing time for each particular cloud or complex of clouds can be of the order of years or tens of years and that this value defines the time scale for variability of scattered flux (even if the primary source was bright only during a much shorter period of time). Even when the primary photons have already left the cloud, multiply scattered photons can be observed. The variability of the primary radiation flux and cloud parameters will affect the appearance of the 6.4 keV line. All important features can be seen considering the simplest case of a point source of the continuum X-ray emission located in the center of a spherically symmetric cloud of neutral gas. In the presence of the steady primary source the expected equivalent width of the 6.4 keV line (with respect to the primary continuum flux) is about $1 \times \tau_T$ keV, where τ_T is the Thomson depth of the medium, surrounding the source[1]. If the primary source radiation diminishes, then the equivalent width immediately rises to the value of ~ 1 keV (e.g. Fabian, 1977, Vainshtein and Sunyaev, 1980) – the value, which is rather insensitive to the cloud parameters like optical depth (as soon as it is low) or particular distribution of scattering matter. This happens because both 6.4 keV and scattered continuum fluxes are proportional to the optical depth of the cloud and intensity of the primary radiation. However later (after the diminishing of the primary source flux) the value of the equivalent width can change considerably.

We briefly discuss below how the effects of time delay, large opacity and scattering by bound electrons affect the appearance of the 6.4 keV line. For simplicity we are considering the simplest case of a point source of continuum X-ray emission in the center of a spherically symmetric cloud of neutral gas.

One of the most important parameters determining the possibility of the

---

[1] Calculating τ_T we take into account all electrons bound in atoms and molecules; see details below

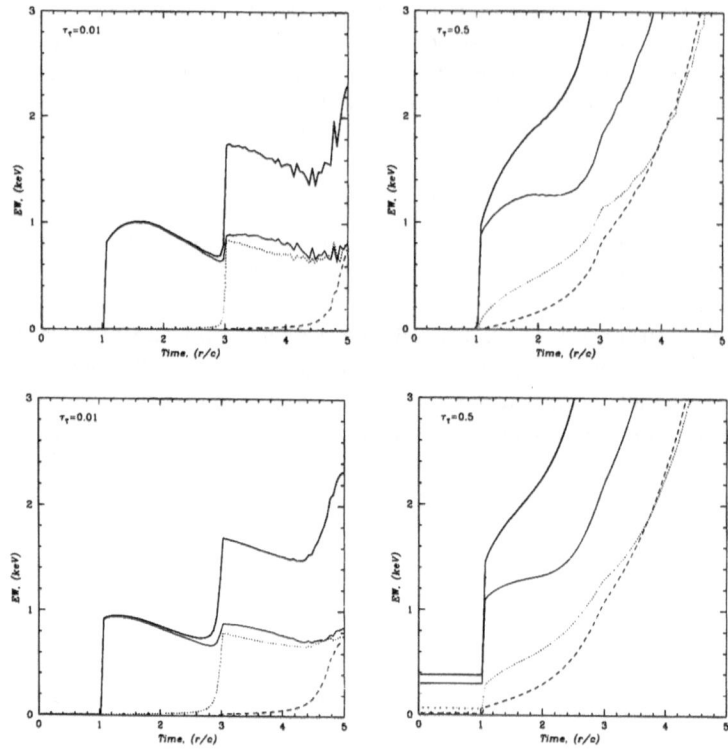

Fig. 2. The equivalent width versus time for the compact source of continuum emission at the center of the uniform spherical cloud. The upper row of figures corresponds to the short flare of the compact source flux, the lower row of figures corresponds to the "switch–off" of the steady source. The thin solid line shows the contribution of the 6.4 keV photons, which left the cloud without further interactions. The dotted line shows the contribution of 6.4 keV photons which have been Thomson scattered once and dashed line shows the 6.4 keV photons which undergo more than one scattering before escape from the cloud. The thick solid line shows total contribution of all these components.

line detection is it's equivalent width (EW), i.e. the ratio of the flux in the line to the spectral density of the continuum flux in the vicinity of the line. We use Monte–Carlo simulations in order to generate spectra, emerging from a spherically symmetric cloud with the point source of the continuum radiation (power law with photon index $\Gamma = 1.8$) in the center. Initial seed photons were generated in the 1–18 keV energy range. Since temporal behavior of the 6.4 keV line flux is in question we kept record of escape time for photons, emerging from the cloud. Shown in Fig.2 is the evolution of the line equivalent width with time for the case of a short flare at $t = 0$ (upper row of figures) and theta–function-like behavior of the primary source flux (lower row of figures). In the latter case the flux of the source was stable before the moment of time $t = 0$ and dropped to zero after this moment. In both cases $t = 1$ (time is measured in $\frac{r}{c}$ units) corresponds to the moment of time, when direct radiation (towards observer) leaves the cloud. Two values of the cloud radial Thomson depths $\tau_T = 0.01$ (a) and 0.5 (b) were used. The thin solid line shows the contribution of the 6.4 keV fluorescent photons, which left the cloud without further interactions. The dotted line shows the contribution of the 6.4 keV photons which have been Thomson scattered once and the dashed line shows the 6.4 keV photons which undergo more than one scattering before they escaped from the cloud. At last thick solid line shows total contribution of all these components. Note that scattered 6.4 keV photons may have an actual energy lower than 6.4 keV (we discuss the shape of the line profile below). Obviously the continuum is also scattered with intensity being proportional to τ^n after n scatterings (in the limit of small optical depth).

The first thing apparent from Fig.2 is the sharp change of EW at the $t = 1$. This is obviously related to the disappearance of the direct component (escaping from the cloud without interaction), which contributes to the continuum emission. Immediately after $t = 1$ EW jumps to the level of ~ 1 keV for both values of τ_T (Fig.2 a,b). Indeed (see e.g. Vainshtein and Sunyaev, 1980) both flux in the line and the level of the scattered continuum emission are (to the first order) proportional to the amount of matter over the radius of the cloud. As the results, the EW (ratio of the line flux to the continuum spectral density) happens to depend only weakly on the actual parameters of the cloud.

For large values of optical depth (Fig.2b) the behavior of the EW is very different – its value starts to rise from early beginning. The reason for such behavior is the large photoabsorption depth. Continuum photons (at about 6 keV) have a high probability to be absorbed before they can escape from the cloud. On the other hand high energy photons (at the energy 10–20–30 keV) can reach peripheral regions of the cloud without being absorbed, ionize an electron from the iron K–shell and thus produce a 6.4 keV photon, which can escape further.

Even apart from the photoabsorption effect the total EW calculated as the sum of unscattered and scattered 6.4 keV photons (Fig.2 a,b) grows with time. For low values of optical depth this growth is seen as the sharp jump at $t = 3R/c$. This jump corresponds to the moment when all photons, which undergo only one interaction (absorption + fluorescence or single Thomson scattering) are leaving the cloud. After that moment of time the spectrum is dominated by the photons passed through more than one interaction. One can see that the EW of the 6.4 keV complex increases at that moment. This is because the continuum is now dominated by twice Thomson scattered photons, while the line is constructed from two kinds of photons: (a) photons which first were Thomson scattered, then ionized an electron from the iron K–shell and produced new 6.4 keV quantum and (b) Thomson scattered 6.4 keV photons. The EW for each of these components is about 1 keV. After each additional scattering (when the spectrum is dominated by the photons passed through n interactions) the total equivalent width will increase further (roughly proportional to n). In other words one can say that with each additional "interaction" Thomson scattering does not change the existing EW of the line, while ionization of iron K-shells adds more "fresh" 6.4 keV photons.

Finally one can note that although the EW of the 6.4 keV complex rises with time the absolute flux declines with time. The decline of the flux is caused by the escape of the photons from the cloud and photoabsorption of photons. Note that for the Sgr B2 complex ASCA detected the 6.4 keV flux at the level of few $10^{-4} phot/s/cm^2$ (Koyama, 1996) and even a 1000 times weaker line will still give a few hundred counts for 100 ks HTXS observation. It is interesting that during the period of time $\sim$ 10-20 years (the value not incomparable with the life time of best modern space observatories) the parameters of the line may substantially evolve.

In the previous discussion the fact that photons can change their energy during scattering was not taken into account. However for the 6.4 keV line even single scattering causes significant blurring of the line, which is especially important for future instruments with high energy resolution. Since we are considering scattering by the neutral matter, then all complications related to bound electrons (discussed in Sunyaev and Churazov, 1996; see next section) are to be taken into account. In particular for the 6.4 keV photons after each scattering (by a hydrogen atom) about 14% of the photons will maintain the initial energy unchanged (Rayleigh or elastic scattering). Some small fraction photons will cause excitation of electrons in the hydrogen atoms, leading to the appearance of Raman satellites of the line at the energies $h\nu_0 - 13.6 \times (1 - 1/n^2)$ eV, where n is the principle quantum number of excited level. But the largest fraction of photons will be Compton scattered by hydrogen atoms, leading to the ionization of an electron and decrease of the photon energy by $\sim$ 13-200 eV. Due to the motion of the electrons bound in hydrogen atoms, the backscattering peak will be smeared out.

Shown in Fig.3 are the spectra, emerging from the spherical cloud (with the radial optical depths of 0.2) after a given time from the switch-off of the compact source. We assume here that fluorescence produces a monochromatic line at the energy of 6.4 keV. One can see that after multiple scatterings complex a shoulder forms on the low energy side of the line. As discussed above the total equivalent width of the whole complex increases with time. However the photons are now distributed over a rather broad energy range, which increases with time. The equivalent width of the unshifted 6.4 keV line itself will still increase with time. This is related in part with the more rapid decline of the continuum at 6.4 keV due to photoabsorption (see above) and in part with Rayleigh scattering, which leaves $\sim$14% of the photons at the initial energy.

Thus the value of the equivalent width and the shape of the 6.4 keV iron K_α line, emerging from the neutral matter, illuminated from inside or outside by the X-ray continuum spectrum, contains information on the time history of the illuminating continuum flux. Assuming normal abundance of iron in the scattering media the value of an equivalent width of about 1 keV indicate that we are dealing with a scattered component. Values in excess of 1 keV can be

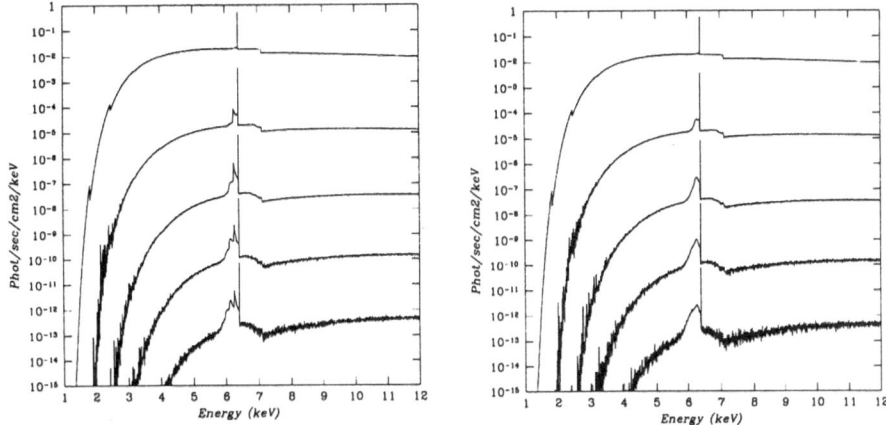

Fig. 3. Spectra emerging from the uniform cloud with radial Thomson depth $\tau_T = 0.2$ at different moments of time after a short flare of the continuum emission from the point source in the center of the cloud. For the left figure the recoil effect was calculated as for free cold electrons at rest. For the right figure effects of bound electrons were taken into account (but only for the 6.4 keV line). Spectra correspond to the intervals of time: 0–1, 1–2, 2–3, 3–4, 4–5 (in units R/c) from top to the bottom curves correspondingly. Each subsequent spectrum was multiplied by 0.05 for clarity. Spectra are plotted with an energy resolution of 5 eV.

due to strong photoabsorption or multiple Thomson scattering. Spectroscopic analysis of the 6.4 keV line profile and the shape of the 7.1 keV absorption edge can help to distinguish between these possibilities. When combined with broad band spectroscopic measurements it can be used to determine the position and flux history of the primary continuum source. A new generation of X–ray instruments should be capable of achieving the required sensitivity and energy resolution to accurately measure the detailed structure of the 6.4 keV line in the Galactic Center region. Comparing the flux and shape of the line from different molecular clouds in the region, one can reconstruct the date and duration of the flare, responsible for observed scattered emission. Note also that observations, over a period of 5–10 years, may show the variability of the line flux and shape.

More detailed discussion is given in Churazov & Sunyaev, 1997.

3   Scattering by electrons bound in neutral atoms

As was mentioned in the previous section the shape of the lines in the scattered spectra is affected by the properties of a scattering media. Compton scattering of X-rays by free electrons is known to play an important role in a number of astrophysical objects: early Universe, hot gas in clusters of galaxies, accretion disks around neutron stars and black holes, plasma clouds surrounding compact

X-ray sources. In many cases X-rays are scattered by neutral gas rather than free electrons. One can mention the reflection of X-rays by the solar photosphere (X-rays are produced by the flares above the solar surface and a considerable fraction of them will be reflected by the photosphere, having low degrees of ionization $\frac{e}{H} \leq 10^{-3}$); reflection of X-rays by the photospheres of cold flaring stars (T-Tauri, late stars coronas); outer regions of the extended accretion disks in binary systems, having low degree of ionization; molecular tori around central objects in the active galactic nuclei and QSOs. Compact X-ray sources, embedded into the dense molecular clouds are of special interest.

As is well known change of a photon energy after scattering by free electron at rest unambiguously related with the scattering angle due to momentum and energy conservation laws. For the scattering by the electron, bound in a hydrogen atom, additional factors complicate the process: finite binding energy of the electron and motion of the electron in the atom. Because of discrete energy levels of the electron in the exited state not all values of photon energy change are possible. Because of the "random" motion of the electron within the atom change of the photon energy is no longer an unambiguous function of the geometry.

Usually the scattering process (by electrons bound in atom) is divided into three branches:

A) Rayleigh (coherent) scattering: $\gamma_1 + H = \gamma_2 + H$. The energy of the photon remains the same, only direction changes. The recoil effect is $\sim m_p/m_e$ times smaller than in the case of scattering by free electron.

B) Raman scattering: $\gamma_1 + H = \gamma_2 + H(n,l)$, where $H(n,l)$ denotes one of the excited states of the hydrogen atom. The energy of the photon decreases by the value of excitation energy: $h\nu_2 = h\nu_1 - E_{n,l}$ – Raman satellites of the line appear.

C) Compton scattering: $\gamma_1 + H = \gamma_2 + e^- + p$, accompanied by ionization of atom. The energy of the photon decreases by the value of ionization potential and kinetic energy of electron after scattering: $h\nu_2 = h\nu_1 - 13.6eV - E_e$.

For Compton scattering the final state of the electron corresponds to the continuum. Unambiguous relation between scattering angle and change of the photon frequency breaks even if the atom or molecule are not moving before scattering. This is because the electron is not "at rest" in the atom, but has certain distribution over momentum. This ambiguity of energy transfer does not violate conservation laws, since heavy nucleus can carry the necessary momentum, having negligible kinetic energy. One can estimate the influence of the motion of an electron in an atom simply considering scattering of a photon by a free electron having velocity comparable to the velocity of the electron bound in atom. Due to the Doppler effect energy transfer may vary by a factor of $\sim h\nu_1 \times \frac{v}{c}$ for scattering by a large angle. The broader the distribution of the electron initial momentum, the large are the deviations of energy change can be compared to pure recoil in the case of scattering by free a electron at rest. This is especially important for scatterings by a large angle, causing the smearing of the backscattered peak (Fig.4a). One can mention the similarity of the distortions of the left wing of the line for scattering by hydrogen with distortions appearing for scattering by free electrons with a temperature ~ 10 eV. Shown in Fig.4b is the spectrum of 6.4 keV photons, scattered by free electrons with Maxwellian distribution of momenta with temperatures 0.1, 1 and 10 eV. One can see that with the increase of temperature "blurring" of the left wing also increases. The right side of the line does not change much. In astrophysical conditions the line profile will be always blurred: at low temperatures electrons are bound in atoms and the low frequency wing is blurred due to the distribution of electron momentum in the atom, while at high temperatures electrons are free and blurring is caused by the Maxwellian distribution of electron momenta. Note that in the astrophysical condition (interstellar media, stellar atmospheres, accretion disks) hydrogen is ionized at the temperatures of the order of 1 eV. Therefore "blurring" of a line may be minimal at the gas temperatures $T \sim 1\text{--}5$ eV, when hydrogen is already ionized, but the thermal velocities of electrons are still lower than the typical velocity of a electron in a hydrogen atom.

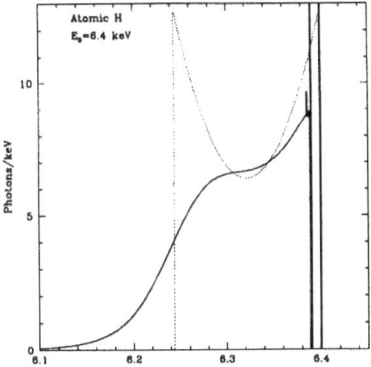 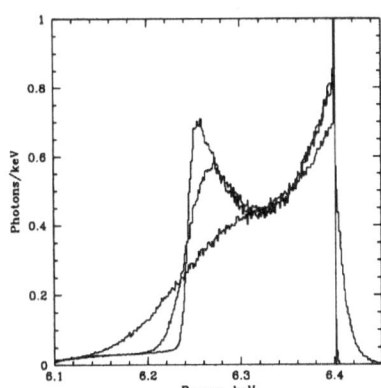

Fig. 4. a) Neutral hydrogen: Spectrum of the scattered 6.4 keV photons averaged over all scattering angles (solid line). About 13.5 % of photons undergo elastic scattering, 2.3 % of the photons change energy by 10.2 eV – Raman satellite of the line appears, corresponding to the excitation of the second level in the hydrogen atom. The same spectrum calculated for scattering by free cold electrons is shown by the dotted line. b) Emission spectrum emerging from a cloud of free electrons with small optical depth (electron temperatures are 0.1, 1 and 10 eV for three curves respectively). Monte-Carlo simulations. From Sunyaev & Churazov, 1996.

Rayleigh scattering does not change the energy of the line. Thus in scattered spectra some fraction of photons will keep their initial energy. Due to Raman scattering a set of satellites at the left (low energy side) of the line will appear (Fig.4a). The energy gap between the unshifted line (due to Rayleigh scattering) and first Raman satellite as well as intensities and energies of Raman satellites are unique for given type of atoms. This opens the principle possibility to distinguish contributions of different elements to the scattering cross section and thus to determine the abundance of elements in the scattering media. The most interesting is the case of helium because of it's large abundance. For the scattering by a helium atom along with the increase of Rayleigh scattering (by a factor of 2 due to coherent scattering) the structure of the lines, corresponding to Raman scattering, changes strongly. In particular the gap between ground and first excited state in the helium atom is ~20 eV, compared to 10.2 eV for hydrogen. Note, that for ~ 6 keV photons the wavelength $\lambda \sim 2\mathring{A}$ is comparable with the size of the atom and the parity selection rule is not strict. Due to the stronger binding of the electron in the helium atom, electron momentum distribution is substantially wider than that in atomic or molecular hydrogen. As a result (see Churazov, Sunyaev, Vainshtein, 1997) the left wing of the line will be more "blurred".

The energy gap is about twice as large as that for the hydrogen atom and significant differences in the recoil profile allow one to hope that studying the recoil spectrum might become a method for helium abundance determination. Note that even after multiple scatterings photon does not appear in this energy gap.

Of course study of the recoil profile of the 6.4 keV line requires high energy resolution and large effective area of the spectrometer. We discussed above the effects which can be observed by the instruments of future missions like Spectrum-X-Gamma with Bragg spectrometer, AXAF and XMM with gratings, ASTRO-E with X-ray bolometers and especially HTXS. Among the most promising objects for such studies are the molecular clouds in the Galactic Center region. Present day telescopes (e.g. ASCA) clearly detect very bright fluorescent lines from this region (e.g. Koyama, 1994), which flux is proportional to the Thomson optical depth of the cloud. One can hope that with the new generation of X-ray spectrometers it will be possible to observe second order effect - recoil effect due to the scattering of iron K_α photons, born in the cloud, by the molecular hydrogen. This effect is proportional to the square of the cloud optical depth: e.g. for a cloud with $\tau_T \sim 0.2$ about 20% of the photons in the fluorescent line will be scattered causing a decrease of the photons energies. The variability of the primary X-ray continuum source (or specific geometry – e.g. illumination from the side) can greatly improve the observational conditions, since the direct component can be absent. Further improvement in energy resolution (e.g. HTXS) will make possible helium abundance determination.

Acknowledgement: This work was supported in part by the grant RBRF 96-02-18588.

References

Bally J. & Leventhal M., 1991, Nature, 353, 234

Bambinek W. et al. 1972, Rev. of Modern. Phys., 44, 716

Chen W. Gehrels N., Leventhal M., 1994, ApJ, 426, 586

Churazov E. et al., 1993, ApJ, 407, 752

Churazov E., Gilfanov M., Sunyaev R., 1996, ApJ, 464, L71

Churazov E., Sunyaev R., Vainshtein L., 1997, in preparation

Churazov E., Sunyaev R., 1997, in preparation

Cook W.R., Grunsfeld J.M., Heindl W.A., Palmer D.M., Prince T.A., Schindler S.M., Stone E.C., 1991, ApJ, 372, L75

Hertz P., Grindlay J., 1984, ApJ, 278, 137

Fabian A.C. Nature, 269, 672

Koyama K, 1994, New Horizon of X-ray Astronomy, FSS-12, 181, Univ. Acad. Press, Tokyo

Koyama K. et al., 1996, Publ. Astron. Soc. Japan, 48, 249

Markevitch M, Sunyaev R., Pavlinsky, 1993, Nature, 364, 40

Ramaty R., Leventhal M., Chan K.W., Lingenfelter R.E., 1992, ApJ, 392, L63

Mirabel I.F., Morris M., Wink J., Paul J., Cordier B., 1991, A&A, 251, L43

Mirabel I.F. et al., 1992, Nature358, 215

Morrison R., McCammon D., 1983, ApJ, 270, 119

Pavlinsky M., Grebenev S., Sunyaev R., 1994, ApJ, 425, 110

Skinner G.K. et al., 1991, A&A, 252, 172

Sunyaev R. et al., 1991, ApJ, 383, L49

Sunyaev R., Markevich M., Pavlinsky M., 1993, ApJ, 407, 606

Sunyaev R. & Churazov E., 1996, Astronomy Letters, 22, 648

Vainshtein L. & Sunyaev R., 1980, Soviet Ast. Letters, 6, 673

Radiative and advective accretion disks around black holes

M.A. Abramowicz

Department of Astronomy & Astrophysics, Chalmers University of Technology, 412 96 Göteborg, Sweden

Abstract:

Observations suggest that in several sources (SXTs, some low luminosity AGNs, the Galactic Centre) accretion disks around black holes are of the standard Shakura-Sunyaev type (SSD) at large radii, and of the advectively dominated (ADAF) type at small radii. The transition between the two types occurs at thousands or tens of thousands of gravitational radii away from the central black hole. Both SSDs and ADAFs are present during long periods of the quiescent state of SXTs, as well as during the rapid transient phase. The physical reason for the transitions is not yet known. In this article, I argue that the transition should occur due to a (yet unknown) *global* interaction between the ADAF and the SSD. A few simple possibilities that come to mind do not work, however. I show that the transition region between the ADAF and the SSD should rotate at super Keplerian orbital speeds and that its hydrodynamical structure is seriously two dimensional.

1 Introduction

The newly discovered solutions which describe stable, optically thin, hot, advectively dominated accretion flows (ADAFs) have attracted a lot of attention because of their remarkably successful applications to X-ray transients, some active galactic nuclei, and the Galactic Centre. (See Narayan 1996 for a review.) Observations of spectra and variability of these objects can be explained by black hole accretion disks which in their inner parts are ADAFs and in the outer parts are standard Shakura-Sunyaev disks (SSDs). Observations indicate that SSD-ADAF transitions do occur, that the transition region is probably rather narrow, and that is located rather far from the central black hole, from hundreds to tens of thousands gravitational radii away. There is no satisfactory theoretical understanding of the nature of these transitions. Many researchers are puzzled by why such transitions should occur at all. They argue that if a disk is of the SS type at some large radius, there is no obvious reason why it should not continue to be of the same SS type all the way down to the black hole. For this reason, the presently prevailing idea is that only an unknown instability, determined by properties of the SS disk at a fixed radial location, could trigger the SS-ADAF transition. It is often argued that the instability must be connected with the local vertical structure of the disk, and thus should depend on details of the vertical radiative transfer, and possibly on the disk – corona interaction. For example, if the corona heats the SS disk sufficiently strongly, the disk will lose thermal balance and evaporate (Meyer and Meyer-Hofmeister, 1994).

However, despite a lot of effort, the very existence of such an instability has not been proved in the distant region where the SSD-ADAF transition is likely to occur. It may be that the SSD plus corona system *is* locally stable there. In addition to this physical difficulty there is a mathematical one: Regev and Lasota, (1996) *rigorously proved* that general mathematical conditions for SSD-ADAF transitions are impossible to meet if the transition region is narrow and the processes governing its structure are local.

The above physical and mathematical difficulties may suggest that the presently prevailing idea that the SSD-ADAF transitions are due to a *local instability of the standard SS disks*, may be only a part of the story, and something quite relevant may still be missing. Thus, one should try to look for alternative explanations and possibilities.

2 Types of accretion disks: linear branches

Today's state of the art in accretion disk modelling is quite demanding. Both the local and global models that are being constructed are non linear and rather complex. This reflects the non linear complexity of the physics of accretion — a difficulty that cannot be avoided. Figure 1, taken from Björnsson et al. (1996), illustrates this point. It shows families of *local* accretion disk models, at the fixed radial location $r = 10r_G$, around a black hole with mass $M = 10M_\odot$. The models accurately describe the radial advection of heat, and relevant radiative and plasma processes in the vertical structure: black body radiation when the disk is optically thick, and two temperature plasma, bremsstrahlung, synchrotron radiation, Comptonization and electron-positron pairs when the disk is optically thin. Figure 1 shows just one particular aspect of the models: the value of surface density Σ. The lines in the Figure correspond to families of models with a fixed value of the viscosity parameter α. Each model in each family corresponds to a single point on a particular curve, i.e. to a particular choice of the pair $\dot{M}$, α. Chen et al. (1996) demonstrated that although for different radial locations the picture is different, *the basic topology of the solutions is always the same.*

Figure 1 looks complicated, but I am convinced that there is a rather simple way to gain an intuitive insight into its deep physical meaning. My intuition is governed by the remark that what the Figure *really* shows, is a network of simple straight-line "branches" connected by very short non-linear joints. The very reason for the existence of linear branches is rather simple. On each particular branch, exactly one physical process dominates the cooling, and thus the equations describing accretion are linear (in their logarithmic version). The dominant cooling processes are indicated by labels in Figure 1: Pairs, Advection, Bremsstrahlung, Compton, and Optically Thick Black Body. I will now describe the most relevant physical properties of the six physically most important sets of branches: SS(g), SS(r), SLIM, SLE, ADAF, THICK.

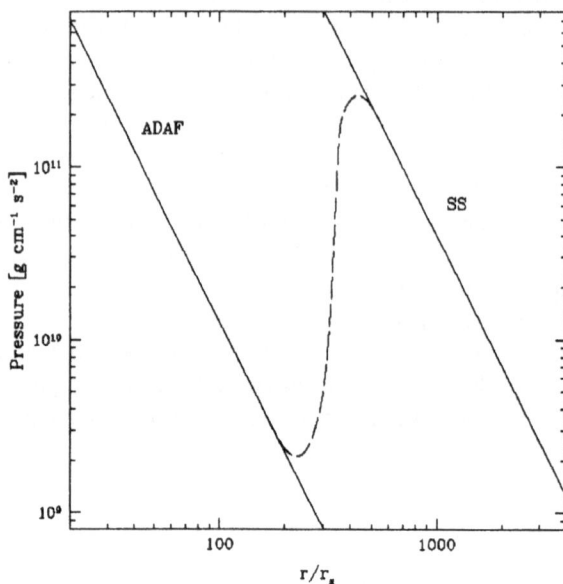

Fig. 1.

SS(g): These disks are very thin geometrically, very optically thick, and gas pressure supported. Radial advection of heat is negligible. Solutions of this type exist for all values of α, but only for $\dot{M} \lesssim 10^{-3} \dot{M}_{\mathrm{Edd}}$. The disks are thermally and viscously stable. They were first calculated by Shakura and Sunyaev (1973).

SS(r): These disks are very thin geometrically, optically thick and radiation pressure supported. Radial advection of heat is negligible. Solutions exist for all values of α and for $10^{-3} \dot{M}_{\mathrm{Edd}} \lesssim \dot{M} \lesssim \dot{M}_{\mathrm{Edd}}$. The disks are thermally and viscously unstable. They were first calculated by Shakura and Sunyaev (1973).

SLE: These disks are very thin geometrically and optically. They are gas pressure supported. Radial advection of heat is negligible. Solutions exist for all values of α and for $\dot{M} \lesssim \dot{M}_{\mathrm{Edd}}$. The disks are viscously stable, but thermally unstable. They were first calculated by Shapiro, Lightman and Eardley (1973).

SLIM: These disks are moderately thin geometrically, optically thick and they are radiation pressure supported. Radial advection of heat is important. Solutions exist for $\alpha < \alpha_{\mathrm{crit}}$ and for $\dot{M} \gtrsim \dot{M}_{\mathrm{Edd}}$. The disks are thermally and viscously stable. Their radiative efficiency is small. They were first calculated by Abramowicz et al., (1986, 1988). Global models have been constructed by several teams, and I will discuss them later.

ADAF: These disks are not very geometrically thick, optically thin and gas pressure supported. Radial advection of heat is important. Solutions exist for $\alpha < \alpha_{\mathrm{crit}}$ and for $\dot{M} \lesssim 10^{-1} \dot{M}_{\mathrm{Edd}}$. The disks are thermally and viscously stable.

They look faint and hot because their radiative efficiency is very small and their temperature very high — close to the virial temperature. They were first calculated by Narayan and Yi (1996) and Abramowicz et al., (1996). Very recently, accurate *global* models of ADAFs have been constructed. They employ differential equations in the pseudo-Newtonian potential (Chen et al., 1996b, Narayan et al. 1996, Matsumoto et al. 1996) or fully relativistic differential equations in Kerr geometry (Chen et al. 1996c). These solutions go all the way from large radii down to the horizon, and pass smoothly through the sonic point. Collectively, these solutions take into account non-Keplerian angular momentum distribution, radial pressure gradient, radial advection of heat, and radiative cooling by bremsstrahlung and Comptonization in a two-temperature plasma. They fully confirm the topology of the solutions shown in Figure 1. (Note: the value of α_{crit} is very sensitive to details. The most accurate recent calculations point to high values of $\alpha_{crit} > 1$. Thus, it looks as though only the solutions corresponding to small, sub-critical, α are astrophysically relevant.)

THICK: For these, the large vertical thickness makes it impossible to use the 1D+1D method of constructing disk models. Approximate analytic 2D models were constructed a long time ago by the Warsaw group (see Abramowicz, Calvani and Nobili, 1980). Later, several authors constructed more accurate numerical models (see Igumenshchev et al., 1996 for references).

3   Transitions between branches

Combined solutions, which for different ranges of radii belong to different types of disk, are of course possible. Radial transitions between different types could obviously be smooth if the models belong everywhere to one of the four families which do not cross the critical solution, corresponding to $\alpha = \alpha_{crit}$. In Figure 1, the critical solution (not shown) locates just between the short-dashed and dot-dashed lines.

Two such combined solutions with smooth radial transitions between different types of accretion disk have been explicitly constructed.

SS(g)-SS(r)-SLIM: For this solution, $\alpha < \alpha_{crit}$, $\tau > 1$. Global ($10^4 r_G \geq r \geq r_G$) accretion disk models of this family, consisting of the three different types in different radial ranges, have been constructed by Abramowicz et al. (1988) and later by other authors. These disks are everywhere optically thick. They always have an unstable SS(r) part for some range of radii and for this reason they cannot be stationary. However, the instability does not destroy these disks: instead, they undergo a limit-cycle behaviour which may explain temporal behaviour of AGNs on time-scales ranging from a few years to a few tens of years. Their spectral properties fit the BBB component in the observed AGN spectra (Szuszkiewicz et al., 1996).

These combined models provide an example of radial SSD-ADAF transitions because they are always of the SSD type very far away from the black hole, and of the ADAF (optically thick) type very close in. In the transition region, the angular momentum is super-Keplerian and the pressure has two extrema, one maximum and one minimum (Abramowicz et al., 1988). I argue that these properties are genuine: they must be shared by all SSD-ADAF transitions including, in particular, also the recently discussed SSD-ADAF transitions which involve optically thin ADAFs.

SS(g)-SS(r)-SLE: For these solutions, $\alpha > \alpha_{\mathrm{crit}}$, τ continuously changes between $\tau \gg 1$ and $\tau \ll 1$. Wandel and Liang (1991) and Liang and Luo (1994) have constructed *local* models of disks of this family covering a large range of radii by combining the three previously known linear branches through a non-linear "bridging" solution based on a phenomenological approximation of radiative transfer for $\tau \approx 1$. In these solutions, advective cooling is negligible. In fact, there is strong advective *heating* present at the SLE part, but it does not change any of the general properties of the solution and thus is not physically interesting. The SLE part of these models is violently thermally unstable, and for this reason these disks cannot exist in reality. However, they do provide a purely formal solution of some academic interest: they are optically thick at large radii and optically thin at smaller radii and so they experience optically thick-thin transitions. Disks of this type, and their optically thick-thin transitions, have recently been discussed by Artemova et al. (1996) who used the same local approach as that used by previous investigators, and fully confirmed the previously obtained results. Their "bridging formula" was slightly different from that used previously, but this introduced no new effects.

4   A few general remarks on SSD-ADAF transitions

Figure 1 shows that for a fixed value of α and a fixed value of $\dot{M} \lesssim 0.1\dot{M}_E$, both ADAF and SSD solutions are possible. They are both stable (Kato et al., 1996ab). Observations tell us that Nature knows how to pick up the ADAF solution at small radii and join it with the SSD at large radii. We do not know how to theoretically explain and numerically model such a transition. Indeed, some recent claims that these transitions have been numerically modelled, are based on a misunderstanding. What was found, was only a rediscovery of the previously known, *unstable*, solution with optically thick-thin transitions, derived long ago by Liang and his collaborators (described here in Section 2). These trivial transitions have nothing to do with the SSD-ADAF transitions which are the subject of the recent interest.

It is unlikely that the SSD-ADAF transition occurs smoothly and is governed by the same equations as those describing the SSD and ADAF. In this case the equations describing the accretion disk structure would "know" about the possibility of the transition. In all of the numerical simulations constructed to date, there is no clue that such a transition may occur. Thus, it is more likely that the transition occurs because of some as yet unrecognized effects or processes that are not included in the standard equations. What could these processes be? The most natural answer is that they could be governed by some global interaction between the ADAF and the SSD. Indeed, the location of the transition region, $r_{\text{trans}} \approx (10^3 - 10^4) r_G$, when expressed in units of the gravitational radius (a natural unit of length here), is a large number of roughly the same magnitude as the ratio of the efficiencies of the ADAF and the SSD, $\eta_{\text{SSD}}/\eta_{\text{ADAF}} \approx 10^2 - 10^4$. To me, this suggests an explanation of why the transition radius is so enormously large. Suppose that the transition is due to some balance of the global effects of the ADAF and the SSD and suppose that these effects are mediated by radiation. Then what comes to mind is that $r_{\text{trans}}/r_G \approx \eta_{\text{SSD}}/\eta_{\text{ADAF}}$ as a natural consequence. Of course, the transition could be triggered by a purely local phenomenon, for example evaporation of the SSD caused by it becoming overheated by the local hot corona. There has been a lot of recent work proceeding in this direction. To my mind, such a possibility is less natural than a globally triggered transition, because it does not give a role for the ADAF. In this interpretation, the ADAF is seen as occurring as a local disease of the SSD. But then, why should the transition occur at a distant location so well tuned to the ADAF efficiency? Of course, this may just be an insignificant coincidence. Two global processes have been discussed in the context of SSD-ADAF transitions: either the ADAF overheats the SSD and "evaporates" it, or the SSD overcools the ADAF (by providing very soft photons) and collapses it (Esin, 1997).

Here, I estimate whether the global heating and evaporation of the SSD due to the integrated ADAF radiation could work. Let us consider an annulus with radial width Δr, located at the transition radius r_{trans}. The relative width $\delta = \Delta r / r_{\text{trans}}$ is most probably $\delta \lesssim 1$ (Regev and Lasota, 1996). Consider a layer with effective optical depth $\tau = 1$ at the top of the annulus. It has surface density $\Sigma = \xi m_p / \sigma_T$, where ξ is a dimensional scaling parameter, and $\xi = 1$ corresponds to pure electron scattering. The power needed to "evaporate" the annulus in the thermal timescale $t_T = 2\pi \Omega_K^{-1} \alpha^{-1}$ is

$$L_{\text{EVAP}} = \frac{(GM)^{3/2} \delta \xi m_p}{r^{1/2} \sigma_T}. \tag{4.1}$$

On the other hand, the power supplied by the ADAF to the annulus is,

$$L_{\text{ADAF}} = \dot{M} c^2 \eta_{\text{ADAF}} \epsilon. \tag{4.2}$$

Here $\epsilon < 1$ describes how much of the ADAF radiation ends up in the annulus. It is a geometrical factor which can be found from the solid angle integration of the ADAF as seen from the annulus. Let us denote $\dot{m} = \dot{M}/(L_{Edd} c^2)$. I assume that when the annulus is overheated by the ADAF radiation, it expands *faster* that it cools by means of its own black-body radiation. The evaporation condition is therefore, $L_{\text{ADAF}} > L_{\text{EVAP}}$, and it takes the form,

$$\frac{\dot{m} \eta_{rmADAF}}{\xi \delta \alpha} > \left(\frac{r_G}{r}\right)^{1/2}. \tag{4.3}$$

With the typical ADAF values of the dimensionless parameters, and $\xi \approx 1$, the left hand side cannot be larger than about 10^{-7}, while the right hand side could not be much less than about 10^{-2}. Thus, this version of global evaporation cannot work. Practically the same conclusion, only in a slightly different context but with a much more accurate treatment of the physics than here, has been reached in a very important recent work by Esin (1997).

However, could some similar process work? To see how difficult this would be to achieve, one should note that in the case of a "modified evaporation", either $\xi < 10^{-5}$, or the power of the radial function on the right hand side should be -3 or less. Evaporation is not the only global alternative. It could be, for example, that the ADAF exerts a damaging radiative torque on the SSD (we have checked that this would not work), or that there is a local SSD instability triggered by global influence of the ADAF. I am convinced that some such process will eventually be found. A *purely* local reason for the transition makes no sense to me.

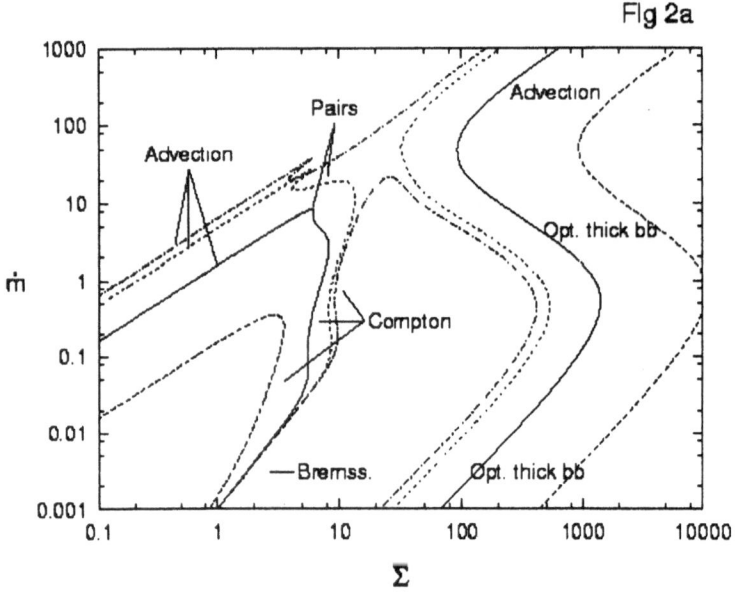

Fig. 2.

5   A necessary condition for the SSD-ADAF transition

Figure 2 shows typical radial dependence of the pressure for SSD and ADAF solutions (the heavy lines), corresponding to a particular choice of accretion rate $\dot{M}$ and viscosity parameter α. In the example shown in the Figure, $\dot{M} =$

$10^{-3} \dot{M}_{\mathrm{Edd}}$ and $\alpha = 10^{-3}$. The ADAF solution is approximated by the self-similar solution of Narayan and Yi (1996). For a different choice of $\dot{M}$ and α, the location of the SSD and ADAF lines would be different, but three features of the Figure will not change: (1) both the SSD and ADAF lines have logarithmic slopes close to $-5/2$, (2) the SSD line is above and to the right of the ADAF line, (3) the relative horizontal distance between the SSD and ADAF lines depends only *very weakly* on the radius, accretion rate, viscosity parameter, and black hole mass,

$$D = \frac{r_{\mathrm{SS}} - r_{\mathrm{ADAF}}}{r_{\mathrm{ADAF}}} \approx 10. \tag{5.1}$$

t is quite clear from Figure 3 that only if $D \geq \delta$, could the slope of $P = P(r)$ be negative everywhere in the transition region. $D = \delta$ is the limiting case for which the slope could be zero everywhere. For transitions with $D < \delta$ the slope must be *positive* in at least part of the transition region. Thus, because $D \approx 10$, for a proper SSD-ADAF transition the slope $P = P(r)$ must be positive somewhere in the transition region. This is equivalent to pressure having at least one maximum and one minimum in the transition region, as shown schematically in Figure 3 by the broken line. Thus, for a proper SSD-ADAF transition, $dP/dr = 0$ in at least two different points within the transition region.

To see what this implies for the angular momentum distribution, we write the condition of balance of forces in the equatorial plane of the disk in the form,

$$\frac{1}{\rho}\frac{dP}{dr} = \frac{\ell^2(r) - \ell_{\mathrm{K}}^2(r)}{r^3} + v_r \frac{dv_r}{dr}. \tag{5.2}$$

Here, ρ is the density of matter, $\ell(r)$ is the angular momentum distribution in the disk, $\ell_{\mathrm{K}}(r)$ is the Keplerian angular momentum distribution, and v is the radial velocity. The inertial force $v_r dv_r/dr$ is usually very small for sub-sonic flows in comparison with ℓ_K^2/r^3 and may be neglected. Thus, from equation (5.2) and the fact that pressure must necessarily have two extrema in the transition region (one maximum and one minimum), it follows that the angular momentum distribution $\ell(r)$ must cross the Keplerian curve ℓ_{K} twice. Between these two crossing points the angular momentum must be super-Keplerian.

A necessary condition for the existence of a proper ($\delta < 1$) SSD-ADAF transition is that in the transition region, the rotation of the disk has a super-Keplerian orbital speed.

The above result is a strict mathematical statement which needs no additional support. However, it is interesting to note that the existence of the crossing point on the ADAF side, i.e. the one which is closer to the black hole, could have been anticipated from the paper of Chen et al. (1996a). They discussed an interesting fact, noticed by Abramowicz et al. (1995), that *local* ADAF solutions exist only below a certain radius r_{out}. The value of r_{out} depends on the accretion rate $\dot{M}$ and the viscosity parameter α. Topologically, the limiting outer radius is similar to the "cusp" at the inner edge of accretion disks (Abramowicz, Calvani and Nobili, 1980). At the cusp $dP/dr = 0$, and the topological similarity suggests that one should have $dP/dr = 0$ also at r_{out}. Equation (5.1) implies that in this case $\ell(r_{out}) = \ell_K(r_{out})$. Indeed, Narayan et al. (1996) constructed models of ADAFs which have exactly Keplerian angular momentum at r_{out}, and the slope of the $\ell(r)$ line is steeper there than that of $\ell_K(r)$. Thus, these two lines do obviously cross at r_{out}.

References

Abramowicz M.A., Calvani M., & Nobili L., 1980, ApJ, 242, 772

Abramowicz M.A., Lasota J.-P. & Xu C., 1986, in *Quasars*, eds. G. Swarup & V.K. Kapachi, (D. Reidel Pub. Co.: Dordrecht)

Abramowicz M.A., Czerny B., Lasota J.-P., & Szuszkiewicz E., 1988, ApJ, 332, 646

Abramowicz M.A., Chen X., Kato S., Lasota J.-P., & Regev O., 1995, ApJ, 443, L61

Abramowicz M.A., Chen X., Granath M., & Lasota J.-P., 1996, ApJ, 471,762

Artemova I.V., Björnsson G., Bisnovatyi-Kogan G.S. & Novikov I.D., 1996, ApJ, 456, 119

Björnsson G., Abramowicz M.A., Chen X., & Lasota J.-P., 1996, ApJ, 467, 99

Chen X., Abramowicz M.A., Lasota J.-P., Narayan R., & Yi I., 1995, ApJ, 443, L61

Chen X., Abramowicz M.A., & Lasota J.-P., 1997, ApJ, in press

Chen X., & Taam R.E., 1993, ApJ, 412, 254

Ensin A.A., 1997, preprint astro-ph/9701039

Igumenshchev I.V., Chen X. & Abramowicz M.A., 1996, MNRAS 278, 236

Luo C. & Liang E.P., 1994, MNRAS, 266, 386

Matsumoto R., Honma F. & Kato S. (1989) in *Theory of Accretion Disks*, eds. F. Meyer & al. (Dordrecht: Kulver)

Meyer F. & Meyer-Hofmeister E., 1994, AA 288, 175

Narayan R., 1996, preprint astro-ph/9611113

Narayan R., & Yi I., 1994, ApJ, 428 L13

Narayan R., & Yi I., 1995a, ApJ, 444, 231

Narayan R., & Yi I., 1995b, ApJ, 452, 710

Narayan R., Honma F., & Kato S., 1996, ApJ, submitted

Regev and Lasota, 1996 private communication

Shakura N.I., & Sunyaev R.A., 1973, A&A, 24, 337

Shapiro S.L., Lightman A.P., & Eardley D.M., 1976, ApJ, 204, 187

Szuszkiewicz E., Malkan M.A. & Abramowicz M.A., 1996, ApJ, 458, 474

Wandel A. & Liang E.P., 1991, ApJ, 380, 84

Observational progress on accretion disks: a turning point

M.-H. Ulrich

European Southern Observatory

Abstract:
A turning point has been reached in the observations of AGN for several reasons: (1) the development of a method to derive the mass of the central black hole (2) the identifications of several subsets of AGN which cover a wide range of accretion rates (3) the observations of the appearance of broad lines in galaxies which had previously displayed moderate signs or no signs at all of activity and (4) the growing evidence for a central massive black hole in normal galaxies.

This opens new avenues of investigations such as the exploration of the entire parameter space defined by the black hole mass and the accretion rate, and the observational exploration of the duty cycles of nuclear activity.

The most recent observations of the broad emission lines strengthen the disk plus wind model of the BLR, a model which appears at present, to be the most plausible.

1 Introduction

There are several independent lines of evidence for the presence of an accretion disk. In three energy ranges (mid X-rays, UV and optical) there is evidence for a dense cool medium which scatters, or absorbs and re-emits radiation emitted at higher energy by the central source (Pounds & al 1990; for a review, Ulrich, Maraschi & Urry 1997). Various considerations, in particular relative to the solid angle of this cool medium as viewed from the central source strongly suggest that this medium has a flat geometrical structure. An attractive possibility, entirely consistent with the observations, is to identify this flat structure with the accretion disk surrounding the black hole assumed to be present in AGN.

The disk is probably magnetized and magnetically accelerated outflows are the likely sources of the broad line region (BLR) gas clouds.

Recent and current work on this topic is reviewed. An agenda is drafted for AGN variability studies in the next decades and beyond.

2 Different types of continuum variations in the UV and X-ray ranges, and evidence for different emission mechanisms

2.1 The rapidly variable UV component and the irradiation model The UV continuum flux varies on all measurable time scales, minutes to decades. The fastest variations are known to be incompatible with models of variable fueling in standard optically thick, geometrically thin, accretion disks (Clarke 1987). What could then cause the fast variations in low luminosity AGN ? An answer is offered by the good correlation and the simultaneity observed between the variations of the UV continuum flux and the variations of the mid X-ray flux on time scales of weeks to months (Perola & al 1986, Warwick & al 1996, Clavel & al 1992).

This is the basis of the irradiation model in which the variable part of the UV continuum flux is thermal emission by a dense cool medium whose temperature is modulated by the incident variable X-ray flux. A consistency check for this model is that the energy in the variable X-ray component is larger than that in the UV variable component. This is indeed verified in low luminosity AGN, the only ones for which simultaneous UV/X-ray observations have been performed.

It is not known whether the irradiation model applies to high luminosity AGN for lack of enough simultaneous UV/X-ray observations (these AGN are in general are rather faint). These observations should be the priority of the next space missions capable of measuring the UV and mid X-ray flux of relatively faint objects.

Explosive reconnections probably dissipate significant power via magnetic flares within the disk corona (Galeev & al 1979, Blandford & Payne 1982) and X-ray emission is produced via inverse Compton emission in the hot corona surrounding the colder accretion disk (Haardt & Maraschi 1991). Turbulence and magnetic reconnections can produce a variable X-ray emission even if the accretion rate is constant.

2.2 The slow varying UV component On time scales of years the proportionality of the UV and medium X-ray fluxes, which is so remarkable on time scales of weeks to months, breaks down. Figure 1 shows the UV vs X-ray flux of NGC 4151 during campaigns of observations separated by several years. During each campaign, the UV continuum flux is correlated with the mid X-ray flux (though not in a detailed way) but between the two campaigns of 1983/1984 and 1993, the UV flux alone underwent a large increase. A similar situation is observed in NGC 5548 and F9 where on long time scales, the UV flux varies with a much larger amplitude than the X-ray flux (Clavel & al 1992; Morini & al 1986).

Figure 2 shows the variations of the flux of NGC 4151 at 1440 A during the life time of IUE 1978-1996 (for clarity not all data points are plotted). There is a slow large amplitude variation on which is superposed episodes of fast variations which are those which can be expained by the irradiation process described in section 2.1. The mechanism for longer term (years) variations is unknown.

2.3 The Narrow Line Seyferts 1: Eddington accretion rate? The most extreme soft-X-ray variability occurs in Narrow-Line Seyfert 1 galaxies (NLS1), a subset of AGN with very steep X-ray spectra and narrow optical emission

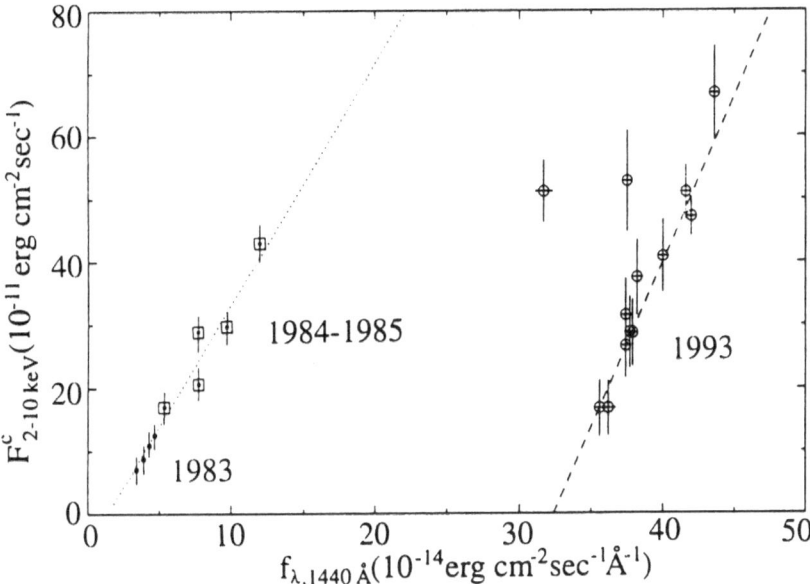

Fig. 1. X-ray flux (absorption corrected, 2-10 keV) versus UV flux (1440 Å) for NGC 4151. *Right:* ASCA observations in 1993 November 30-December 13, with best-fitting linear correlation. *Left:* EXOSAT observations in 1984 December 16 - 1985 January 28 (large crosses) and 1983 November 7-19 (small crosses) with best-fitting linear correlation. The good correlation of the UV and X-ray flux on time scale of weeks and months breaks down at long (years) and short (days) time-scales. (Adapted from Warwick & al 1996 and Perola & al 1986)

lines (FWHM less than 2000 km s^{-1}; Osterbrock & Pogge 1985, Boller & al 1996). In some NLS1, the energy in the soft X-ray range exceeds the energy in the medium/hard component and, moreover, the medium X-ray spectrum is exceptionally steep —characteristics shared with Galactic black hole candidates. This suggests that in NLS1, or at least in some of them, the accretion rate is close to the Eddington limit and the soft X-rays represent viscous heating of the accretion disk (Pounds et al 1995). Alternatively, a small black hole with an accretion rate of $L/L_{Edd} \sim 0.1$ could also emit a very hot spectrum with such an intense soft X-ray component.

3   The broad emission lines region and the disk plus wind model

3.1 The composite model of the BLR   The emission line spectrum of AGN is best explained in a "composite model" (Kwan & Krolik 1981, Collin-Souffrin & al 1986, Netzer 1987) in which the low ionization lines (LIL; Balmer lines and

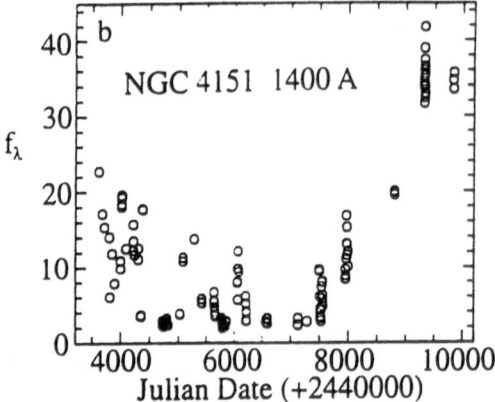

Fig. 2. Long and short term UV continuum variations in NGC 4151 over the 17 years of IUE. The passage through the deep minimum was interrupted by short excursions to medium bright level. The vertical groups of points, unresolved on this scale, are IUE campaigns with an adequate sampling interval of typically three days (Ulrich & al 1997)

FeII multiplets)) come from a very dense medium with $N_e \geq 10^{11}$ cm^{-3}, an ionization parameter much less than 0.1, and a column density exceeding 10^{24} cm^{-2}. This medium must have a flat geometry and could be the accretion disk. The high ionization lines, HIL of e.g. CIV, NV, HeII, come from a more dilute medium (van Groningen 1987, Collin-Souffrin & Lasota 1988) or an ensemble of clouds in a broad cone above and below the disk, with the gas density in the clouds not larger than a few 10^9 cm^{-3} and an ionization parameter of the order of 0.3.

In addition, spectroscopic observations show the gas velocity and the degree of ionization to be correlated: among the high ionization lines, the broadest FWMH and the most extensive wings are those of the most highly ionized species (e.g. Antonucci & Cohen 1983, Ulrich & al 1984; Krolik & al 1991).

3.2 The crucial role of the variability studies: the derivation of the mass of the central black hole Current variability studies have produced crucial results on the BLR and foremost among them is the derivation of the central mass as outlined below:

a) The linear scale of the emission line region can be derived from variability studies. It is derived from the time delay with which the intensity variations of a certain line follows the flux variations of the UV continuum. This delay is interpreted as the light travel time between the UV continuum source (a point source at the AGN center) and the gas emitting this line.

b) The broad line region is stratified. The wings of the lines respond faster than the core to the continuum variations and among the HIL the highest ionized lines respond faster and with a larger amplitude to the continuum variations.

The picture that emerges is that of a stratified highly ionized set of clouds with the most highly ionized and fastest moving gas closest to the center. The degree of ionization and velocity of the gas decrease outwards. To first order the intensity and profile variations of the emission lines fit the model of a variable central continuum source illuminating a stratified BLR. In detail, the situation is complex with the profiles displaying transient asymmetries, peaks and shoulders.

c) The absence of dominant radial motions is consistent with the gas clouds being in virialized chaotic motions. These results on the gas velocity field, and the determination of the gas radial distance from the time delay as described in (a) allow derivation of the central mass. This represents a fundamental step forward in AGN studies.

d) The 3-D architecture and velocity field of the gas clouds can, in principle, be reconstructed from the analysis of well-sampled, high S/N light curves of the continuum and of the intensity and profiles of the emission lines (See "Reverberation Mapping of the Braod Line Region in Active Galactic Nuclei", Gondhalekar & al 1994). The data presently on hand suggest that the LIL and the HIL do not have the same geometry: The LIL variations point to a near absence of matter along the line-of-sight, as expected from a disk at small inclination, while in contrast, the CIV wings respond with a very short time delay consistent with the HIL coming from a broad cone or cylinder, in which case some matter lies along the line of sight (e.g. Horne 1994).

e) The blueshift of the HIL: evidence for a wind. In radio-quiet AGN, the HIL are blueshifted (by 0 to ~ 1500 km s^{-1}) with respect to the LIL, which themselves are at the host galaxy redshift (e.g. Wilkes 1984, Marziani & al, 1996), The blueshift of the HIL is a key information to decipher the structure of the BLR as it indicates that (i) to first order, the HIL and the LIL have different velocity fields and thus cannot not come from exactly the same region (ii) the HIL clouds have a preferred direction of motion.

3.3 The "disk plus wind" model: the best model at present The properties of the BLR gas clouds, described in section 3.2, are best understood in the "disk plus wind" model in which the LIL come from the surface of the accretion disk and the HIL come from a wind originating from the disk (van Groningen 1987, Collin-Souffrin & Lasota 1988). At present, this appears to be the best model.

The question as to how far the accretion disk (which is opaque) extends outwards before becoming self -gravitating is far from being resolved. In recent calculations (Huré & al 1997) the disk is found to have a radius of 10^{17} cm. As a consequence, within this radius what looks like virialized chaotic motions of the highly ionized gas clouds are in fact rotation motions, or more precisely helicoidal motions (still reflecting the disk rotation), from which the central black hole mass can be estimated.

Magnetically accelerated outflows from accretion disks and radiatively driven winds are promising models for the formation and evolution of the BLR clouds (Blandford & Payne 1982, Königl & Kartje 1994, Bottorff et al 1996, Murray

& Chiang 1996). The combination of the acceleration of the gas along the field lines (inside the Alfven radius), of the imperfect confinement and of the intense ionizing field is likely to result in a inhomogeneous hot medium with a roughly ordered velocity field. The densest coolest inhomogeneities form the BLR clouds emitting the prominent UV/optical lines. The hottest phase is detected as the fully ionized component of the warm absorber.

Many features of these promising models, however, remain unspecified and can be adjusted to accommodate the observations.

The complex variations of the line profiles (Eracleous & Halpern 1994) suggest the presence of transient inhomogeneities on the surface of the disk which locally enhance the line emission thus altering the profile of the lines emitted by the disk such as Hβ. These inhomogeneities could also be the sites of increased gas extraction from the disk surface, lifting additional clouds above the disk and producing shoulders and other features observed in the HIL profiles. The double peaked Balmer lines (Eracleous & Halpern 1994, Gaskell 1996) present in some radio galaxies may have their origin in the (magnetized) disk structural properties associated with jet formation (The HIL lines never show double peaks).

In summary, the disk plus wind model provides a plausible explanation for the main properties of the BLR clouds, in particular for the blueshift of the HIL which would be difficult to understand in other models. In that sense it appears to be the best model at present.

4   An exciting future for AGN research

4.1 Exploring the M - M$_\odot$ parameter space As stated above a major advance in AGN research has been the development of a method to derive the mass of the central black hole (Section 3.2). At the same time, in another major development, several subsets of AGN have been recognized, which have widely different accretion rates.

As mentionned in section 2.3 the recently identified subclass of NLS1 contains objects which may be accreting at the Eddington rate. At the other extreme in accretion rate are the newly identified subclass of the very weak AGN found from a search for very weak broad lines or unresolved weak UV sources at the nucleus of nearby galaxies (Ho & al 1996, Maoz & al 1995). These objects have a very low intrinsic accretion rate but their mass is not known. Through a systematic study of the line and continuum variability of significant samples in these subclasses the central mass present in the members of these classes can be determined. These observations form one of the most promising observational areas in the near future. Similarly, monitoring high-luminosity AGN will allow the measurement of the emission line and continuum variability and the measurement of the mass of the central black hole in the brightest objects in the universe.

With these data in hand one will be in the position to explore the entire parameter space defined by the black hole mass,M and the accretion rate, M$_\odot$.

4.2 Duty cycles and Families of AGN The identifications of subclasses of AGN lead naturally to the question as to whether the different subsets of AGN represent successive phases, possible repetitive phases in the life of a galaxy nucleus (Eracleous & al 1995) or whether they are in fact several families of AGN evolving along separate scenarios.

An argument in favor of the passage of one subset to another is provided by the unexpected appearance of broad emission lines in galactic nuclei which had earlier displayed only weak signs of activity (NGC 1097 Storchi-Bergmann & al 1995) or no activity at all (NGC 4552 Renzini & al 1995). In the elliptical galaxy NGC 4552 a fairly bright unresolved UV source appeared at the center between two HST observations separated by two years. Subsequent spectroscopy has revealed the presence of broad emission lines (Renzini & al 1995). Was it accretion of a passing star or of an interstellar cloud by a dormant central black hole?

Closely related to this question is the increasingly strong evidence for the presence of a supermassive black hole in nuclei of galaxies.

The data necessary to pursue this line of investigation which is related to the duty cycle of AGN and to the frequency of accretion of stars or small gas clouds, are of two types: 1) statistics of the number of AGN in each subset in the general galaxy population, and 2) long time monitoring of members of the various subsets of AGN, and of quiescent nuclei.

5 Summary: The next millenium of AGN variability studies ?

AGN variability studies provide several absolutely critical types of information: (1) The separation of the different components of the continuum and of the lines through their different variability characteristics (2) The derivation of the mass of the central black hole (3) Evidently, they provide information on the variations of the accretion rate but the interpretation of this information is a domain still largely unexplored.

It is clear that our agenda for the near and distant future should contain the following items:

1- Measure the mass of the black hole from line and continuum variability studies in the newly identified subsets of AGN such as the very low luminosity AGN and the NLS1. Similarly, long-term campaigns have to be started to monitor the continuum and lines of the the very high luminosity AGN, and so derive the central mass of the brightest objects in the universe. One or two decades of assiduous work should suffice to get interesting answers.

2- Establish whether or not the irradiation model applies to high luminosity AGN, by simultaneous monitoring of the UV and mid X-ray fluxes.

3- Confirm, or invalidate, the disk plus wind model from new higher quality data - starting with the low luminosity AGN because they are the easiest to observe.

4- A task less simple and of uncertain outcome is the study of the long term variations. We know that they appear in at least two forms: (i) slow large amplitude variations like that of NGC 4151 between 1978 and 1996, and (ii) the unexpected brightening of the nuclear continuum flux and/or broad lines in nuclei known previously to be "passive" or only weakly active (e.g. NGC 1097, NGC 4552). The frequency of such events is entirely unknown and can be estimated only through observations of a large number of weakly active nuclei and of normal galaxies over many years, decades, centuries or longer.

The on-going development of calibrated electronic archives is the only way to secure the data on variable phenomena on time scales 10, 100, 1000 times longer than has been observable up to now.

References

Antonucci R. R. J. & Cohen R. D. 1983, Ap.J. 271, 564

Blandford R. D. & Payne D.G., 1982, MNRAS. 199, 883

Boller T. & al 1996, Astron. & Astrophys. 305, 53

Bottorff M. & al 1997 preprint

Clarke C. J., 1987, Bull. Am. Astron. Soc. 19, 732

Clavel J. & al 1992, Ap.J. 393, 113

Collin-Souffrin S. & al 1986 Astron. & Astrophys. 166, 27

Collin-Souffrin S. & Lasota J.-P. 1988, Publ Astron. Soc. Pac. 100,1041

Eracleous M. & Halpern J. P. 1994, Ap.J. Suppl 90, 1

Eracleous M. & al 1995, Ap.J. 445, L1

Galeev A. A., Rosner R. & Vaiana G. S., 1979, Ap.J. 229, 318

Gaskell C. M. 1996 in Jets from Stars and Galactic Nuclei Editor W Kundt 165, Berlin: Springer-Verlag

Gondhalekar P. M., Horne K. H. & Peterson B. M. 1994 Astron. Soc. Pac. Conf. Series Vol. 69

Haardt F. & Maraschi, L. 1991, Ap.J. 380, L51

Ho L. C. Sargent W. L. W. & Filippenko A. V. 1996 Ap.J. Suppl 98, 477

Horne K. H. 1994 in Reverberation Mapping Editor & Co-Editor Gondhalekar P. M., Horne K. H. & Peterson B. M. 1994 Astron. Soc. Pac. Conf. Series Vol. 69, 23

Huré J.-M. & al 1997, preprint

Königl A. & Kartje J. F. 1994 Ap.J. 434, 446

Krolik J. H. 1991, Ap.J. 371, 541

Kwan J. & Krolik J. H. 1981, Ap.J. 250, 478

Maoz D. & al, 1995, Ap.J. 440, 91

Marziani P. & al, 1996, Ap.J. Suppl 104, 37

Morini M. & al, 1986, Ap.J. 306, L71

Murray N. & Chiang J. 1997, Ap.J. 474, 91

Netzer H. 1987, MNRAS., 225,55

Osterbrock D. E. & Pogge R. 1985, Ap.J. 297, 166

Perola G. C. & al 1986, Ap.J. 308,508

Pounds K.A. & al, 1990, Nature 344, 132

Pounds K.A. & al, 1995, MNRAS., 277, L5

Renzini A. & al, 1995, Nature, 378,39

Storchi-Bergmann T. & al, 1995, Ap.J. 443, 617

Ulrich M.-H., & al, 1984, MNRAS., 206, 221

Ulrich M.-H., Maraschi, L. & Urry M. 1997, Annual Review of Astron. & Astrophys. preprint

van Groningen E. 1987, Astron. & Astrophys 186, 103

Warwick R. S. & al 1996, Ap.J. 470, 349

Wilkes B. J. 1984, MNRAS, 207, 73

Magnetic fields and precession in accretion disks

F. Meyer

Max-Planck-Institut für Astrophysik, Karl-Schwarzschild-Str. 1, 85740 Garching, Germany

Abstract. We discuss tentative evidence for magnetic flux concentration in accretion disks. It derives from analysis of magnetically driven winds, missing radiation from inner disk parts and dwarf nova oscillations in cataclysmic variable systems and the precession period of the galactic black hole binary GRO 1655. We show that the evidence supports the notion that magnetic flux imposed on the disk from the surrounding magnetosphere of the secondary star becomes advectively concentrated to values that allow the acceleration of strong winds, significantly extract energy from the disk and could lead to a tilting instability of inner parts of the accretion disk.

1 Introduction

Cataclysmic binaries have been a fertile observational ground for accretion disk physics. Here we look at possible manifestations of magnetic flux concentration in the inner parts of their accretion disk. These are (1) the winds observed in UV resonance lines (Drew 1987, 1991, Mauche & Raymond 1987, Verbunt 1991, Woods et al. 1992). They may be too strong to be explained by radiation driving (Verbunt 1991). (2) Missing radiative energy flux in the inner parts of stationary accretion disks revealed by eclipse mapping for a number of nova-like systems (Rutten et al. 1992, Baptista, Steiner & Horne 1996). It can be taken as evidence for magnetic energy extraction (Pringle 1993). (3) Tentative interpretation of dwarf nova oscillations (Warner 1995) as illumination of a tilted disk by a rotating spot on the white dwarf (this contribution) and the anomalous precession period of 3^d of the galactic black hole binary GRO 1655 (Mirabel & Rodriguez 1996) interpreted as a magnetically tilted and precessing inner part of an accretion disk.

In the following sections we discuss the magnetospheres of the secondary stars, the flux disconnection and advective concentration in the accretion disk, and the consequences for wind driving, energy extraction and tilting of the inner disk.

2 Magnetospheres of the secondaries

In cataclysmic binary systems the secondary stars are convective and in rapid rotation. One expects effective dynamo action and strong magnetic fields. Evidence for this is activity of flare stars and star spots on BY Draconis stars, and

X-ray emission of low mass main sequence stars. The magnetosphere of such stars reaches far out as indicated by large scale magnetic interaction in RS Can Ven stars. Thus the accretion disks around the primary white dwarf in cataclysmic variable systems probably are imbedded in the secondaries' magnetospheres.

The field strength near the accretion disk depends on the distribution of magnetic flux on the surface of the secondary. Following the indications of large spots on the surface of BY Draconis stars we assume, e.g. two large circular magnetic spots of opposite polarity each covering $1/8$ of the stellar surface and one stellar radius away from each other. This provides a magnetic field strength B_d near the accretion disk

$$B_d = \frac{1}{8} \left(\frac{R_*}{a} \right)^3 B_s = 30\text{gauss} \tag{1}$$

assuming a spot magnetic field strength of $B_s = 4000\text{gauss}$ in approximate equilibrium of magnetic pressure in the spot and photosphere gas pressure outside. Numbers are for a typical 1 $M_\odot$ white dwarf primary and Roche lobe filling $0.3M_\odot$ secondary (R_* secondary's radius, a binary separation).

Finite magnetic diffusivity of the accretion disk will allow outside magnetic flux to file through the accretion disk. Two effects then determine the further fate of such flux. The systematic inward accretion carries flux inward and concentrates it at the center, and the Kepler motion will wind it up and exert torques on the flux tubes. A flux tube can only sustain a limited amount of torque which roughly corresponds to the kink instability limit of equal longitudinal and toroidal field strength. Further winding eventually opens up the flux tube and releases tension to infinity. This can lead to reconnection to the original unwound flux configuration (as in solar flares). Reconnection can only occur on the Alfvén travel time scale. If winding is rapid the flux tubes presumably stay open to infinity persistantly.

The non-dimensional number

$$S = \frac{\ell \Omega_K}{V_A} \tag{2}$$

relates the wind-up time to the (dynamical) reconnection time (ℓ length of flux tube, $\approx a$, Ω_K Kepler frequency at disk foot point, V_A Alfvén speed). From the conservation of magnetic flux and torque along a flux tube one derives that the total turn-around angle between footpoint and largest flux tube diameter at height ("45° angle of field") is

$$\phi = \ell/r_m \tag{3}$$

where r_m is the tube radius at maximal extent. Thus the amount of winding is of order one turn and the Kepler frequency is the effective winding frequency if r_m is of order of the disk radius r_d.

The Alfvén speed

$$V_A = \frac{B}{\sqrt{4\pi\rho}} \tag{4}$$

contains the density within the flux tube extending between secondary and disk. A lower limit neglecting other possible feed-in processes may be obtained from the theory of Shmeleva and Syrovatskii (1973) who calculated the coronal density for given thermal conduction flux F. We take

$$F = \frac{B_\ell B_\phi}{4\pi} r_m \Omega_K = \frac{B_d}{4\pi} r_d \Omega_K \tag{5}$$

(B_ℓ, B_ϕ longitudinal and toroidal field components) as the input of magnetic work per cm^2 and sec and estimate the coronal temperature T from

$$F = -\kappa_0 T^{5/2} \frac{dT}{d\ell} \simeq \frac{2}{7} \kappa_0 \frac{T^{7/2}}{\ell} \tag{6}$$

($\kappa_0 T^{5/2}$ electron thermal conductivity, Spitzer 1962).

We apply their analysis to determine the equilibrium pressure inside the flux tube and from this and the temperature the density ρ. For a one solar mass white dwarf, a radius of expanded flux tube of approximatively disk radius $r_d = 10^{10.4}$cm, a field strength $B_d = 30$gauss the thermal flux becomes

$$F = 10^{11.1} g/sec^3 r_{9.5}^{-3/2} \tag{7}$$

the coronal temperature

$$T = 10^{8.2} r_{9.5}^{-3/7} K \tag{8}$$

the particle number density

$$n = 10^{11.4} r_{9.5}^{-3/2} / cm^3 \tag{9}$$

and the Alfven speed

$$V_A = 10^{7.1} r_{9.5}^{3/4} cm/sec \quad . \tag{10}$$

$r_{9.5}$ is the flux tube radius at the concentrated foot point in the disk in units of $10^{9.5}$cm. This low speed yields for the number S

$$S = 10^{2.7} r_{9.5}^{-9/4} \tag{11}$$

This large value indicates that even with less than 100% thermalization efficiency any magnetospheric flux that diffuses into the accretion disk will open up towards infinity and become disconnected from the secondary star.

3 Flux concentration

Magnetic flux which has diffused into the accretion disk is subject to inward transport by accretion and, with increasing concentration, to outward magnetic diffusion

$$\frac{\partial B}{\partial t} + \frac{1}{r}\frac{\partial}{\partial r}\left[r(\eta\frac{\partial B}{\partial r} + v_r B)\right] = 0 \tag{12}$$

where v_r is the inward radial velocity and η the magnetic diffusivity. In stationary accretion flow

$$v_r \approx \alpha V_s^2/v_K = \alpha v_K (H/r)^2 \tag{13}$$

($v_K = (GM/r)^{1/2}$ Kepler velocity, H scale height of the disk $\approx$ half disk thickness, α viscosity parameter). One may parametrize the magnetic diffusivity by a magnetic α-parameter

$$\eta = \alpha_M H V_s = \alpha_M r v_K (H/r)^2 \tag{14}$$

and might expect α and α_M to be of similar order of magnitude though their ratio is unknown. We argue below that α_M/α might be $\leq 1/2$ from observational evidence.

The diffusion equation has two particular solutions. In the stationary state inward advection and outward diffusion are in balance. This leads to the distribution (Pringle 1993)

$$B = B_0 (r/r_0)^{-\alpha/\alpha_M} \tag{15}$$

We note that such a distribution contains more flux in the interior than present at outer disk radius r_0 if $\alpha/\alpha_M > 2$. Though the secondary star's magnetosphere contains only a limited amount of magnetic flux this would not prevent the accumulation of flux values surpassing it in the disk. This becomes possible by the opening up of any flux that has diffused into the disk. The magnetospheric field lines no longer subject to the twisting by the disk motion will then reconnect to dynamical equilibrium and always restore the original magnetosphere. This sheds additional light on observed differences between dwarf novae in outburst and novalikes or UX Uma system which display stationary accretion (Tout, Pringle & La Dous 1993).

In dwarf nova quiescence accretion has practically ceased, $v_r \approx 0$. Magnetic diffusion if it persists then lets the flux concentration fade away on a timescale

$$t_M = \frac{r^2}{\eta} \simeq \frac{1}{\alpha_M}\left(\frac{r}{H}\right)^2 \frac{1}{\Omega_K} \tag{16}$$

which is of the order of magnitude of the quiescence time. Thus the next outburst can only concentrate the smoothed out magnetospheric field B_d again. The outburst does not last long enough to accumulate more and more flux from the outside as is possible for the stationary accreting systems discussed above. In order that this difference arises one clearly needs $\alpha/\alpha_m \geq 2$ as stated above. Then nova-like systems will have stronger fluxes in their inner accretion disk than dwarf novae in outburst.

4   Magnetically accelerated winds

Winds with speeds of several thousand km/s have been observed in UV resonance lines from dwarf novae and nova-like systems. Though these were generally attributed to radiation driving from hot inner disk (like O-star winds) it
was noted (Verbunt 1991) that density estimates indicated mass outflow surpassing the photon pressure capability. Further, eclipse mapping has shown a
number of nova-like systems in which the radiation of the inner disk surface falls
significantly below that expected from stationary accretion (Rutten et al. 1992,
Baptista et al. 1996). There seems to be no acceptable alternative to magnetic
extraction of energy. As shown above this extraction has to occur via a wind
driven magnetically since the disk field lines are not connected to the secondary
anymore. This requires fields of sufficient strength.

Extraction of magnetic energy on open flux tubes depends on the mass flow
rate and the terminal speed that a wind reaches. Magnetically accelerated winds
have been studied by many authors in particular for jet formation (Blandford
& Payne 1982, Sakurai 1985, 1987, Camenzind 1986, Nitta 1994). We show that
the wind parameters are related to the shape $s(z)$ with which the flux tube
concentrated in the inner disk at radius $s(0) = r_0$ expands upward with height
z (s, ϕ, z cylindrical coordinates centered on the white dwarf). The mass flux is
determined by the density and the flux tube crosssection at the point where the
rising gas reaches sound velocity. This speed is so small compared to the Kepler
speed with which the flux tube rotates around that even for finite values of
B_ϕ/B_z the gas "frozen" to the field lines must corotate with the Kepler frequency
$\Omega_K = (GM/r_0^3)^{1/2}$. For ds/dz finite this exerts centrifugal forces which have a
finite component along the magnetic field and thus centrifugally lift the gas
upward. We include the gas pressure, for simplicity in isothermal approximation

$$P = \rho V_s^2 \tag{17}$$

(V_s = isothermal sound speed).

To derive the equation of motion for the poloidal velocity v_p one projects centrifugal, gravitational, and pressure gradient forces first on the inclined field direction (B_p poloidal, B_ϕ toroidal components) and takes the poloidal component
of that vector. If one further uses the equation of continuity in the approximate
form

$$\pi s^2 \rho v_p = \frac{\dot{M}_w}{2} = \text{const} \tag{18}$$

($\dot{M}_w$ mass loss rate for both disk sides) one obtains

$$\left(v_p - \frac{B_p^2}{B_p^2 + B_\phi^2} \frac{V_s^2}{v_p} \right) \frac{dv_p}{dz} = -\frac{GM}{r_0^3} z \frac{B_p^2}{B_p^2 + B_\phi^2} \cdot \left[\frac{1 + \frac{s}{z}\frac{ds}{dz}}{(\frac{s^2+z^2}{r_0^2})^{3/2}} - (1 + 2\frac{V_s^2}{v_k^2}\frac{r_0^2}{s^2}) \frac{sds}{zdz} \right]$$

$$\tag{18}$$

The first term in the angular bracket derives from the downward pull of gravity, the other two terms from centrifugal acceleration and the pressure gradient. The latter term proportional to $V_s^2/v_K^2 \approx 0(10^{-3})$ can be neglected.

Sound transition occurs for

$$v_p^2 = \frac{B_p^2}{B_p^2 + B_\phi^2} V_s^2 \tag{19}$$

equivalent to sound speed along the inclined field lines. This occurs where the acceleration on the right hand side changes sign. Since below this point the solution is subsonic and approximately in hydrostatic equilibrium with an exponential decrease of density upwards, in order to reach significant wind mass loss this transition must occur at sufficiently low heights. We see that this is mainly determined by balance between centrifugal and gravitational forces, pressure plays a negligible role. Therefore the rate of widening $s(z)$ of the fluxtubes is the one determining factor. We parametrize $s(z)$ by curvature of field lines at $z = 0$,

$$s^2 = r_0^2 + \kappa z^2$$

$\kappa \equiv r_0/r_c$ with r_c the radius of curvature at $z = 0$. The expression in angular brackets becomes

$$\frac{1 + \kappa}{[1 + (1+\kappa)z^2/r_0^2]^{3/2}} - \left(1 + 2\frac{V_s^2}{v_K^2}\frac{1}{(1 + \kappa z^2/r_0^2)}\right)\kappa \tag{20}$$

Neglecting $V_s^2/v_K^2 \ll 1$ this expression passes through zero (sound transition) at

$$z_s^2/r_0^2 = \frac{(\frac{1+\kappa}{\kappa})^{2/3} - 1}{1 + \kappa} \tag{21}$$

which yields $z_s/r_0 \to \sqrt{2/3}\frac{1}{\kappa}$ for $\kappa \to \infty$, $\to \kappa^{-1/3}$ for $\kappa \to 0$, and $= 0.54$ for $\kappa = 1$.

The wind mass loss rate then becomes

$$\frac{1}{2}\dot{M}_w = \pi r_0^2 \rho_0 \frac{B_p}{\sqrt{B_p^2 + B_\phi^2}} V_s \exp\left[-\frac{1}{2} - \frac{GM}{r_0 V_s^2}[1 - \frac{3}{2}(\frac{\kappa}{1+\kappa})^{1/3} + \frac{1}{2}\frac{\kappa}{1+\kappa}]\right] \cdot \cos\chi \tag{22}$$

where the integration over the cross section at $z = z_s$ was replaced for simplicity by multiplication with the crosssectional area.

$$\cos\chi = \sqrt{\frac{\frac{1}{1+\kappa} + (\frac{\kappa}{1+\kappa})^{1/3}}{1 - \kappa + (1 + \kappa)(\frac{\kappa}{1+\kappa})^{1/3}}} \tag{23}$$

is the vertical projection cosine of the poloidal velocity, $\cos\chi \to 1$ for $\kappa \to 0$, $\to \sqrt{3/5}$ for $\kappa \to \infty$, and $= 0.903$ for $\kappa = 1$.

The factor $\frac{GM}{r_0 V_s^2} = \frac{v_K^2}{V_s^2} \simeq (r/H)^2$ is of order $\geq 10^{2.8}$ which cuts off significant wind mass loss if the square bracket in the exponent does not become small. For this we have

$$\left[1 - \frac{3}{2}(\frac{\kappa}{1+\kappa})^{1/3} + \frac{1}{2}\frac{\kappa}{1+\kappa}\right] \to 1 \quad \text{for} \quad \kappa \to 0, \to 1/(6\kappa^2) \quad \text{for} \quad \kappa \to \infty,$$

$$\text{and} \quad = 0.059 \quad \text{for} \quad \kappa = 1 \qquad (24)$$

Taking the whole accretion disk and its atmosphere as isothermal (which over-estimates the wind losses) one obtains

$$\frac{\dot{M}_w}{\dot{M}_d} = \frac{2\pi r_0^2 \rho_0 V_s v_K}{4\pi \alpha r_0^2 \frac{H}{r}\rho_0 V_s^2} \cdot f(\kappa) = \frac{1}{2\alpha H/r}\frac{v_K}{V_s} f(\kappa) = \frac{1}{2\alpha}(\frac{r}{H})^2 \cdot f(\kappa) \qquad (25)$$

where

$$f(\kappa) = \frac{\cos \chi(\kappa)}{\sqrt{e}} \exp\left\{-(\frac{r}{H})^2[1 - \frac{3}{2}(\frac{\kappa}{1+\kappa})^{1/3} + \frac{1}{2}\frac{\kappa}{1+\kappa}\right\} \qquad (26)$$

For values $\alpha = 1/3$, $H/r = 10^{-1.4}$ in the disk this gives

$$\frac{\dot{M}_w}{\dot{M}_d} = \frac{1}{10} \quad \text{for} \quad \kappa = 2.85, \quad \text{but} \quad = 10^{-13.4} \quad \text{for} \quad \kappa = 1 \qquad (27)$$

This illustrates that sufficient curvature of the fieldlines is necessary to achive significant mass loss. Taking the asymoptotic opening angle ϑ,

$$\tan \vartheta = \frac{s}{z} \to \sqrt{\kappa}$$

as a guide for the enclosed flux fraction we find that $\kappa = 3$ corresponds to half the hemisphere above the disk filled with the expanding conical flux tube which should then contain roughly $\frac{1}{2}$ of the total expanding flux. This supports again the flux distribution

$$B \sim B_0(r/r_0)^{-n}$$

with n about 2 as argued before. We now change to spherical coordinates (r, ϑ, ϕ). Higher up the flow passes the Alfvén critical point

$$v_r = \frac{B_r}{\sqrt{4\pi\rho}} \qquad (28)$$

Up to that height the magnetic field is dominant and enforces co-rotation

$$v_\phi \simeq r\Omega_p \sin \vartheta \qquad (29)$$

(ϑ opening angle of flux tube).

Together with continuity of mass

$$\frac{1}{2}\dot{M}_w = 2\pi(1 - \cos \vartheta)r^2 \rho v_r \qquad (30)$$

and flux conservation

$$B_r 2\pi r^2 (1 - \cos\vartheta) = B_0 \pi r_0^2 \tag{31}$$

(B_0 vertical field component in the disk) one derives

$$v_r = \frac{B_0^2 r_0^4}{4(1 - \cos\vartheta)\dot{M}_w} \frac{1}{r^2} \tag{32}$$

at $r = r_A$, $v_r = v_A$. The magnetic field is dominant below this point and near the Alfvén-critical point supplies the work that accelerates the wind through this region. In rough order of magnitude

$$\frac{B_r B_\phi}{4\pi} \pi r^3 \sin^3\vartheta \Omega_K = \frac{\dot{M}_w}{2} \frac{1}{2} (v_r^2 + v_\phi^2) \tag{33}$$

(On the right handside we may also include additional terms of similar magnitude, as the Poynting flux.) We assume that in the critical point the magnetic field becomes significnatly inclined, $B_\phi \approx B_r$ as it no longer is able to inforce co-rotation and that v_ϕ falls below the co-rotational speed. From the condition of "frozen-in" magnetic field we have

$$\frac{B_\phi}{B_r} = \frac{v_\phi - r\Omega_K \sin\vartheta}{v_r} \tag{34}$$

fulfilled with $v_\phi \approx v_r \simeq \frac{1}{2} r\Omega_K \sin\vartheta$. From this and equation (32) one obtains

$$\left(\frac{r}{r_0}\right)^3 = \frac{B_0^2 r_0}{2\dot{M}_w \Omega_p} \frac{1}{\sin\vartheta(1 - \cos\vartheta)} \tag{35}$$

$$\left(\frac{v_r}{v_K}\right)^3 = \frac{B_0^2 r_0}{2\dot{M}_w \Omega_p} \frac{\sin^2\vartheta}{(1 - \cos\vartheta)} \tag{36}$$

for the position and speed at the critical point. It shows the dependence on fieldstrength and $\dot{M}_w$, the latter being crucially dependent on the field topology near the disk surface, i.e. the flux distribution in the disk.

An upper limit to the strength $\dot{M}_w$ comes from the available accretion power in the underlying disk,

$$\frac{\dot{M}_w}{\pi r_0^2} \frac{v_r^2 + v_\phi^2 + GM/r_0}{2} \leq \frac{3}{4\pi} \frac{GM\dot{M}_d}{r_0^3} \tag{37}$$

($\dot{M}_d$ the disk mass accretion rate) which gives a limit on the amount of mass loss $\dot{M}_w$ that a disk can sustain,

$$\frac{\dot{M}_w}{\dot{M}_d} \leq \frac{1}{64[\frac{B_0^2 r_0}{2\dot{M}_d \Omega_K} \frac{\sin^2\vartheta}{1-\cos\vartheta}]^2} = \frac{1}{16[\frac{B_0^2 r_0}{4\pi P\alpha \frac{H}{r}} \frac{\sin^2\vartheta}{1-\cos\vartheta}]^2} \quad . \tag{38}$$

The latter expression measures the magnetic energy density in terms of the thermal pressure P in a standard accretion disk. The tilt of the magnetic field at the disk surface is derived from the magnetic work that energizes wind and magnetic energy outflow at the Alfvén point. One obtains

$$\left(\frac{B_\phi}{B_z}\right)_0 = 3 \left(\frac{2\dot{M}_w \Omega_K}{B_0^2 r_0}\right)^{1/3} \left(\frac{\sin^2 \vartheta}{1 - \cos \vartheta}\right)^{2/3} \tag{39}$$

The above limit on $\dot{M}_w/\dot{M}_D$ inserted yields an upper limit on the magnetic stress,

$$\frac{B_r B_\phi}{4\pi} \leq \frac{3}{2} \alpha \frac{H}{r} P \tag{40}$$

and near this limit a significant fraciton of the accretion energy is given to the magnetized wind. As derived before, whether this limit is reached depends on $\dot{M}_w$ and thus on the geometry of the magnetic field configuration as parametrized by κ, and thus on the flux distribution in the disk.

We note that equation (38) does not constrict the value of the magnetic field strength but only the product $\dot{M}_w B_0^4$. Thus one cannot exclude magnetic fields that become as strong as the disk pressure.

Winds as discussed in this section presumable are the cause of cataclysmic variable winds and the phenomenon of missing radiative flux in these systems. With the field B_d provided by the secondary's magnetosphere at outer disk radius r_d one obtains

$$\frac{B_0^2}{4\pi P \alpha \frac{H}{r}} = \frac{B_0^2 r_0}{\dot{M}_d \Omega_K} = \frac{B_d^2 r_d^4}{\dot{M}_d (GM r_0^3)^{1/2}} = 3 \tag{41}$$

for $B_d = 30$ gauss, $r_d = 10^{10.4}$cm and $\dot{M}_d = 10^{-9} M_\odot$/yr. This is the value required for strong winds from the interior parts of accretion disks and is a surprizing coincidence. The observed wind speeds are of the order of the Kepler speed as required for strong winds.

Further, "tilt" of the wind column (Verbunt 1991) would be a natural consequence of the unconnected wind magnetic field which is pushed ("diamagnetically") away from the secondary's magnetic field. A phase shift of maximal absorption by $\Delta\phi \simeq 0.2 - 0.3$ (YZ Cnc, Woods et al. 1992) away from $\phi = 0.5$ expected for a symmetric magnetosphere might then be due to asymmetry in the spot distribution on the surface of the secondary.

5   Tilt instability of a disk with a magnetized wind

A magnetized wind exerts breaking stress

$$\tau = \frac{B_z B_\phi}{4\pi} \tag{42}$$

on the two surfaces of the accretion disk. If an inner part of the accretion disk is tilted by a small angle ϑ away from the vertical but its magnetic fields are confined by the surrounding fields of the untilted disk around like in a vertical

magnetic "channel" or tube, then the braking torque becomes unevenly distributed over the surface of the inclined disk as the magnetic field tries to approach a
force-free but torque transporting configuration in this vertical channel. It does
this when the magnetic equilibrium speed V_A is larger than the speed v_K of
moving fluxlines around the disk. This is the case below the Alfven critical point
of the wind.

To a first approximation this exerts a torque per unit surface that increases
the tilt, of

$$2 \frac{B_z B_\phi}{4\pi} \vartheta \, r \cos^2 \phi$$

(from both sides together) where ϕ is the azimuthal angle in the disk counted
from the line of nodes, i.e. the intersection of tilted disk plane with that of the
surrounding untilted one.

On a ring of surface $2\pi r \Delta r$ we obtain the torque ΔT_m

$$\Delta T_m = 2\pi r^2 \frac{B_z B_\phi}{4\pi} \vartheta \Delta r \tag{43}$$

Given the angular momentum of the disk ring

$$\Delta I = 2\pi \, \Sigma r^4 \Omega_K \tag{44}$$

one obtains a growth time for the tilt

$$t_m = \frac{\vartheta \Delta I}{\Delta T_m} = \frac{2}{\Omega_K} \left(\frac{r}{H} \right)^2 \frac{4\pi P H / r}{B_r B_\phi} \tag{45}$$

Growth only occurs if viscous damping does not exceed this excitation torque.
The viscous evolution of tilted accretion disks is investigated in a series of papers
by Pringle and other authors. Here we give a simple estimate.

The viscous damping arises from vertical shear between Kepler velocities of
tilted and untilted parts. This shear is proportional to $\cos \phi$,

$$\frac{\partial v_{sh}}{\partial r} = v_K \frac{\partial \vartheta}{\delta r} \cos \phi \quad . \tag{46}$$

With viscosity μ the viscous torque working against the tilt on the circumferen-
nce becomes

$$T_v = 2\pi r^2 \frac{H}{r} \mu v_K \frac{d\vartheta}{dr} \quad . \tag{47}$$

We parametrize the viscosity μ of the vertical shear

$$\mu = \frac{2}{3} \alpha_m \frac{P}{\Omega_K}$$

and take the torque difference over the ring width Δr to obtain

$$\Delta T_v = \frac{dT_v}{dr} \Delta r = \Delta r \frac{d}{dr} \left[\frac{4\pi}{3} \alpha_m \frac{H}{r} r^4 P \frac{d\vartheta}{dr} \right] \quad . \tag{48}$$

For stationary accretion disks,

$$\frac{H}{r}P \approx \frac{\dot{M}_d v_K}{4\pi\alpha r^2} \sim r^{-5/2} \quad , \tag{49}$$

the bracket is proportional to $r^{3/2}\frac{d\vartheta}{dr}$. If we take $\frac{d\vartheta}{dr} = \frac{\vartheta}{r}$ we obtain

$$\Delta T_v = 2\pi r^2 \Delta r \alpha_m \frac{H}{r}P\vartheta \tag{50}$$

Excitation occurs for $\Delta T_m > \Delta T_v$, or

$$\frac{B_z B_\phi}{4\pi\alpha_m \frac{H}{r}P} > 1 \quad . \tag{51}$$

Though our estimates are rough they indicate that for sufficently strong magnetic tension disks may become tilt unstable. We note comparing with the estimated upper limit on magnetic tension by exhaustion of accretion power,

$$\frac{B_z B_y}{4\pi\alpha \frac{H}{r}P} < \frac{3}{2} \quad .$$

excitation would occur for

$$\frac{\alpha_m}{\alpha} < \frac{B_z B_\phi}{4\pi\alpha \frac{H}{r}P} < \frac{3}{2} \quad . \tag{52}$$

Since we do not know the ratio of the α-parameters nor the numerical reliability of our estimates the question of whether magnetized disks become tilt unstable must remain open. We will further on assume that this is the case. It is interesting to note that one is again in the strong wind range.

6   Dwarf nova oscillations

Dwarf novae in outburst and stationary nova-like systems are observed to display small amplitude photometric oscillations in visual light and in X-rays (for a review see Warner 1995). These oscillations have periods from 10 to 100 sec and intermediate Q value of 10^4 to 10^6. This makes them different form the low Q quasi periodic oscillations and high Q oscillation of DQ Her and intermediate polar systems. We suggest that they result from the illumination of a tilted part of an accretion disk by accretion spots that rotate on the surface of the white dwarf.

   A characteristic variation of oscillation frequency during the outburst of AH Her has been observed (Hildebrand et al. 1980) with shortest period near light maximum. This indicates that the rotation speed of the presumed accretion spots varies with accretion rate. This would fit to the concept that the accretion spot rotation rate is different from that of the white dwarf but is frictionally coupled and thus lags behind Kepler motion by an amount depending inversely on the accretion (spin-up) rate. It also would place the accretion spot rotation away from the tilted direction of the disk and between this and the white dwarf rotation (which should be normal to the orbital plane).

This seems necessary in order to obtain a varying illumination of the tilted disk as the visibility angle of the spot from the disk varies with the spot rotation. A spot rotating at the white dwarf surface in the tilted disk plane would not produce a varying illumination. It is noteworthy that the oscillations were not visible around maximum of the visual light where the spot rotation plane presumably is most aligned with the tilted disk.

The observed variation of the oscillation period during an outburst is not compatible with a rotating spot of a white dwarf magnetic field. One would instead suspect magnetic spots formed by the magnetized accreting matter.

The precession of a tilted magnetized accretion disk within the fields of the surrounding untilted disk is discussed below. In the dwarf nova case it is too slow to affect the rapid dwarf nova oscillations.

7   Precession of GRO 1655

The galactic black hole binary GRO J1655-40 shows superluminal motion and jet precession with a period of 3 days (Tingay et al. 1995, Hjellming & Rupen 1995, Mirabel & Rodriguez 1996). This precession period is remarkable in view of the orbital period of 2.6 days (Baileyn et al. 1995). The system contains a $7M_\odot$ black hole and a secondary of $2.34M_\odot$ (Orosz & Bailyn 1997).

An inclined accretion disk filling 0.7 of the primary Roche lobe would precess in the gravitational field of the secondary with the much longer period of 17 times the orbital period. Thus the precession must be internal to the disk-black hole system.

The secondary star has evolved off the main sequence and displays a F3IV to F5IV spectral type (Orosz & Bailyn 1997). It is convective and like in the late-type secondaries one expects dynamo magnetic fields. We do not know their strength nor surface distribution but assume again magnetic spots of pressure equipartition field strength. We estimate an atmospheric pressure of $P \approx 10^{2.5} \mathrm{dyn/cm^2}$ for this star and a starspot magnetic field of

$$B_* = \sqrt{4\pi P} = 60 \mathrm{gauss} \quad . \tag{53}$$

The same scaling as for the cataclysmic binary system (same mass ratio) yields a magnetospheric fieldstrength at the accretion disk of

$$B_d = 0.5 \mathrm{gauss} \quad . \tag{54}$$

We now assume that such fields become concentrated as in the case of cataclysmic variables and ask whether the tilting instability of the inner accretion disk can be reached. We then investigate what precession period would result.

The rapid evolution of the secondary toward the red giant state sustains a large mass overflow rate of order $10^{-7} M_\odot/\mathrm{yr}$ which we take as scaling parameter.

The disk around the $7M_\odot$ black hole then becomes radiation pressure dominated for

$$r < 10^{9.9}\dot{M}_{-7}^{16/21}\kappa_{.4}^{6/7}\alpha^{2/21}\,\text{cm}\quad,\tag{55}$$

where $\dot{M}_{-7}$ is the mass accretion rate in units of $10^{-7}M_\odot/\text{yr}$ and $\kappa_{.4}$ the opacity in units of the electron scattering opacity. This derives from the standard accretion disk theory and α-friction.

From the same considerations of hydrostatic support and radiative energy transport together with α-parametrization for the frictional stress

$$\tau = \alpha P\tag{56}$$

(Shakura & Sunyaev 1973) one obtains the following estimates for scaleheight

$$H = \frac{3}{4\pi}\frac{\kappa\dot{M}}{c}\quad,\tag{57a}$$

surface density

$$\Sigma = \frac{16c^2}{9\alpha\kappa^2\,\Omega_K\dot{M}}\quad,\tag{57b}$$

temperature T

$$T^4 = \frac{c^2\Omega_K}{4\alpha\sigma\kappa}\tag{57c}$$

(σ Stefan-Boltzmann constant), radiation pressure

$$P_r = \frac{c\Omega_K}{3\alpha\kappa}\tag{57d}$$

and density

$$\rho = \frac{32\pi c^3}{27\alpha\kappa^3\dot{M}^2\Omega_K}\quad.\tag{57e}$$

A radiation pressure supported disk is unstable to viscous and thermal instability ("Lightman-Eardley" instability) if friction is coupled to total ($\approx$ radiation) pressure but stable for coupling to gas pressure. The disk in this binary system seems to extend inward beyond $r = 10^{9.9}\text{cm}$ and irradiation from central energy release would keep the outer disk temperature at values above 10^4K preventing dwarf nova type instability. We take this as indication that α-coupling to the gas pressure describes this disk appropriately.

We may then use the above estimates for the disk quantities with

$$\alpha = \alpha_{eff} = \alpha_0 P_g/P_r\tag{58}$$

where $\tau = \alpha_0 P_g$ is the assumed frictional stress.

$$P_g = \frac{\Re}{\mu}T\rho\tag{59}$$

is the gas pressure ($\Re$ gas constant, μ molecular weight). Thus α_{eff} is a small quantity.

We then obtain the following values

$$\alpha_{eff} = \left(\frac{32\pi}{9\sqrt{2}}\right)^{4/5} \frac{c^2(\Re/\mu)^{4/5}}{\kappa^{9/5}\alpha_0^{1/5}\sigma^{1/5}\Omega_K^{7/5}\dot{M}^{8/5}} \quad , \tag{60}$$

$$P_r = \frac{c\Omega_K}{3\kappa}\frac{1}{\alpha_{eff}} \tag{61}$$

We can now evaluate the precession time t_p of tilted disk in the surrounding fields. The disk ring of surface $2\pi r\Delta r$ and thickness $2H$ permeated by a magnetic field B constitutes a magnetic dipole of moment

$$m = BHr\Delta r \tag{62}$$

which, inclined to a surrounding field, also of strength B by angle ϑ feels the torque

$$T_p = B \cdot m = B^2 Hr\Delta r\vartheta \tag{63}$$

and precesses with period

$$t_p = \frac{2\pi I\vartheta}{T_p} \tag{64}$$

where $I = 2\pi r^3 \Omega_K \Sigma \Delta r$ is the angular momentum. After further algebraic transformation this becomes

$$t_p = \frac{16\pi}{\Omega_K\alpha_{eff}} \left(\frac{r}{H}\right)^3 \left(\frac{4\pi P\alpha_{eff}\frac{H}{r}}{B^2}\right) \tag{65}$$

The last expression with B^2 replaced by $B_z B_\phi$ is the term that must be of order unity if the tilting instability is to work and for efficient wind acceleration B_ϕ becomes comparable to $B_z = B$.

Evaluation of the last equation, taking this factor as unity, gives

$$r = 10^8 \text{cm} \tag{66}$$

for a resulting precession period of 3 days.

At this radius the value of B must be about

$$B = (4\pi P\alpha_{eff}\frac{H}{r})^{1/2} = (\dot{M}\Omega_K/r)^{1/2} = 10^{6.2}\dot{M}_{-7}^{1/2}\text{gauss} \tag{67}$$

which translates to a value of the secondary's magnetosphere near the disk rim of

$$B_d = B\left(\frac{r}{r_d}\right)^2 = \frac{1}{6}\text{gauss} \tag{68}$$

which is close to the value $B_d \approx \frac{1}{2}$ gauss estimated for the magnetosphere.

In view of the assumptions that had to be made one cannot consider these estimates as a proof for a magnetic precession of a tilted inner accretion disk in GRO 1655. But the numbers at least to not seem to exclude this possibility.

8   Conclusion

We have argued for the advective concentration of magnetic flux in the inner accretion disk which entered the disk from the surrounding magnetosphere of the companion star. This appears to account for speed and mass loss rate of observed winds and missing energy in cataclysmic variable systems. We show the possibility of a magnetic disk tilting instability and suggest a relation to the dwarf nova oscillation phenomenon. Finally we show that the same instability may account for the anomalous precession period of GRO 1655.

References

Baptista R., Steiner J.E., Home K., 1996, MNRAS **282**, 99

Blandford R.D., Payne D.G., 1982, MNRAS **199**, 883

Camenzind M., 1986, A&A **156**, 137

Drew J.E., 1987, MNRAS **224**, 595

Drew J.E., 1991, in *Structure and Emission Properties of Accretion Disks*, eds. C. Bertout, S. Collins, J.-P. Lasota, J. Tran Tkanh Van, Editions Fronteères, Gif sur Yvette, 331

Hildebrand R.H., Spillar E.J., Middleditch J., Patterson J., Stiening R.F., 1980, ApJ **238**, L145

Hjellming R.M., Rupen M.P., 1995, Nature **375**, 464

Mauche C.W., Raymond J.C., 1987, ApJ **323**, 690

Mirabel I.F., Rodriguez L.F., 1996 in *Röntgenstrahlung from the Universe*, Internat. Conf. on X-Ray Astronomy and Astrophysics, eds. H.U. Zimmerman et al. (MPE Report 263), p.111

Nitta S., 1994, PASJ **46**, 217

Orosz J.A., Bailyn C.D., 1997, ApJ **477**, 876

Pringle J.E., 1993, in *Astrophysical Jets*, eds. M. Livio, C.P. O'Dea, Cambridge University Press

Rutten R.G.M., van Paradijs J., Tinsbergen J., 1992, A&A **260**, 213

Sakurai T., 1985, A&A **152**, 121

Sakurai, T., 1987, PASJ **39**, 821

Shakura N.I., Sunyaev R.A., 1973, A&A **24**, 337

Shmeleva O.P., Syrovatskii S.I., 1973, Solar Physics **33**, 341

Spitzer L., 1962, Physics of fully ionized gases, 2nd ed., Interscience, New York

Tingay S.J. et al., 1995, Nature **374**, 141

Tout C.A., Pringle J.E., la Dous C., 1993, MNRAS **265**, 25

Verbunt F., 1991, Adv. space Res II, (11)57

Warner B., 1996, Cataclysmic Variable Stars, Cambridge Astrophysics Series Vol. **28**, p.42

Woods J.A., Verbunt F., Collier Cameron A., Drew J.E., Piters A., 1992, MNRAS **255**, 237

Subject and Object Index

Author Index

New Series m: Monographs

Lecture Notes in Physics

For information about Vols. 1–455
please contact your bookseller or Springer-Verlag